THE HISTORY OF
MODERN ASTRONOMY
AND ASTROPHYSICS

BIBLIOGRAPHIES OF THE HISTORY
OF SCIENCE AND TECHNOLOGY
(Vol. 1)

GARLAND REFERENCE LIBRARY
OF THE HUMANITIES
(Vol. 304)

Volume 1

Bibliographies of the
History of Science and Technology

Editors

Robert Multhauf
Smithsonian Institution, Washington, D.C.

Ellen Wells
Smithsonian Institution, Washington, D.C.

THE HISTORY OF MODERN ASTRONOMY AND ASTROPHYSICS
A Selected, Annotated Bibliography

David H. DeVorkin

GARLAND PUBLISHING, INC. • NEW YORK & LONDON
1982

Library of Congress Cataloging in Publication Data

DeVorkin, David H., 1944–
   The history of modern astronomy and astrophysics.

   (Bibliographies of the history of science and
technology ; v. 1) (Garland reference library of the
humanities ; v. 304)
   Includes index.
     1. Astronomy—History—Bibliography. 2. Astrophysics
—History—Bibliography. I. Title. II. Series.
III. Series: Garland reference library of the humanities ;
v. 304.
Z5154.H57D48 [QB15]   016.52   81-43349
ISBN 0-8240-9283-X   AACR2

Printed on acid-free, 250-year-life paper
Manufactured in the United States of America

GENERAL INTRODUCTION

This bibliography is one of a series designed to guide the reader into the history of science and technology. Anyone interested in any of the components of this vast subject area is part of our intended audience, not only the student, but also the scientist interested in the history of his own field (or faced with the necessity of writing an "historical introduction") and the historian, amateur or professional. The latter will not find the bibliographies "exhaustive," although in some fields he may find them the only existing bibliographies. He will in any case not find one of those endless lists in which the important is lumped with the trivial, but rather a "critical" bibliography, largely annotated, and indexed to lead the reader quickly to the most important (or only existing) literature.

Inasmuch as everyone treasures bibliographies, it is surprising how few there are in this field. Justly treasured are George Sarton's *Guide to the History of Science* (Waltham, Mass., 1952; 316 pp.), Eugene S. Ferguson's *Bibliography of the History of Technology* (Cambridge, Mass., 1968; 347 pp.), François Russo's *Histoire des Sciences et des Techniques. Bibliographie* (Paris, 2nd ed., 1969; 214 pp.), and Magda Withrow's *ISIS Cumulative Bibliography. A bibliography of the history of science* (London, 1971–; 2131 pp. as of 1976). But all are limited, even the latter, by the virtual impossibility of doing justice to any particular field in a bibliography of limited size and almost unlimited subject matter.

For various reasons, mostly bad, the average scholar prefers adding to the literature, rather than sorting it out. The editors are indebted to the scholars represented in this series for their willingness to expend the time and effort required to pursue the latter objective. Our aim has been to establish a general framework which will give some uniformity to the series, but otherwise to leave the format and contents to the author/compiler. We have urged that introductions be used for essays on "the state of the field," and that selectivity be exercised to limit the length of each volume to the economically practical.

Since the historical literature ranges from very large (e.g., medicine) to very small (chemical technology), some bibliographies will be limited to the most important writings while others will include modest "contributions" and even primary sources. The problem is to give useful guidance into a particular field—or subfield—and its solution is largely left to the author/compiler.

In general, topical volumes (e.g., chemistry) will deal with the subject since about 1700, leaving earlier literature to area or chronological volumes (e.g., medieval science); but here, too, the volumes will vary according to the judgment of the author. The topics are international, with a few exceptions (Greece and Rome, Eastern Asia, the United States), but the literature covered depends, of course, on the linguistic equipment of the author and his access to "exotic" literatures.

Robert Multhauf
Ellen Wells

Smithsonian Institution
Washington, D.C.

CONTENTS

ILLUSTRATIONS

INTRODUCTION

The period covered by the citations in this bibliography begins with the invention and application of the telescope to astronomy, though there are a few studies that examine previous periods. A strong emphasis is placed upon the literature of recent astronomy, dating from the middle of the 19th to the middle of the 20th century.

Selection criteria include several factors. First, citations are to both popular and scholarly works, but in every case the works must either provide adequate documentation or represent the opinions and recollections of those directly involved in the astronomical problem area. Second, most of the cited works must be reasonably accessible to nonspecialists through access to a university library or major public library. Preference is for works in English, although important or unique studies in all languages are included where necessary for proper coverage and completeness. Since the author is not a specialist in languages, many of the foreign-language citations are annotated based upon secondary source information including reviews and citations. High priority is given to all identified and reasonably accessible bibliographies that include astronomy, the history of astronomy, or material of interest to studies in the history of astronomy.

The selection process began with a complete search through all volumes of major journals (both historical and astronomical) likely to have significant amounts of astronomical history. Articles, book reviews, and notices of books and publishing projects were all examined for inclusion. The following journals were so examined:

> *Advances in Astronomy and Astrophysics*
> *Annals of Science*
> *Annual Review of Astronomy and Astrophysics*

Archive for History of Exact Sciences
Archives Internationales d'Histoire des Sciences
British Journal for the History of Science
Centaurus
History of Science
Isis
Journal for the History of Astronomy
Minerva
Popular Astronomy
Quarterly Journal of the Royal Astronomical Society
Technology and Culture
Vistas in Astronomy

While the above were examined in a systematic manner, other journals and periodicals were certainly not ignored; citations to articles in many of them were collected prior, during and after the systematic searches. In some cases, these were citations which were encountered through the course of examining articles in the journals that were systematically searched. In these and other cases, serendipity was the agent for encounter.

Virtually all monographs initially identified by book reviews in these and other journals were eventually examined through access to a number of major libraries including the libraries of the astronomical departments at Yale and Harvard, the science and history of science libraries at Yale, the Yale University Library, and the Naval Observatory Library in Washington, D.C. A general search for additional monographs was conducted at Yale and Harvard. Several specialist booksellers in the United States and Britain as well as a small number of very helpful colleagues in astronomy and the history of science aided the selection process, and I am indebted to them for this. Finally, several major, exhaustive bibliographical sources were examined, including the *Isis Cumulative Bibliography* (item 68), the *Astronomischer Jahresbericht* (item 69), and D.A. Kemp's *Astronomy and Astrophysics, a Bibliographical Guide* (item 28). These three bibliographical sources, along with Houzeau and Lancaster (item 24) and Poggendorff (item 47), constitute a fundamental base of information. Of these five sources, Kemp's is the most accessible. Though he lists only 32 references as historical works in his appendix, in fact, a large fraction of his 3642 citations provide invaluable historical material in the form of comprehensive reviews, listings of star catalogues of all types, and chronological tables for astronomical dating and calendar systems. Kemp's collection does not, however, contain much information on biographical studies, nor does his collection include strictly historical sources. Kemp should therefore be used in conjunction with the *Isis Cumulative Bibliography* and the remaining three sources indicated above, to gain complete knowledge of the literature. It is hoped that this present bibliography, with its annotation and selected coverage, will provide a more efficient means of finding articles and monographs of specific interest to those who are unfamiliar with the five sources cited above. Then, once this source is used as an entry device to a specific historical problem area, the other five bibliographical sources may be used if complete coverage is desired.

There are many important sources for the study of the history of astronomy that could not be properly highlighted within this bibliography. Therefore, while the incomparable *Dictionary of Scientific Biography* is well known to historians of science, its inclusion (item 1290) in the section on Biographical, Autobiographical and Collected Works does not provide it with proper prominence, and it might be overlooked by those not in history, especially astronomers, who do not know that it is a true fundamental source of general information. Likewise, in the journal section only a few astronomical journals have been identified as being of historical interest, whereas in fact there are many series that the historian should know about. These include *Ciel et Terre* (1880–), *Himmel und Erde* (1889–), *Bulletin Astronomique* (1884–), *The Astronomical Register* (1863–1886), *L'Astronomie* (1882–), and Hermann J. Klein's *Jahrbuch der Astronomie und Geophysik* (1890–1913) and his earlier *Wochenschrift für Astronomie, Meteorologie und Geographie*, a yearly almanac of events, begun by G.A. Jahn in 1847. Similarly, *Die Sterne* began in 1927 as the successor to *Sirius* (1868–) and continues today to be an excellent source of both contemporary and historical information, as does the publication *Sky and Telescope* (item 387) which grew out of the association of two earlier popular reviews in 1941. While it was extremely tempting to include many of the historical articles that have appeared in the pages of this last magazine in its 40 years of publication, it was decided not to do so, since without exception these otherwise excellent studies and sketches do not provide detailed scholarly citations. This is doubly unfortunate because a general cumulative index to the magazine does not, as yet, exist; however, a representative sampling of the articles that have appeared has been published (item 143). Other excellent popular reviews, including the *Irish Astronomical Journal*, *The Griffith Observer*, and the *Leaflets* of the Astronomical Society of the Pacific, now partially superseded by *Mercury*, were not systematically examined, but a number of important citations from these sources do appear. These too have been limited in number, however, for they usually do not provide adequate documentation.

Another extremely useful source of information not explicitly cited here is the many observatory publications that, over the past century, have chronicled the research activities of several hundred astronomical observatories and institutions. Astronomical observatories traditionally cooperate in an exchange program of research publications and notices of activity. It is common for each observatory to periodically discuss its history, instrumentation, and staff changes. This was true of David Gill's Cape Observatory *Annals* (item 836), as it was for the Cambridge Solar Physics Laboratory (item 336a), for the Harvard College Observatory (items 261, 266, 334), and for the Lick, Yerkes and Mount Wilson Observatories. Each year the Harvard College Observatory director and many of his counterparts at other observatories would prepare annual reports to their governing bodies, and many of these are incorporated into their observatory publications. Throughout most of its history the *Monthly Notices of the Royal Astronomical Society* would report on the activities of observatories within the British Empire (see item 50), as would the *Vierteljahrsschrift der astronomischen Gesellschaft* (item 356) for member observatories within that major organization. In the United States, *Popular Astronomy*, from the 1890s through 1951, reprinted observatory reports, as did its predecessors, *The Sidereal Messenger* and *Astronomy and Astrophysics* (item 48). This activity was gradually taken over by *The Astronomical Journal* and is now incorporated into the American Astronomical Society's *Bulletin*. In like manner, reports of British departments of astronomy and observatories are published in the *Quarterly Journal of the Royal Astronomical Society*.

In addition to observatory reports, we have also slighted unpublished sources; listings are provided here where titles of M.S. and Ph.D. dissertations might be found (items 20, 35a), but general sources on dissertations must also be examined for comprehensive coverage. The British Society for the History of Science publishes an annual report of dissertations completed and in progress in Great Britain. It is compiled from questionnaires sent to history, history of science, history of education, sociology, and social relations of science departments.

The growing availability of archival sources for the study of the history of astronomy has only been hinted at within this work. While we have listed reports of the existence of various archival holdings worldwide (see, for example, items 5, 24, 25, 26, 35, 44, 48a, 63, 64, 65, 67), and have included notice of several major finding aids (items 5, 25a, 28a, 41, 56), the reader should be aware that vast and largely untapped archival resources do exist and, through organization and microfilming, are slowly becoming generally available. The most comprehensive project has been conducted recently by the Center for History of Physics of the American Institute of Physics (335 East 45th Street, New York, NY 10017). A preliminary report on their work, prepared by Spencer R. Weart and the author, will be appearing in a forthcoming issue of *Isis*. Earlier reports of direct interest have been listed here (items 41, 65, 67). It should also be noted here that the Center for History of Physics maintains a constantly updated card file locating all known resources and, wherever possible, includes descriptions and finding aids of many archival sources for both physicists and astronomers, as well as for their institutions. Another important aspect of their work in the recent past has been to conduct a major oral history program in modern astronomy. To date well over one hundred astronomers have been interviewed and the interviews have been transcribed, edited and prepared with abstracts, tables of contents and indexes. These have been consulted by a number of astronomers and historians, and, with the publication of a general finding aid to the program materials, forthcoming possibly by 1982, it will become a central source of scholarly research. The Center issues a *Newsletter* reporting on its activities and related activities in the histories of physics and astronomy communities. It is available by writing to the Center.

Several centers of archival activity exist in Britain. In London, the Royal Commission on Historical Manuscripts (Quality House, Quality Court, Chancery Lane, London WC2 1HP) is a standard source for updated finding aids to the papers of British scientists, and in Oxford the Contemporary Scientific Archives Centre (10 Keble Road, Oxford 0X1 3QG), founded by Margaret Gowing, takes in papers for organization and preservation. They publish periodic reports of their activities, available upon request. Finally, a generally available source for archival information is the microfiche edition of *Archives of British Men of Science*, consisting of 58 fiche cards identifying the location of many collections. The microfiche edition was published by Mansell of London, and was prepared by R.M. MacLeod and James Friday of the Science Policy Research Unit at the University of Sussex in 1972. Other major sources in the United Kingdom include the Royal Society, where all correspondence relating to their publications is housed; the Royal Astronomical Society (item 5); the Cambridge University Observatory Library; the Crawford Library of the Royal Observatory, Edinburgh; and the Royal Greenwich Observatory, Herstmonceaux, where all records pertaining to the history of the Greenwich Observatory are kept with the exception of records referring to the determination of longitude, which are housed at the National Maritime Museum Naval Archives. Finally, all meteorological records and all records pertaining to anyone in the scientific civil service in Britain are kept in the Public Record Office in London.

In Europe and the U.S.S.R. relatively little in the way of coordinated work has been done. Kulikovsky (item 35) reports on progress in the Soviet Union. There is Ernst Zinner's classic compilation of European manuscript repositories (items 70 and 176), and K.R. Biermann's survey of mid-19th-century material in German-speaking countries (item 164). Genevieve Feuillebois reviews (item 283) the holdings of the Paris Observatory, which has just been chosen as the official repository of the records of the International Astronomical Union. It is also believed that several major projects are under development in Europe. A center of activity of this sort is the Archenhold-Sternwarte, 1193 Berlin-Treptow, Alt Treptow 1, GDR (D.B. Herrmann, Director).

The most general sources of archival information in the United States are the *Directory of Archives and Manuscript Repositories in the*

United States and the *National Union Catalog of Manuscript Collections*. The first is maintained by the National Historical Publications and Records Commission of the National Archives and Records Service, Washington, D.C., and lists some 2675 institutional repositories in the United States, giving brief annotations on the scope of the holdings, types of records held, and the approximate volume of material. It should be understood that unless the repository is primarily or significantly of astronomical interest, there will probably not be explicit mention of astronomical material. The second general source is maintained by the Library of Congress and is a collection of contributed paragraph descriptions of the contents of collections held in the United States.

The cut-off date for inclusion of citations in this bibliography was December 31, 1980. Since that time, issues of *Isis*, the *Journal for the History of Astronomy*, and many other journals have contained articles of interest here. Similarly, a major project, the *Cambridge General History of Astronomy*, under the general editorship of M.A. Hoskin with four volume editors, is well under way, but has no definite completion date.

Since the production of any bibliography is necessarily a project that has time value and quickly becomes dated, it was decided at the outset to produce this bibliography in a form suitable for processing by computer. The general literature search was first done using standard 5″ × 8″ file cards, with clearly identified addressable fields for author(s), title, journal or publisher, publication date, category classification, subject index, subclassification categories, and annotation text. These cards were then edited and typed in machine readable format using a standard IBM OCR-A typing element. The typewritten copy was then optically read and placed onto magnetic tape, and from there onto disk for further editing and the eventual automatic sorting by category, author, and date. The software for this activity was kindly provided by Kunie F. DeVorkin, and additional support was generously given by Jack F. Staff and Carol Cross. I am deeply indebted to them for their generosity, perseverance and understanding.

The classification system developed for this bibliography was synthesized from the systems employed by D.A. Kemp (item 28) and by the *Astronomischer Jahresbericht* (item 69), with a number of alterations dictated by the nature of astronomical progress since the invention of the telescope and by the nature of the published historical record. Except for general studies it was rare to find either a historical paper or review that covered three hundred years of activity in any one problem area. Usually, specific time periods were examined; in discussing the historical problem of the nature of the nebulae, for example, studies tended to center either upon the era of William Herschel or the post-spectroscopic era of Keeler, van Mannen, and Hubble. The classification thus created represents the major periods of astronomical activity. The categories, and their subdivisions, can be seen in the table of contents and in the detailed breakdown of contents at the end of this introduction.

It is painfully evident that in producing a classification of this type, many historical reviews overlap two or more categories. But few have real strengths in more than one area. The fully annotated citation with full documentation appears only in that category where it makes the most marked contribution. Unnumbered cross-references (marked with "*") then appear in secondary categories identifying only author and title, with cross-reference to the location of the original citation. It was therefore necessary to include about 600 cross-references for the some 1400 primary citations.

The documentation for each citation is straightforward, but some comments on the annotation are necessary. First, the active verb that usually introduces the annotation implies the overall character of the contribution: "Lists" is clearly different in treatment than "Examines" or "Reviews" or "Analyzes." Next, especially in reviews written by astronomers, references to "modern" studies or to "work to date" refer to the date of the original cited publication, and not to the date of the compilation of this bibliography. Finally, when it was convenient to do so, the number of citations utilized in the historical work is indicated. This provides some means of assessing what might be a valuable reference leading to deeper study. In many cases, especially in older monographs with unnumbered footnote citations, an estimate of the number of references is indicated by the words "few," "some," "numerous," "extensive," etc. "Indirect" or "textual" citations refer to incomplete references provided within the text.

The value of a comprehensive author, name and subject index cannot be overestimated in a bibliographical work of this type. In fact, no matter what classification is used, or how many cross-references are provided, most users of this bibliography will probably start in the index. The index provided at the end of this bibliography indicates authors by an "*". Other proper names are included when the citation provides significant information about them, or when their names appear in titles or as editors and translators. Subject headings are not comprehensive, however, and are included only for historically coherent topics not general enough for the major organizational classification of this bibliography.

Finally, a remark or two about getting familiar with the history of modern astronomy should be made. For broad introductions to facts, figures and trends, J.L.E. Dreyer's *A History of Astronomy from Thales to Kepler* (item 89) could be followed by Robert Grant's *History of Physical Astronomy* (item 91), Agnes Clerke's *A Popular History of Astronomy during the 19th Century* (item 82) and finally by Otto Struve's and Velta Zebergs' *Astronomy of the 20th Century* (item 113). While these are still, admittedly, the best reviews in English in their field, they are all purely scientific narratives, and do not reflect the character and trends of the modern study of the history of science. To gain some insight into how modern history is done, *New Aspects in the History and Philosophy of Astronomy* (item 123) might be helpful, as might M.A. Hoskin's and O. Gingerich's short essay (item 169). Similarly, the expected volumes of the Cambridge *General History of Astronomy* will have editorial prologues that provide useful historiographical information. But the only sure way to see how it is done—what problems are addressed, and how they are answered—is to read as many selections of direct topical interest as possible. The annotations in this bibliography are designed with this in mind, to aid both the exploratory and directed reader to material that is likely to be of direct interest and value.

I would like to extend my appreciation to the many people and institutions who aided in this project. In addition to those mentioned above, H. DeVorkin, P.M.E. Erwood, E. Forbes, B. Goldstein, D.B. Herrmann and K. Hufbauer provided both encouragement and aid, as did the following organizations: Cognitronics

Corporation, Shah Services, and the Departments of the History of Science and of Astronomy at Yale University. I am indebted to the series editors and to the publisher for their good criticism and suggestions.

But even with this support, errors persist for which I take full responsibility. Those identified in the final stage of proofreading include:

p. 42	item 157	should be in Monographs
p. 72	item 264	should be in Compendia
p. 98	item 362	should be in Compendia
p. 179	item 650	should be in Computational Methods
p. 180	item 651	should be in Computational Methods
p. 180	item 653	should be in Theoretical Astronomy (General)
p. 190	item 683	should be in Positional Astronomy (General)
p. 191	item 686	should be in Positional Astronomy
p. 222	item 807	should be in Computational Methods
p. 268	item 952	should be in Astrophysics (General)
p. 338	item 1196	should be in Extra-Galactic Astronomy and Cosmology

The abbreviation "JHA" used frequently in citing book reviews refers to the *Journal for the History of Astronomy.*

The suffixes "a" and "b" in some citations refer to late insertions.

David H. DeVorkin
National Air and Space Museum
February, 1982

DETAILED CONTENTS

BIBLIOGRAPHIES: pp. 3–20
Bibliographies include both comprehensive and selective series; special listings and indexes; manuscript archives finding aids; library listings; educational resource letters; listings of theses and dissertations; selective subject matter lists; cumulative journal indexes.

GENERAL (MONOGRAPHS): pp. 21–34
General histories of astronomy; major popular and technical reviews that either include historical material or represent landmark studies of the astronomical field; major reviews within standard encyclopedic volumes; major reviews of large areas of the history of astronomy.

GENERAL (COMPENDIA): pp. 34–40
Collections of review papers, both contemporary and historical, in continuing series; special issues of periodicals and annuals devoted to astronomical history; monographs with major historical papers on different topics; multi-author volumes; encyclopedias with volumes or chapters on astronomy or the history of astronomy; source books reprinting primary papers of historical importance; symposium proceedings.

GENERAL (ARTICLES): pp. 40–44
Articles appearing in journals or as chapters in monographs that cover major periods in the history of astronomy, or major themes; general commentary on astronomy or the history of astronomy; many citations linked to the section on Historiography.

HISTORIOGRAPHY: pp. 44–48
Monographs and papers containing material of use in the methodology of history of astronomy including classification of subject areas, examination of data growth, applications of computers, problems in editing, applied history of astronomy.

REGIONAL HISTORIES: pp. 49–67
Monographs and papers devoted to the history of astronomy within an identified locale—region, state, nation, continent or hemisphere; the growth of observational facilities, teaching institutions, interest in and

support for astronomy in general; amateur astronomy; national or regional bibliographies; comparative studies of different regions; national or regional archival resources.

INSTITUTIONAL HISTORIES: pp. 67–71
Studies of institutions linked with astronomical activity, other than observatories and universities; histories identifying the role of institutions in astronomy.

OBSERVATORIES: pp. 71–94
Histories of specific observatories, observing stations or expeditions; the development of observatories; reviews of observatory activity; studies of major astronomical instrumentation; manuscript collections dealing with specific observatory histories; descriptions of major observatory programs and instrumentation.

UNIVERSITIES AND COLLEGES: pp. 94–95
Programs of study; the development of educational resources in specific regions.

SOCIETIES (NATIONAL AND INTERNATIONAL): pp. 96–103
Annual reviews of national and international societies; internationalism in astronomy; histories of societies; international cooperative ventures in astronomy and geophysics; the influence of societies upon astronomy.

JOURNALS: pp. 103–106
The development of purely astronomical journals; major astronomical journals and magazines of interest to historians of astronomy; indexes to major journals of historical interest.

INSTRUMENTATION (GENERAL): pp. 107–113
Monographs and articles dealing with general aspects of astronomical instrumentation; major works that include general information; museum catalogues; the role of instrumentation in astronomy.

PRE-TELESCOPIC INSTRUMENTATION: pp. 114–115
Sighting devices; major non-optical observatories; astrolabes and sundials.

OPTICAL INSTRUMENTATION AND TECHNIQUE: pp. 115–126
The origins and development of telescopic instruments and equatorial mountings; techniques of manufacturing optical equipment for astronomical use; major optical manufacturers; manufacture of optical glass.

PHOTOGRAPHIC INSTRUMENTATION AND TECHNIQUE:
    pp. 126–130
The development of photographic techniques for astronomy; photosensitive detectors; photographic photometry.

POSITIONAL ASTRONOMY (STELLAR POSITIONS AND
MOTIONS—CATALOGUES): pp. 229–237
The determination of stellar positions and motions; star globes, charts,
maps and atlases; stellar distances by astrometric means; stellar bright-
nesses; double star observations; constellation boundaries; the Solar
Parallax.

POSITIONAL ASTRONOMY (ALMANACS AND EPHEMERIDES):
pp. 238–241
Studies in the production of tabulated information on celestial phenomena, for use both in positional astronomy and in astronomical navigation (see also the section on Post-Newtonian theoretical astronomy);
Tobias Mayer's tables; nautical almanacs; theories of the motions of the
Moon and Sun; solar, lunar and planetary positions from antiquity to
the present.

POSITIONAL ASTRONOMY (ASTRONOMICAL CONSTANTS):
pp. 241–249
Transits of Venus and the quest for the Solar Parallax; the determinations of the velocity of light; stellar aberration; the reckoning of time;
the gravitational constant; Ether-Drift experiments.

GEODESY: pp. 249–251
Astronomical determinations of the shape of the Earth; astrogeodetic
observatories; the Earth's motions.

ASTROPHYSICS (GENERAL): pp. 253–264
General monographs reviewing broad aspects of astrophysics; studies of
theoretical physics relating to general astrophysics; laboratory spectroscopy; origins of astrophysics.

ASTROPHYSICS (TERRESTRIAL AND ATMOSPHERIC):
pp. 264–267
Modern studies of geomagnetism and solar-terrestrial relationships; atmospheric physics; theory of radiative transfer; particles and fields; time
scales; high-altitude atmospheric research.

ASTROPHYSICS (PLANETS AND PLANETARY ATMOSPHERES):
pp. 267–271
Modern studies of the physical nature of planetary surfaces and
planetary atmospheres based upon spectroscopic and photographic
techniques; history of the nature of Mars; the origins of lunar features.

ASTROPHYSICS (COMETS, METEORS, AND THE
INTERPLANETARY MEDIUM): pp. 272–277
Chemical and physical studies of terrestrial impact craters and meteori-
tic material; studies of the Zodiacal Light and other components of the

interplanetary environment; the origin and formation of comets; cometary orbits and meteor swarms.

ASTROPHYSICS (SOLAR PHYSICS): pp. 277–287
Modern spectroscopic, visual and photographic studies of the Sun's surface and atmosphere; solar magnetism; origin and nature of sunspots; the solar wind; solar eclipses and gravitational red shift; theoretical studies of the structure of the Sun; chemical composition of the solar atmosphere.

ASTROPHYSICS (MAINTENANCE OF SOLAR HEAT AND
 ORIGIN OF SOLAR SYSTEM): pp. 288–296
Theories of the origin of the Solar System from Laplace to the present, interpreted in light of modern knowledge; theories of the source and maintenance of the heat of the Sun from the mid-19th century to the present; time scales.

ASTROPHYSICS (THE STARS): pp. 296–300
Reviews of knowledge about single stars: sizes, distances, colors, brightnesses.

ASTROPHYSICS (BINARY AND MULTIPLE STARS): pp. 300–305
General reviews and articles on binary star systems; the determination of orbits; stellar masses and dimensions; evolution of binary systems; evolutionary steps in the understanding of the period/luminosity relationship.

ASTROPHYSICS (VARIABLE STARS, NOVAE AND
 SUPERNOVAE): pp. 305–310
Monographs and papers on the history of study of intrinsic variable stars; historical records of novae and supernovae; theories of pulsating variable stars.

ASTROPHYSICS (STELLAR SPECTROSCOPY AND SPECTRAL
 CLASSIFICATION): pp. 310–315
Monographs, texts and articles on the interpretation of stellar spectra for information about stellar atmospheres; the classification of stars by their spectra; spectral atlases.

ASTROPHYSICS (THE HERTZSPRUNG-RUSSELL DIAGRAM):
 pp. 316–319
The origins and evolution of the diagram; the contributions of Ejnar Hertzsprung and H.N. Russell.

ASTROPHYSICS (STELLAR STRUCTURE AND EVOLUTION):
 pp. 319–325
Stellar energy sources; the internal structures of stars; nucleosynthesis

and nuclear physics; theories of stellar evolution and energy; the conservation of energy.

THE HISTORY OF
MODERN ASTRONOMY
AND ASTROPHYSICS

BIBLIOGRAPHIES

1. Adams, G.H., and H.R. Calvert. *Special Exhibition of Historic Astronomical Books*. London: Science Museum Book Exhibition No. 3, 1954. Pp. 29.

   Lists 75 books dating from 1474 to 1837 of importance to the progress of all aspects of astronomy, with annotations on the importance of each work. Includes bibliography of secondary sources.

2. Baldet, M.F. "Liste générale des comètes de l'origine à 1948." *Annuaire publié par le Bureau des Longitudes 1950*, B1-B86.

   Lists 1619 comets from 2315 B.C. to 1948 including designation; discovery date; part of the sky within which comet was first seen; place of discovery and discoverers. Numerous bibliographical citations and notes.

3. Barr, E. Scott. *An Index to Biographical Fragments in Unspecialized Scientific Journals*. Alabama: University of Alabama, 1974. Pp. x + 294.

   Citations for over 7700 scientists taken from eight English-language general science journals (i.e., *The American Journal of Science, Nature*). Includes birth and death dates, major field, and listing of biographical notices. Time span is from 1792 to 1932, concentrating on 1870 to 1920. 1500 portrait references.

4. Barr, E. Scott. "Biographical Material in the First One Hundred Volumes of *The Astrophysical Journal*." *Isis*, 65 (1974), 513-515.

   Covers the period 1895-1944 and includes 73 names.

5. Bennett, J.A. "The Manuscript Archives of the Royal Astronomical Society." *Quarterly Journal of the Royal Astronomical Society*, 18 (1977), 459-463.

Provides descriptive summary of the manuscripts held by
the RAS dating from approximately 1750. The collection is
in three sections: RAS Letters; RAS Papers; RAS MSS. The
first consists of incoming general correspondence from 1820;
in the second are documents such as minute books, collections
of papers submitted to the Society, records of Council discussions; the third has collections donated by astronomers
or their descendents and consists of 35 named collections.

6. Besterman, T. *A World Bibliography of Bibliographies and of
    Bibliographical Catalogues, Calendars, Abstracts, Digests,
    Indexes and the Like.* 3rd Edition. 4 volumes. Geneva:
    Societas Bibliographica, 1955-1956.

    Volume 1, published in 1955, contains partially annotated
    citations to over 60 astronomical bibliographies covering
    all periods.

*   Biermann, Kurt-R. "Attempt at a Classification of Unpublished
    Sources in the More Recent History of Astronomy in German-
    speaking Countries."

    Cited herein as item 164.

*   Bok, Bart J. "Report on Astronomy."

    Cited herein as item 150.

*   Brown, Harrison, ed. *A Bibliography on Meteorites.*

    Cited herein as item 971.

*   Bruhns, C., ed. *Vierteljahrsschrift der astronomischen
    Gesellschaft.*

    Cited herein as item 356.

7. [Brussels Observatory]. *Bibliography of Astronomy, 1881-
    1898.* England: University Microfilms Ltd., 1970. 18 reels.

    Fills the time gap between Houzeau and Lancaster's
    Bibliographie générale and the *Astronomischer Jahresbericht.*
    Produced under the auspices of the IAU. The original is on
    some 50,000 citation slips at the Royal Observatory in
    Brussels, and this microfilm is organized into two-year
    sections arranged by a subject classification derived from
    the *Jahresbericht* as used in *Volume 38.* A short history of
    the project that produced this bibliography is contained in:
    JHA, 2 (1971), 53-54. A *Guide* is available, written by
    J.B. Sykes: item 59.

8. Collard, Auguste. *L'astronomie et les Astronomes*. Brussels: G. van Oest, 1921. Pp. viii + 119.

   Cites 758 items in three major sections: dictionaries, biographies, specific subject areas. Items arranged chronologically within each subject area. Some brief annotation.

9. Collier, Beth A., and Anthony F. Aveni. *A Selected Bibliography on Native American Astronomy*. Hamilton, N.Y.: Colgate University Department of Astronomy, 1978. Pp. 148.

   Cross-referenced bibliography of about 1500 articles and books on the astronomy of native cultures in the Americas.

10. Collinder, Per. "Astronomical Books and Papers Printed in Sweden between 1881 and 1898: Bibliography and Historical Notes." *Arkiv för Astronomi*, 4, No. 19 (1966), 323-339. Reprinted: *Meddelanden fran Uppsala Astronomiska Observatoriums*, No. 159.

    Sources arranged by topic with an author index and detailed historical commentary.

11. Conner, Elizabeth. "The Mount Wilson Observatory Library." *Publications of the Astronomical Society of the Pacific*, 62 (1950), 98-104.

    Description of the holdings, including antiquarian texts from the 16th century through Newton. At the date of the article, library contained between 17,000 and 18,000 volumes including bound periodicals, observatory publications and astronomical monographs.

12. Corbin, Brenda G., ed. *Catalog of the Naval Observatory Library*. 6 volumes. Boston: G.K. Hall, 1976. Pp. 3500+.

    Lists the contents of the library through photoreproduction of card catalog of authors, titles, subjects, serials and cross-references. Includes over 75,000 items on secondary historical literature. Arranged by subject and alphabetical within each each subject range.

13. Dickson, Katherine Murphy. *History of Aeronautics and Astronautics, a Preliminary Bibliography*. Washington, D.C.: National Aeronautics and Space Administration, 1968. Pp. viii + 420.

    Emphasizes other bibliographical sources. Includes 920 titles with annotation. Reviewed in: *Technology and Culture*, 11 (1970), 485-487.

*   Dunkin, Edwin. *Obituary Notices of Astronomers*.

       Cited herein as item 1273.

14.   Emme, Eugene M. "Aeronautics, Rocketry and Astronautics."
       Technology and Culture, 9 (1968), 436–455.

       Reviews present historical research programs in aero-
       space technology, the history of Project Mercury, rocket
       technology. 15 citations.

*   Feuillebois, Genevieve. "Les Manuscrits de la Bibliothèque
       de l'Observatoire de Paris."

       Cited herein as item 283.

15.   Forbes, Eric G. "The Crawford Collection of Books and
       Manuscripts on the History of Astronomy, Mathematics,
       etc., at the Royal Observatory, Edinburgh." *British
       Journal for the History of Science*, 6 (1972–73), 459–461.

       Brief description of the contents and highlights of this
       major collection which includes over 2000 books printed
       before 1770.

16.   Gray, George J. *A Bibliography of the Works of Sir Isaac
       Newton*. Second Edition. London: Dawsons, 1966.

       Updated and greatly expanded by P. and R. Wallis as item
       62.

16a.   Hamel, Jürgen. *Bibliographie der astronomiehistorischen
       Veröffentlichungen in der Deutschen Demokratischen
       Republik*. Teil III (1974–1979). Berlin-Treptow: Archen-
       hold-Sternwarte, 1980. Pp. 49.

       A continuation of the general bibliographical project
       identified in item 66. No annotations. 593 citations.

*   Heinemann, K. *Astronomische Bibliographie: zum 50. Jubilaüms-
       band des 'Astronomischen Jahresberichts.'*"

       Cited herein as item 381.

17.   Heinemann, K. *Verzeichnis von Sternkatalogen 1900-1962*.
       Heidelberg: Astronomischen Rechen-Instituts, 1964.
       Number 16. Pp. 40.

       Indexes star catalogues published between 1900 and 1962
       for individual stars. Includes information on sources of
       data, compilers and bibliography.

*   Hoffleit, Dorrit. "The Library of Harvard College Observatory."

    Cited herein as item 299.

18. Hoffleit, Dorrit. "Bibliography of Meteoric Dust with Brief Abstracts." *Harvard Reprint Series II, No. 43.* (1952), 1-45.

    Selected annotated alphabetical (by author) listing of papers, 1819-1952, dealing with aspects of interplanetary dust in the Earth's atmosphere. Based on large card catalogue maintained, circa 1952, at Harvard College Observatory. 505 citations. A continuation was published in: *Smithsonian Contributions to Astrophysics*, 5 (1961), No. 8. Pp. 85-111.

19. Holton, Gerald. "Resource Letter SRT-1 on Special Relativity Theory." *American Journal of Physics*, 30 (1962), offprint, 8 pp.

    References major papers and books with annotations. 99 citations to work after 1923.

20. Hoskin, M.A., and V.E. Thoren, eds. "Theses and Dissertations on the History of Astronomy." *Journal for the History of Astronomy*, 1(1970), 91-92, 167-168; 2(1971), 71; 3(1972), 73, 149; 4(1973), 143; 6(1975), 147; 8(1977), 73.

    Includes titles, authors and institutions for M.Sc., M.Phil. and Ph.D. studies completed or in progress. Summarized in: *JHA*, 11 (1980), 148-151.

21. Houghton, Walter E., et al., eds. *The Wellesley Index to Victorian Periodicals 1824-1900.* Vol. 1. Toronto: University of Toronto Press, 1966. Pp. xxiv + 1194.

    Includes contributions by late 19th-century astronomical authors. Reviewed in: *Isis*, 58 (1967), 251-253.

22. Houzeau, J.C. "Repertoire des constantes de l'astronomie." *Brussels Annals, New Series*, 1 (1878). Pp. xiii + 271.

    Bibliographical listing of accepted astronomical data with citations to major historical works that mark significant steps in the accumulation of that data. Organized by subject category.

23. Houzeau, J.C. *Vade-Mecum de l'astronomie.* Bruxelles: F. Hayez, 1882. Pp. xxviii + 1144.

Expansion of Houzeau's original bibliography on fundamental astronomical data. Includes sections on the general
study of astronomy; the history of astronomy; spherical,
theoretical, practical and physical astronomy; celestial
mechanics; and topical sections on the Sun, the Solar
System, stars, and star groups. A final section on observatories and observational techniques provides brief annotations on the histories of major institutions. General annotations are also provided in each subject category. Includes some 3447 citations.

24. Houzeau, J.C., and A. Lancaster. *Bibliographie générale
    de l'astronomie jusqu'en 1880* [1887]. New Edition with
    introduction and index by David Dewhirst. 2 vols. in 3
    parts. London: Holland Press, 1964. Pp. xxiv + viii +
    858; vii + 859-1727; lxxxviii + 1728-2225.

    Comprehensive and primary source to 1880 covering all
    aspects of astronomy. Volume 1 includes some 16,000 mono-
    graph sources and some 8,000 authors; Volume 2, containing
    periodical literature, contains some 25,000 citations.
    Dewhirst added commentary on the location of important
    manuscript materials, and an author index. Originally
    published in 1887.

25. Howse, Derek. "Greenwich Register of Source Material on
    the History of Astronomy." *Quarterly Journal of the Royal
    Astronomical Society*, 12 (1971), 335.

    Announces general program to collect information on source
    material for historians of astronomy. Worldwide coverage
    is intended for five areas of information: large instru-
    mentation, collections of smaller instruments, manuscript
    collections, relevant illustrative material, bibliography
    of other indexes.

25a. Hoyt, William Graves, ed. *Early Correspondence of the
     Lowell Observatory, Microfilm Edition 1894-1916*. Flag-
     staff, Ariz.: Lowell Observatory, 1973. Pp. 112.

     General description of the contents of the collection,
     made available on 10 rolls of 35mm microfilm. Includes
     the papers of Percival Lowell, A.E. Douglass, V.M. Slipher,
     and miscellaneous Lowell Observatory Papers and photo-
     graphs. The index is arranged by collection, and alpha-
     betically by correspondent within each collection.
     Available from the Lowell Observatory.

26. Jarrell, Richard A. "Astronomical Archives in Canada."
    Journal for the History of Astronomy, 6 (1975), 143-147.

    Describes archives and collections that contain or possess
    at least preliminary finding aids or indices to content.
    Identifies major collections relating to the history of
    the Dominion Observatory, Queen's University Observatory,
    Quebec Observatory, Royal Astronomical Society of Canada,
    Dominion Astrophysical Observatory, McGill Observatory
    and a few others. Briefly describes each institution.
    3 citations. Supplementary notice in *JHA*, 8 (1977), 71-72.

27. Kemp, D.A. "The Crawford Library of the Royal Observatory,
    Edinburgh." *Isis*, 54 (1963), 481-483.

    Description of the contents and development of this major
    astronomical library "second only to the library of Poulkovo
    Observatory in the U.S.S.R."

28. Kemp, D.A. *Astronomy and Astrophysics, A Bibliographical
    Guide*. London: MacDonald and Co., 1970. Pp. xxiii + 584.

    A standard for contemporary astronomy. Covers the whole
    of modern astronomy and astrophysics listing 3642 articles,
    books, monographs, atlases, compendia, dictionaries, and
    other astronomical bibliographies. Includes both primary
    and secondary published literature. Notes the number of
    references cited by each work and their year span. Brief
    annotations and extensive cross-indexes by author and
    subject. In many cases includes papers announcing major
    discoveries or classic defining papers and reviews.
    Arranged chronologically within subject areas. Most cita-
    tions post-World War II. Glossaries of abbreviations and
    journal names.

28a. Kevles, Daniel J., ed. *Guide to the Microfilm Edition of
    the George Ellery Hale Papers, 1882-1937*. Pasadena:
    Carnegie Institution of Washington and the California
    Institute of Technology, 1968. Pp. 47.

    General description of the contents of the collection,
    made available on 100 rolls of 35mm microfilm. Includes
    alphabetical listing by correspondent with date ranges
    and frame number ranges on each roll. The *Guide* is available
    from: The Director, Mount Wilson Observatory, 813 Santa
    Barbara Street, Pasadena, CA 91101, and the microfilms
    from The Microfilm Company of California, 1977 South Los
    Angeles Street, Los Angeles, CA 90011.

29.   Kleczek, J., J.L. Leroy, and F.Q. Orrall. *A General
      Bibliography of Solar Prominence Research, 1880-1970.*
      Prague: Astronomical Institute of the Czechoslovak
      Academy of Sciences, 1972. Pp. 151.

      Includes approximately 800 citations to primary literature
      organized alphabetically by author with subject cross-
      references. Notes existence of other major bibliographies.

30.   Knight, David. *Sources for the History of Science 1660-
      1914.* Ithaca, N.Y.: Cornell University Press, 1975. Pp.
      223.

      One of a series dealing with the location and exploitation
      of historical evidence. Includes six headings: Histories
      of science, manuscripts, journals, scientific books,
      nonscientific books, surviving physical artifacts. Examines
      major secondary sources and provides a critical analysis.
      Reviewed in: *Isis*, 68 (1977), 299-302.

*   Knobel, E.B. "The Chronology of Star Catalogues."

      Cited herein as item 845.

31.   Korzeniewska, O.I. *The Bibliography of Publications of
      Polish Astronomers, 1923-1963.* Warsaw: Polish Astronomical
      Society, 1964. Pp. 221.

      Lists 2025 contributions by Polish astronomers including
      books, scientific papers, review articles omitting purely
      popular literature. Listed chronologically by year and
      alphabetically within each year. Includes name and subject
      indices and list of abbreviations. Short introduction in
      English.

32.   Kuchowicz, B. *Nuclear Astrophysics: A Bibliographical
      Guide* (4 parts). Warsaw: Nuclear Energy Information
      Center, 1965. Reprinted, New York: Gordon and Breach,
      1967. Pp. x + 441.

      The original edition had an effective date range of
      1885-1959 listing 640 papers and texts dealing with the
      abundances of the elements, nuclear energy sources in the
      sun and stars, stellar structure and evolution, and re-
      lated topics. The completed work in 1965 extends date to
      1964 with about 1400 additional items. Includes author
      index.

33.   Kuchowicz, B. *The Bibliography of the Neutrino.* New York:
      Gordon and Breach, 1967. Pp. 440.

Lists 2003 publications, arranged chronologically, for
the period 1929-1965. Includes subject and author indices,
with listings of major conference and institutional reports.

34. Kulikovsky, Piotr G. *Istoriko-astronomicheskie issledovaniya*.
    Moscow: Akademiya Nauk. Issued annually (not for all
    years).

    Annual selective bibliography of books and articles in
    the history of astronomy covering primarily Russian litera-
    ture but including foreign citations. Published since 1955,
    the major papers have been identified in two issues of the
    British Journal for the History of Science, 1 (1963), 391;
    2 (1964), 84-89 covering the period through the 1962 issue
    of Volume 8. See also item 42.

35. Kulikovsky, Piotr G. "Sources for the History of Astronomy
    in the Scientific Archives of the U.S.S.R." *New Aspects
    in the History and Philosophy of Astronomy* (item 123),
    245-252.

    Surveys work completed or in progress in the Soviet
    Union. 19 citations.

35a. Kulikovsky, Piotr G. "Bibliography of Dissertations on
    the History of Astronomy in the USSR after the Second
    World War." *Journal for the History of Astronomy*, 11
    (1980), 221-224.

    Provides an annotated list of 19 dissertations completed
    as of 1974.

36. Lalande, J.J. *Bibliographie astronomique: Avec l'histoire
    de l'astronomie, 1781-1802*. Paris: De l'Imprimerie de
    la République, 1803. Pp. viii + 915.

    The bibliography, comprising some 660 pages and including
    all periods since antiquity, is arranged in chronological
    order and includes many biographical references to con-
    temporary literature. Many later items are heavily annotated.
    The history, extending from pages 661 to 880, is a strictly
    chronological narrative of major events outlined without
    direct citation. Includes author index to bibliography.

37. Lavrova, N.B. "An Outline of the History of Astronomical
    Bibliography." *Istoriko-Astronomicheskie Issledovaniya*,
    5 (1959), 83-196.

    Reviews (in Russian) 246 bibliographies published between
    1545 and 1958. Extended in item 38.

38. Lavrova, N.B. *Bibliografia Astronomicheskikh Bibliografii*.
     Moscow: Astronomicheskii Sovet Akademii Nauk SSR, 1962.
     Pp. vi + 110.

     Review of 248 bibliographical sources including bibliog-
     raphies, library catalogues, and scholarly works with ex-
     tensive bibliographical sections from mid-18th century to
     the present. A large fraction of the bibliographies in-
     cludes lists and catalogues of stellar objects and astro-
     nomical phenomena. Preface and table of contents in French,
     annotations in Russian.

39. Lavrova, N.B., ed. *Bibliography of Works on Astronomy
     Performed in the U.S.S.R. from 1917-1957*. Translated by
     Foreign Technology Division, Wright Patterson AFB.FTD-
     63-1149/1+2+4, 1963. Pp. 672.

     Provides transliteration of bibliographic entries without
     transliteration of the original Russian title. Contains
     more than 9500 titles of books and articles by over 1800
     authors.

40. Lavrova, N.B. *Bibliography of Russian Astronomical Litera-
     ture, 1800-1900*. Moscow: Moscow University, 1968. Pp.
     385.

     Cites several thousand books and articles by Russians
     or about Russian astronomy from journals and periodicals
     worldwide. Organized by subject category. No detailed
     annotation.

*   Lecat, Maurice. *Bibliographie de la Relativité*.

     Cited herein as item 788.

*   Ley, Willy. *Rockets, Missiles, and Space Travel.*

     Cited herein as item 505.

41. [Lick Observatory Archives]. *Preliminary Finding Aid to
     the Archives of the Lick Observatory*. New York: American
     Institute of Physics, 1980. Pp. viii + 56.

     Based upon the card catalogue maintained by Lick Ob-
     servatory Archives staff in the McHenry Library, Univer-
     sity of California at Santa Cruz. Includes outgoing letters
     of the Directors, incoming and outgoing correspondence
     from the 1870s through the 1950s, a photographic collec-
     tion, and collections of staff as well as the early records
     of the Astronomical Society of the Pacific. Finding aid
     notes correspondents, date ranges and approximate volume.
     See also item 56.

W.W. Campbell and the old Mills Spectrograph set up for visual observations at Lick Observatory in the early 1890s (see Item 41). *(Photograph
courtesy of the Lick Observatory Archives, Santa Cruz, California.)*

41a. Looney, John L. *Bibliography of Space Books and Articles from Non-Aerospace Journals, 1957-1977*. Washington, D.C.: NASA History Office, 1979. Pp. xv + 243.

Provides over 300 citations arranged by topic and alphabetically by author. Based upon 18 journals, indexes, abstracts and other bibliographies. Includes a section on space science noting numerous sources on space astronomy.

* MacPike, Eugene F. "Doctor Edmond Halley."

Cited herein as item 1337.

42. Meadows, A.J. "Studies on the History of Astronomy." *Journal for the History of Astronomy*, 1 (1970), 160-163.

Abstracts of contents of volumes 9 (1966) and 10 (1969) of bibliographies prepared in Moscow under auspices of Commission 41 of the IAU. See also item 53.

43. Morgan, Julie. "The Huggins Archives at Wellesley College." *Journal for the History of Astronomy*, 11 (1980), 147.

Brief description of six observing notebooks of Huggins and his wife dating from 1856 through 1901.

44. Mourot, Suzanne, and D.J. Cross. "Astronomy Archives in Australia." *Journal for the History of Astronomy*, 4 (1973), 66-68; 211-213.

Description of the holdings of the Mitchell Library, Sydney, and of the Archives Offices of New South Wales from the late 18th century through the 20th century. Second article reviews holdings at Riverview Observatory, founded in 1909 by Pigot, and records of Melbourne Observatory held in the Mount Stromlo Library.

45. Munsterberg, Margaret. "The Bowditch Collection in the Boston Public Library." *Isis*, 34 (1942-43), 140-142.

Brief description of Bowditch's library deposited by his sons which originally included 2500 works on astronomy and mathematics, hundreds of pamphlets and maps, but which has now increased to over 10,000 volumes through a trust fund and donations. Includes mention of manuscript sources.

45a. Nicolson, Marjorie. "Resource Letter SL-1 on Science and Literature." *American Journal of Physics*, 33 (1965), 175-183.

Includes numerous citations to studies of the plurality of worlds.

46. Ordway, F.I. *Annotated Bibliography of Space Science and Technology, with an Astronomical Supplement.* Washington, D.C.: Arfor Publ., 1962. Pp. x + 77.

    Lists 483 titles of books from 1931 through 1961 chronologically by decades in 1930s and 40s, and by years thereafter. Includes annotation, author and title indices.

47. Poggendorff, J.C. *Biographisch-literarisches Handwörterbuch zur Geschichte der exacten Wissenschaften.* Berlin: Akademie Verlag and various other publishers, 1858-present.

    Continuing standard series published by various firms providing biographical information on physical scientists and abbreviated short lists of their major publications. Volumes 1-6 (1858-1940) reprinted by Edmonds Brothers, Ann Arbor, Michigan, 1944.

48. [Popular Astronomy]. *A General Index.* Northfield, Minn.: Popular Astronomy, 1909. Pp. 148.

    Includes complete run to 1909 for *Popular Astronomy*, *Astronomy and Astrophysics*, and the *Sidereal Messenger*. Indexed by author and by subject from 1882. Includes numerous annual reports of observatories, 66 obituary notices for astronomers, detailed cross-indices to reports on astronomical phenomena.

48a. Roland, Alex. *A Guide to Research in NASA History.* Washington, D.C.: History Office, NASA Headquarters, 1980 (5th Edition). NASA HHR-50. Pp. 65.

    Informative research guide and finding aid to resources within NASA for the study of all aspects of space history. Resources center upon NASA history, but many pre-NASA records are identified as preserved in the NASA History Office. The *Guide* also identifies the records of other NASA Centers and specialized holdings in libraries in the Washington, D.C., area. NASA-supported historical programs are listed and described as are the contractural programs the NASA History Office maintains and fosters with free-lance historians.

49. Roller, Duane H.D., and Marcia M. Goodman. *The Catalogue of the History of Science Collections of the University of Oklahoma Libraries.* 2 volumes. London: Mansell, 1976. Pp. xii + 584; 608.

    Reproduction of card catalogue listing some 30,000 printed works and 10,000 items on microfilm. Reviewed in: *JHA*, 8 (1977), 215.

50. [Royal Astronomical Society]. *General Index*. 5 volumes.
     London: Royal Astronomical Society, 1870; 1896; 1911;
     1934; 1953. Pp. 212; 207; 198; 168; 70.

     Cumulative indices published by the Society for its
     journal, the *Monthly Notices*, covering the years 1828 to
     1950. Includes author, subject, celestial object listings.
     First two volumes especially useful for listings of asteroids
     comets and general celestial events. All volumes include
     biographical notices identifying over 1400 names, observa-
     tory activities, and "Reports of the Royal Astronomical
     Society Council," which identify major astronomical achieve-
     ments in each year. Volume 3 includes a general index to
     illustrations appearing in the *Monthly Notices* volumes
     1-70 and in the *Memoirs*, Volumes 1-59.

51. [Royal Astronomical Society]. *Catalogue of the Library of
     the Royal Astronomical Society to June 1884*. London:
     Royal Astronomical Society, 1886. Pp. 408.

     Alphabetical listing by author or institution of some
     7000 titles of offprints, books and journals. Cross-
     references included. Supplements include: 1894-1898 (1900)
     and 1898-1925 (1926), adding another 340 pages of material.

52. [Royal Society (London)]. *International Catalogue of
     Scientific Literature*. 14 volumes. London: Harrison and
     Sons, 1901-1918.

     Each volume of between 175 and 304 pages each lists
     between 1300 and 2400 citations to work published annually,
     but irregularly, during each year from 1900 through 1915.
     Part E contains astronomical literature.

53. [Russian Academy]. *Bibliography of Books and Papers
     Published on the History of Astronomy*. Moscow: Astronom-
     ical Council of the U.S.S.R. Academy of Sciences, 1961- .

     Selected bibliographical listings prepared since 1961
     under the auspices of Commission 41 (History of Astronomy)
     of the International Astronomical Union. Averages 300-400
     citations per year without detailed annotation. See also
     item 42.

54. Russo, François. *Elements de Bibliographie de l'Histoire
     des Sciences et des Techniques*. Paris: Hermann, 1969.
     Pp. xv + 214.

     Within this general, well-documented bibliography are
     topical sections listing secondary works on astronomy and

astronomers. Update of 1954 *Bibliographie Histoire des
Sciences et des Techniques*. Reviewed in: *Isis*, 45 (1954),
204-205; *Centaurus*, 15 (1970), 190-191.

55.   Seal, Robert A. *A Guide to the Literature of Astronomy*.
      Colorado: Libraries Unlimited, 1977. Pp. 306.

      Heavily annotated introduction to contemporary literature
      citing 578 monographs, dictionaries, journals, abstracts
      and technical reports. Includes a brief section on secondary
      historical literature.

56.   Shane, Mary Lea. "The Archives of Lick Observatory."
      Journal for the History of Astronomy, 2 (1971), 51-53.

      Description of the Archives and short sketch of the origin
      of this major collection within the historical context of
      the Lick Observatory. See also item 41.

57.   [Smithsonian Institution]. *Preliminary Guide to the Smith-
      sonian Archives*. Washington, D.C.: Smithsonian Institu-
      tion Press, 1971. Pp. 72.

      Brief descriptive entries provide an overview of the
      sources for historical research at the Smithsonian. Covers
      archives and office records. Reviewed in: *Isis*, 65 (1974),
      256.

58.   Struve, Otto, and Eduardo Lindemanno. *Catalogus Bibliotheca
      Speculae Pulcovensis*. 2 volumes. St. Petersberg: A.
      Eggers, 1860; 1880. Pp. xxx + 970; xvii + 640.

      Catalogues of the Library of the Pulkovo Observatory.
      Volume 1 by the elder Otto Struve; Volume 2 extended by
      Lindemann. Volume 1, to 1858, includes some 4113 titles
      of major works, 7625 volumes of serials, 143 charts and
      14,634 minor papers. Volume 2 adds 1737 major titles,
      4814 serials and 9096 minor titles. Provides comparison
      descriptions to items common in Lalande's *Bibliographie*
      (item 36). Organized by major topic areas: mathematics,
      astronomy, geodesy, physics, and subject areas within
      astronomy. Alphabetical within each subject category. No
      detailed annotation.

59.   Sykes, J.B. *Guide to the Bibliography of Astronomy 1881-
      1898*. Ann Arbor: University Microfilms, 1970. Pp. 16.

      The *Bibliography* itself, consisting of some 18 reels of
      35mm microfilm, fills in the gap between Houzeau and
      Lancaster's *Bibliographie Générale* and the *Astronomischer
      Jahresbericht*. See item 7. Reviewed in *JHA*, 2 (1971), 53.

*   Todd, David P. *Stars and Telescopes*.

    Cited herein as item 1408.

60. Tuckerman, Alfred. *Index to the Literature of the Spectro-scope*. 2 volumes. Washington: Smithsonian, 1888; 1902. Pp. x + 423; iii + 373.

    General survey of spectroscopic literature. Volume 1 includes all sources to July 1887, and Volume 2 extends the date to 1900. Includes section on historical articles and a 70-page section on astronomical spectroscopy in Volume 1. Some annotations. Volume 1 includes 3829 citations to 799 authors; Volume 2 includes similar but unspecified number of citations.

61. [United Nations]. *International Space Bibliography*. New York: United Nations, 1966. Pp. iv + 166.

    Survey of all levels of literature dealing with space flight and space science classified by country, secondary sources and bibliographies, periodicals, texts, and monographs. Approximately 3000 citations.

62. Wallis, Peter, and Ruth Wallis. *Newton and Newtoniana, 1672-1975. A Bibliography*. London: William Dawson and Sons, 1977. Pp. xxiv + 362.

    Contains approximately 4000 citations of translations, reviews, collections, articles and books about Newton and Newtonian science. Includes library locations and detailed bibliographical annotations. Update and expansion of item 16. Reviewed in: *Isis*, 70 (1979), 623.

63. Warner, Brian. "Astronomical Archives in Southern Africa." *Journal for the History of Astronomy*, 8 (1977), 217-222.

    After a brief sketch of the history of astronomy in Southern Africa, major archival resources are identified residing in Johannesburg (Lacaille, J. Herschel, Maclean, Airy, Smyth); Bloemfontein (A.W. Roberts; E.C. Pickering); Pretoria (Union Observatory, Gill, Innes, many major 20th-century names); and Capetown (Maclean, Mann, the Herschels, Cape Observatory, Henderson, etc.).

64. Warner, Brian. "Cape of Good Hope Royal Observatory Papers in the Archives of the Royal Greenwich Observatory." *Journal for the History of Astronomy*, 9 (1978), 74-75.

    Description of major collection that includes ledgers, observations and correspondence of Fearon Fallows, G.B.

Airy, Thomas Henderson, Thomas Maclear, E.J. Stone and David Gill.

(65.) Warnow, Joan Nelson. *A Selection of Manuscript Collections at American Repositories*. New York: American Institute of Physics, 1969. Pp. vi + 73.

Includes information on manuscript collections of astronomers, astronomical institutions and societies noting location, date range, size, state of organization. Oral history interviews and pictorial collections are also listed. Reviewed in: *Isis*, 61 (1970), 535.

* Wasson, John T. *Meteorites*.

Cited herein as item 991.

66. Wattenberg, Diedrich. *Forschungen und Publikationen zur Geschichte der Astronomie in der Deutschen Demokratischen Republik. Eine Bibliographie (1949-1969)*. Berlin-Treptow: Archenhold-Sternwarte, 1969. Pp. 138.

Bibliography of articles and books on the history of astronomy written by authors within the GDR, or on topics pertaining to astronomy within the GDR since its formation. First in a continuing series. Part 2, covering the years 1970-1974, was published in 1975 in 71 pages. Part 3 was published in 1980 as item 16a. Reviewed in *JHA*, 1 (1970), 159.

67. Weiner, C., and Joan N. Warnow. "Source Materials for the Recent History of Astronomy and Astrophysics: A Checklist of Manuscript Collections in the United States." *Journal for the History of Astronomy*, 2 (1971), 210-218.

Short list of about 130 collections including collection title, type of materials, date ranges and approximate size.

68. Whitrow, Magda, ed. *Isis Cumulative Bibliography. A Bibliography of the History of Science Formed from Isis Critical Bibliographies 1-90, 1913-1965*. 3 volumes. London: Mansell, 1971; 1976. Pp. lxx + 664; vi +789; xciv + 678.

Comprehensive and nearly exhaustive partially annotated bibliography compiled from the yearly *Isis Critical Bibliographies* constituting a primary reference to the field. Volume 1 is organized alphabetically by name of personality with some 40,000 entries for over 10,000 names. Volume 2

contains 3800 entries for references to institutions including universities, hospitals, libraries, societies,
etc. Volume 3 contains some 21,500 entries organized by
subject disciplines and objects of study within the history
of science, medicine and technology for all periods. A
fourth volume for civilizations and periods is in preparation, and extension volumes covering the period 1966-1975
are appearing. Volumes 1 and 2 are reviewed in: *Isis*, 66
(1975), 262-263; Volume 3 is reviewed in: *Annals of Science*,
34 (1977), 432-434.

69.   Wislicenus, Walter F. *Astronomischer Jahresbericht*. Volumes
      1-20, Berlin: Georg Reinner, 1900-1920; Volumes 21-68,
      Berlin: W. de Gruyter, 1921-1969.

      Comprehensive abstract and bibliography published annually
      under the auspices of the Astronomisches Rechen-Institut
      of Berlin. Founded by Wislicenus and edited in turn by
      a number of staff members, this exhaustive work includes
      biographical and historical sections. Replaced by *Astronomy
      and Astrophysics Abstracts* in 1969, which, while retaining
      the same basic classification, is now annotated in English.

70.   Zinner, Ernst. *Verzeichnis der astronomischen Handschriften
      des deutschen Kulturgebietes*. München: Beck, 1925. Pp.
      544.

      A catalogue of astronomical manuscripts found primarily
      in European libraries. Includes sources to the mid-19th
      century. Lists over 12,000 items. Reviewed in: *Isis*, 15
      (1931), 193-195.

71.   Zinner, Ernst. *Geschichte und Bibliographie der astronomische
      Literatur in Deutschland zur Zeit der Renaissance*. Leipzig:
      Hiersemann, 1941. Pp. viii + 452.

      Bibliography of astronomical literature published in
      Germany from 1448 to 1630. Over 5000 citations arranged
      by year of publication and alphabetical within each year.
      Includes a separate listing of especially significant
      works, a detailed index and short biographies and sketches
      of the astronomy of the period. Detailed review of this
      work by E. Rosen points to the effect of Zinner's nation-
      alism on the body of the bibliography: *Isis*, 36 (1946),
      261-266.

GENERAL HISTORIES

MONOGRAPHS

72. Abetti, Giorgio. *The History of Astronomy*. New York: Henry Schuman, 1952. Pp. x + 338.

    General descriptive introduction covering all subjects and all periods presented chronologically in three broad areas: "From the Beginning to Copernicus"; "Reformation of Astronomy"; "The Modern Era." Includes an appendix on "development and aims of the world's astronomical observatories" and a concluding chapter on the IAU and the future of astronomy. Emphasizes the modern periods and stresses the importance of the advance of observational astronomy. No direct citations. Reviewed in: *Isis*, 44 (1953), 298-300.

73. Airy, George Biddell. *Abriss einer Geschichte der Astronomie (1800-1832)*. Wien: Gerold, 1835. Pp. 117.

    Brief review of astronomical progress, translated into German by C.L. Littrow. Emphasizes physical astronomy. Existence of English edition not determined.

74. Arago, François. *Popular Astronomy*. 2 volumes. W.H. Smyth and Robert Grant, translators and editors. London: Brown, et al., 1855. Pp. xlviii + 707; xxxii + 846.

    Translation of expanded text of Arago's lectures at the Observatory of Paris ending in 1846. Comprehensive descriptive review of astronomy arranged by subject but including extensive and detailed historical summaries. Topics of historical interest in Volume 1 include: Horology; the history of telescopic equipment; the nature of nebulae; theories of the Milky Way; the discoveries of proper motion and solar motion; the nature of comets. In Volume 2, historically treated areas include: the Earth's figure, structure and motions; cosmical nature of meteors; the

discovery of aberration and the determination of the
velocity of light; the discovery of Neptune; the calendar;
growth of observatories. Numerous editorial annotations
and partial citations.

75. Armitage, Angus. *A Century of Astronomy*. London: Sampson
     Low, 1950. Pp. xi + 255.

     Provides a general non-technical overview of the develop-
     ment of all branches of astrophysics emphasizing astro-
     nomical spectroscopy, stellar evolution from Lockyer to
     Russell, and modern cosmology.

76. Bailly, Jean-Sylvain. *Histoire de l'astronomie moderne
     depuis la fondation de l'école d'Alexandrie, jusqu'à
     l'époque le MDCCXXX*. 3 volumes. Paris: Bure, 1779–1782.
     Pp. xvi + 727; 751; 415.

     General descriptive history from antiquity. Volume 2
     begins with Kepler and emphasizes the development of New-
     tonian celestial mechanics to 1730 highlighting cometary
     and planetary theory and the development of major observa-
     tories at Greenwich and Paris. Volume 3 discusses contempo-
     rary mid-18th-century astronomy. Includes general index
     to all volumes.

77. Bailly, Jean-Sylvain. *Histoire de l'astronomie ancienne
     et moderne*. 2 volumes. Paris: Bernard, 1805. Pp. xviii +
     371; 490.

     General history of astronomy. Volume 1 covers pre-history
     through Greece and the Far East. Volume 2 reviews astronomy
     from the time of Copernicus through Newton. Five additional
     lectures are provided to review 18th-century progress,
     general speculations, and a general summary.

78. Berry, Arthur. *Short History of Astronomy*. London: Murray,
     1898. Reprinted, New York: Dover, 1961. Pp. xxxi + 440.

     Popular introduction covering all periods to the 19th
     century. Short commentary on text and life of author in
     Isis, 28 (1938), 418-420.

79. Bigourdan, G. *L'astronomie, évolution des idées et des
     méthodes*. Paris: E. Flammarion, 1911. Pp. vii + 399.

     General history covering all periods. Major sections on
     the growth of astronomical instrumentation and technique;
     the figure of the Earth and its motions; and positional
     astronomy. Some indirect citations.

80. Boquet, F. *Histoire de l'astronomie*. Paris: Payot, 1925.
    Pp. 507.

    General history based primarily upon secondary works in
    French. Approximately 170 pages devoted to 18th century
    and later including detailed review of 18th century celestial
    mechanics. No index.

81. Bryant, Walter W. *A History of Astronomy*. New York:
    Dutton, 1907. Pp. xiv + 355.

    An elementary descriptive history covering all periods.
    The first 11 chapters are chronological from antiquity to
    Herschel, Bessel and Struve. The latter 22 chapters are
    topical dealing with all general aspects of astronomy and
    astrophysics. No direct citations, but notes dependence
    upon the histories written by Agnes Clerke. Reviewed in:
    Popular Astronomy, 16 (1908), 80-82.

82. Clerke, Agnes M. *A Popular History of Astronomy during the
    19th Century* [1885]. 2nd Edition. Edinburgh: Adam and
    Charles Black, 1887. Pp. xvi + 502.

    Standard work (First Edition 1885) covering most aspects
    of astronomy and early astrophysics. Major omission:
    progress in celestial mechanics during period. Stated period
    of review for discussion often preceded by discussion of
    events of earlier periods. In much of the discussion of
    spectroscopic astronomy the author defers to the opinions
    of William Huggins. Includes a chronological table, 1774-
    1887, and extensive citations. 1902 edition reviewed in:
    Popular Astronomy, 10 (1902), 465.

83. Clerke, Agnes M., and Simon Newcomb. "Astronomy." *The
    Encyclopaedia Britannica, Eleventh Edition, Volume II*.
    Cambridge: Cambridge University Press, 1910. Pp. 800-818.

    Provides a concise review of all aspects of contemporary
    astronomy and astrophysics by Newcomb with an excellent
    sketch of history by Agnes Clerke with references and short
    bibliography. Both the Ninth and Eleventh Editions should
    be sought out for reviews of specific subfields.

84. Costard, George. *The History of Astronomy, with its applica-
    tion to geography, history, and chronology: occasionally
    exemplified by the globes*. London: James Lister, 1767.
    Pp. xvi + 308.

    General history including an account of the 1761 transit
    of Venus.

85.   Delambre, Jean. *Histoire de l'astronomie*. 6 volumes.
      Paris: 1817; 1819; 1821; 1827.

      Volumes in chronological order; Volume 3, "... du moyen
      age," and Volumes 4 and 5, "Moderne," have been reprinted
      by Johnson Reprint Corporation of New York but Volume 6,
      "au dix-huitième siècle," has not. Volumes include anno-
      tated abstracts of works of principal astronomers during
      each period. Works are taken in order without detailed
      intercomparison. Introductory essays provide general views
      of progress of astronomy during each period. The last
      volume, on the 18th century, treats both observational
      astronomy and celestial mechanics. Contrasts strongly with
      Robert Grant's history (see item 91).

*   DeSitter, W. *Kosmos*.

      Cited herein as item 1167.

86.   Dijksterhuis, E.J. *The Mechanization of the World Picture*
      [1950]. Oxford: Oxford Univ. Press, 1961; 1969. Pp. vi +
      537.

      Originally published as *De Mechanisering van het Wereld-
      beeld* in Amsterdam in 1950 and translated by C. Dikshoorn.
      Demonstrates that the tendency of science to mathematize
      reality has been a long-standing drive away from ancient
      but persistent animistic views of the world. Covers history
      from Greece to Newton, using citations from original sources.
      Reviewed in: *Centaurus*, 2 (1951), 86–87.

87.   Doig, Peter. *A Concise History of Astronomy*. London: Chap-
      man & Hall, 1950. Pp. vi + 320.

      Over 70 percent of the text deals with the period 1700–
      1950. The 19th- and 20th-century periods are organized
      by subject. Provides a quick introduction to generally
      received highlights in the history of astronomy. Short
      chapter bibliographies. Review by J. Streeter includes
      survey of general texts and reference material in the
      history of astronomy: *Isis*, 42 (1951), 73–75.

88.   Doublet, Édouard L. *Histoire de l'astronomie*. Paris:
      Gaston Doin, 1922. Pp. 572.

      A general history of astronomy concentrating upon post-
      Renaissance through the 19th century. Includes sections
      on Chinese, Hindu, Mexican and Peruvian astronomy as well
      as classical Greek astronomy. Provides background on
      astronomers, their observatories, and the principal histo-

rians of science who wrote on the history of astronomy
(Jean-Frederic Weidler through Pierre Duhem). Doublet's
work was particularly influenced by Duhem. Reviewed in:
Nature, 110 (1922), 600; *Isis*, 5 (1923), 172.

89. Dreyer, J.L.E. *A History of Astronomy from Thales to
     Kepler* [1906]. 2nd Edition. New York: Dover Publ., 1953.
     Pp. x + 438.

     First published in 1906, this revision with introduction
     by W.F. Stahl remains a standard work reviewing in detail
     the growth of planetary theory. Reviewed in: *Isis*, 44
     (1953), 396-397.

90. Forbes, George. *History of Astronomy*. London: G.P. Putnam,
     1909; 1921. Pp. ix + 150.

     Brief general exposition intended for popular audiences.
     Covers all periods.

91. Grant, Robert. *History of Physical Astronomy, from the
     Earliest Ages to the Middle of the Nineteenth Century*.
     London: H.G. Bohn, 1852. Reprinted by Johnson Reprint
     Corporation, 1966. Pp. xx + xiv + 637.

     Concentrates on post-Newtonian celestial mechanics and
     detailed history of early telescopic astronomy. Though
     largely descriptive throughout, textual discussions of
     progress of celestial mechanics are highly detailed and
     authoritative. Considered the classic general work in
     English for this period. Extensive citations to both
     primary and secondary works.

92. Hall, A. Rupert. *The Scientific Revolution 1500-1800*. The
     Formation of the Modern Scientific Attitude. London:
     Longmans, Green & Co., 1954. Pp. xviii + 390.

     Includes excellent narrative on the transition from
     Aristotelian to Newtonian cosmology. Reviewed in: *Annals
     of Science*, 11 (1955), 99-101.

93. Hall, A. Rupert. *From Galileo to Newton: 1630-1720*. London:
     Collins, 1963. Pp. 380.

     General historical narrative including some discussion
     of descriptive and observational astronomy during the
     period.

94. Herrmann, D.B. *Geschichte der Astronomie von Herschel bis
     Hertzsprung*. Berlin: VEB Deutscher Verlag der Wissen-
     schaften (GDR), 1975. Pp. 282.

Surveys the rise of modern astronomy and astrophysics to
about 1930 including early work on the structure of the
galaxy by William Herschel; the search for interstellar
absorption from Herschel's examination of a finite evolving
Universe, through Olbers in 1823 and to Trumpler in 1930;
the importance of the discovery of asteroids; the rise of
spectroscopic astronomy; the professional organization of
astronomy and how it changed during the period. Other
subjects include a short history of star catalogues; the
early hostilities to astrophysics; the rise of technological
improvements and their direct correlation with astronomical
advances; the importance of professional journals. Extensive
documentation and bibliography. Reviewed in: *JHA*, 10
(1979), 67-69.

95.   Herschel, John F.W. *A Treatise on Astronomy*. London:
      Longman, Rees, et al., 1833; 1837; through 1851. Pp.
      viii + 422.

Famous general review and text forming part 43 of
Dionysius Lardner's *Cabinet Cyclopaedia of Natural Philosophy*. The various editions (1837-1851) remained identical
without revision for new discoveries. An expanded revision
appeared in 1849 as *Outlines of Astronomy* (item 96).

96.   Herschel, John F.W. *Outlines of Astronomy*. London: Long-
      man, Brown, et al., 1849; 5th Edition, 1858. Pp. xxiv +
      714.

Considered one of the major general review works of the
mid-19th century. First published in 1849 as an expansion
of his popular *Treatise* (item 95), this text includes far
more useful commentary and documentary material. Of note
was the habit of appending additional and corrective
commentary in subsequent editions to the original text to
account for progress in the field. Thus each edition retains
the original flavor but includes commentary on progress
during the period 1849-1869. Major editions were the 5th
(1858); 9th (1867) and 10th (1869). Notes are clearly
identified and dated.

97.   Hoefer, Ferdinand. *Histoire de l'astronomie*. Paris:
      Hachette, 1873. Pp. 631.

General history covering the origins of astronomy; its
progress in ancient times through the middle ages in all
countries; a general review of the Greek, Roman, and Arabic
periods; and finally a topical presentation of modern
astronomy. Provides only cursory notice of the application
of spectroscopy to astronomy. Numerous citations.

98.   Jahn, Gustav Adolph. *Geschichte der Astronomie*. Leipzig:
      H. Hunger, 1842; 1844. Pp. x + 308; 292.

      General historical review of major aspects of astronomy
      during the first half of the 19th century. The first volume
      covers planetary astronomy, emphasizing Gauss' studies
      of the newly discovered minor planets, but also includes
      descriptive chapters on the sun, planets, satellites, and
      comets. The second volume examines progress in stellar
      astronomy, noting catalogues, double stars, and the search
      for stellar parallaxes. It also includes a chapter on
      astronomical geodesy. Numerous citations to German litera-
      ture. Two volumes bound as one.

99.   King, Henry C. *The History of the Telescope*. High Wycombe,
      Bucks: Charles Griffin, 1955. New York: Dover, 1979.
      Pp. xvi + 456.

      Definitive study of telescopes and telescopic equipment
      including all forms of techniques and technology for
      astronomical observation from the early Greeks to the space
      age. Includes nineteen chapters detailing lenses and optics
      in the Arab world; Galileo's application of the telescope
      to astronomy; early failures to correct for aberrations of
      lenses; aerial telescopes; reflecting telescopes in the
      age of Newton and their many variations; navigational
      sighting devices; Herschel's telescopes; invention of
      achromatic lenses and improved equatorial mountings; im-
      provement of lens glass and problem of production of large
      glass blanks; Lord Rosse's reflectors; growth of instru-
      mentation at the Paris and Greenwich Observatories; the
      first American observatories; production of silver-on-
      glass mirrors; history of spectrum analysis; large refrac-
      tors; solar instrumentation; major reflectors of the 19th
      and 20th centuries; and amateur telescope making. Approxi-
      mately 1700 citations. Reviewed in: *Annals of Science*, 11
      (1955), 268-269.

100. King, Henry C. *The Background of Astronomy*. New York:
      George Braziller, 1958. Pp. 254.

      Provides a descriptive introduction to the history of
      astronomy from antiquity until the end of the 16th century.

101. Lalande, J.J. *Astronomie*. 3 volumes. Paris: Desaint, 1771;
      1781; 1792. Pp. lvi + 608 + 248 (tables); 830; 850.
      The third edition is reprinted, New York: Johnson Reprint
      Corporation, 1966.

Comprehensive review of 18th-century astronomy, heavily
illustrated, including an exposition of computational and
observing techniques. Emphasizes growth of planetary
theories and cosmology.

*   Lalande, J.J. *Bibliographie astronomique; Avec l'histoire
         de l'astronomie, 1781-1802.*

         Cited herein as item 36.

102.   Laplace, P.S. *Precis de l'histoire de l'astronomie.*
         Paris: Courcier, 1821. Pp. 160.

         Dedicated to Humboldt, forms book five of the fifth
         edition of his *Exposition du système du monde*. Constitutes
         a brief history of physical astronomy and cosmology from
         ancient times through the 18th century, and provides a
         short discussion on hopes for future progress.

103.   Lebon, Ernest. *Histoire Abrégée de l'astronomie.* Paris:
         Gautier-Villars, 1899. Pp. vii + 274.

         Primarily a history of 18th- and 19th-century astronomy
         arranged by topic and by author within each topical
         section. Provides a clear review of the contributions of
         French astronomy from the Cassinis through Poincaré, in-
         cluding many biographical sketches, portraits and a
         biographical dictionary.

104.   Loomis, Elias. *The Recent Progress of Astronomy; Especially
         in the United States*. New York: Harper & Bros., 1850.
         Pp. viii + 257.

         Reviews recent discoveries and events including the
         discovery of Neptune, wherein Loomis examines the reli-
         ability of the predictions; the discoveries of asteroids
         and planetary satellites; recent comets; the measurement
         of stellar parallaxes; observations of variable stars and
         novae; the distribution of the stars; and the structure
         of the visible universe and the resolution of nebulae
         into stars by Lord Rosse and by the Bonds at Cambridge.
         In a second part of the book a general review of progress
         of astronomy in the United States is provided through
         a compilation of the various observatories and their
         activities. Later sections review astronomical expeditions,
         the determination of longitude by telegraphic signals,
         and the manufacture of telescopes in the United States.
         No index or direct citations.

105. Macpherson, Hector. *Modern Astronomy. Its Rise and Progress*. Oxford: Oxford University Press, 1926. Pp. 196.

Brief popular history based upon a set of public lectures in Aberdeen during 1925-26. Includes discussions of recent progress in planetary astronomy, stellar astronomy, cosmology and cosmogony. Brief bibliography and index.

106. Mädler, Johann Heinrich. *Geschichte der Himmelskunde*. Braunschweig: G. Westermann, 1873. 2 volumes. Pp. x + 528; 590.

A comprehensive chronological review from ancient times to the present. Includes biographical footnotes and annotation with chronologies of major events within each era, identified by intellectual period. Volume 2 begins with William Herschel and ends with all aspects of mid-19th-century astronomy.

107. Mineur, Henri. *Histoire de l'astronomie stellaire jusqu'à l'époque contemporaine*. Paris: Hermann, 1934. Pp. 57.

Brief contemporary review of stellar astronomy.

108. Moore, Patrick. *The Development of Astronomical Thought*. Edinburgh: Oliver and Boyd, 1969. Pp. vii + 119.

Brief popular review of the history of astronomy with most technical concepts explained in elementary fashion. No citations.

109. Narrien, John. *Historical Account of the Origin and Progress of Astronomy*. London: Baldwin and Cradock, 1833. Pp. xiv + 520.

General introduction to the history of astronomy emphasizing antiquity but bringing narration through to the 18th century. Useful commentary on the development of celestial mechanics. Some citations.

* Newcomb, Simon. *Popular Astronomy*.

Cited herein as item 1402.

* Newcomb, Simon. *Side-Lights on Astronomy and Kindred Fields of Popular Science*.

Cited herein as item 1403.

110. Pannekoek, A. *A History of Astronomy*. London: George Allen and Unwin, 1961. Pp. 521.

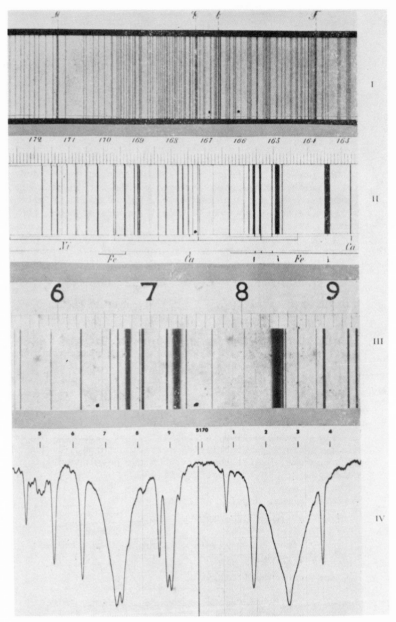

Comparative spectra of the Sun from the 1820s through 1940 showing the improvement in technique. I. Fraunhofer; II. Kirchhoff;
III. Rowland; IV. Utrecht. The wave lengths increase to the left on
I, and to the right on II, III, and IV. *(From A. Pannekoek,* A History
of Astronomy *(Item 110), p. 384.)*

General descriptive history in three parts and 43 chapters originally published as *De Groei van ons Wereldbeeld*
in 1951. Part 1 (pp. 1-172) covers ancient astronomy; Part
2 (pp. 173-310) covers late medieval astronomy through
Newton and the elaboration of Newtonian celestial mechanics
in the 18th century; Part 3 (pp. 311-496) surveys modern
stellar and galactic astronomy and astrophysics. Illustrations of star charts and photographic maps, stellar spectra,
solar spectra, and the planetary surfaces are organized
in chronological sequence demonstrating the growth of
technique and resolution. Includes 214 citations to both
original works and secondary literature. Reviewed in:
Isis, 54 (1963), 494-495.

111.   Ronan, Colin A. *Discovering the Universe. A History of
        Astronomy.* New York: Basic Books, 1971. Pp. vii + 248.

Brief, popular introduction to the history of astronomy
emphasizing 20th-century astronomy. Arranged by topic.
No citations. Reviewed in: *Isis*, 63 (1972), 110.

112.   Sageret, Jules. *Le système du monde de Pythagore à Edding-
        ton.* Paris: Payot, 1931. Pp. 346.

Treats, in five parts, the general aspects of astronomy
since Greek times. The first two parts treat classical
physical astronomy. Parts 3 and 4 examine stellar astronomy
and astrophysics, the Sun and spectroscopy. Only in the
fifth part is there explicit discussion of cosmology, from
the 18th century through Eddington. Numerous citations
and short bibliography.

113.   Struve, Otto, and Velta Zebergs. *Astronomy of the 20th
        Century.* New York: Macmillan, 1962. Pp. 544.

General comprehensive review of all aspects of research
progress from 1900 until the 1950s emphasizing the growth
of observational astrophysics. Chapters include photography of the Milky Way; radial velocities; techniques
of photometry--visual, photographic, and photoelectric;
radio astronomy; the Sun; physics of the Solar System
(planetary atmospheres, the work of Percival Lowell,
comets, and the Moon); the origin of the Solar System
including tidal and collisional theories as well as recent
theories of Kuiper, McCrea, von Weizsäcker, and Oort's
theory on the origin of comets; spectral classification
from the late 19th century through the Henry Draper Catalogue and two-dimensional classification; stellar atmospheres and spectroscopy including mass loss and stellar

populations; stellar structure through the revision of
Russell's theory and the identification of sources of
stellar energy; the Hertzsprung-Russell Diagram and
stellar evolution; double stars of all types, their
discovery and general analysis of observations; forms
of variables including pulsating variables, novae and
supernovae; the discovery of the interstellar medium and
its properties; discrete galactic nebulae; theories of
the structure of the Milky Way; galaxies and the expanding universe. Introductory chapters examine the growth
and nature of the international astronomical profession,
particularly in the U.S.S.R., and the state of astronomy
in 1900. A detailed appendix on the growth of techniques
of stellar spectroscopy, Struve's specialty, was originally
published in 1935. 353 citations, glossary of terms,
brief general bibliography and chronology.

114.   Tauber, Gerald E. *Man's View of the Universe*. New York:
       Crown, 1979. Pp. 352.

       Large-format pictorial history covering all periods in
       introductory fashion. Brief bibliography of secondary
       sources.

115.   Waterfield, Reginald L. *A Hundred Years of Astronomy*.
       London: Duckworth, 1938. Pp. 526.

       General descriptive narrative of the progress of modern
       astronomy and astrophysics from the first measurement
       of stellar parallax and the discovery of Neptune through
       the rise of spectroscopic astrophysics, the Hertzsprung-
       Russell Diagram, galactic structure and cosmology. In-
       cludes a short concluding chapter on astronomical societies
       and sections on time signals by telegraphy, a detailed
       chronological table 1834-1937, and a bibliography listing
       general textbooks, monographs, and review articles with
       approximately 200 citations. Reviewed in: *Isis*, 31 (1939),
       109-112.

116.   Whitney, C.A. *The Discovery of Our Galaxy*. New York:
       Knopf, 1971. Pp. xv + 308 + viii.

       Popular and useful introduction to the history of
       cosmology and cosmogony including three parts: Part 1
       on ancient speculation treats Greek thought through
       Galileo, Newton, and the first discoveries of nebulous
       objects. Part 2 examines the Nebular Hypothesis, the work
       of the Herschels, and of Lord Rosse. Part 3 concludes
       with Hubble and work in the 1950s on the identification

of the spiral form of the Milky Way. Chapter 13 includes a variorum translation of the last chapter of Laplace's *Système du Monde* taken from parts of its six editions. Citations identified within the text. Reviewed in *JHA*, 6 (1975), 141-142; *Hist. of Sci.*, 12 (1974), 299-306.

117. Williams, Henry Smith. *The Great Astronomers*. New York: Simon and Schuster, 1930. Pp. xix + 618.

A popularly written general history centering upon the work of major figures. Includes detailed chapters on the history of recent astrophysics. No direct citations.

118. Wolf, Rudolf. *Geschichte der Astronomie*. Munich: R. Oldenbourg, 1877. Reprinted, 1933. Pp. xvi + 815.

General history covering all periods and aspects. Includes, within a section on "the new astronomy," an extensive review of post-Newtonian celestial mechanics; the development of instrumentation and observations; and commentary on the literature of astronomy and the history of astronomy. Numerous citations. Reviewed in: *Mathematical Gazette*, 17 (1933), 340.

119. Wolf, Rudolf. *Handbuch der Astronomie. Ihrer Geschichte und Literatur*. 2 volumes. Zurich: F. Schulthess, 1890; 1892. Pp. xvi + 712; 658.

Comprehensive reference for the 18th and 19th centuries organized by subject and introduced by a concise history. Includes detailed historical notes throughout both volumes. Extensive citations.

120. Zinner, Ernst. *Die Geschichte der Sternkunde von den ersten Anfängen bis zur Gegenwart*. Berlin: Julius Springer, 1931. Pp. xi + 673.

A general history. Strong emphasis on German contributions. Heavily documented and useful as a general source reference. Abstracted as: *Untersuchungen zur Geschichte der Sternkunde*. Bamberg, 1932. Pp. 62. Reviewed in: *Nature*, 128 (1931), 986-87; *Isis*, 16 (1931), 161-167.

*   Zinner, Ernst. *Geschichte und Bibliographie der astronomischen Literatur in Deutschland zur Zeit der Renaissance*.

Cited herein as item 71.

121. Zinner, Ernst. *Astronomie: Geschichte ihrer Probleme*. Munich: Karl Alber, 1951. Pp. xii + 404.

Extensively annotated series of excerpts from famous astronomical works covering all periods to the 20th century. Organized into 3 broad subject categories: The Solar System (306 pp.); The Stars (61 pp.); Cosmogony (23 pp.). Part 1 outlines in detail the growth of planetary theory, observations of the Sun, Moon, planets and comets; Part 2 reviews theories of the Milky Way and the observations of stellar characteristics; Part 3 is limited to a review of the cosmogonies of Kant and Laplace. Text and translations in German. Extensive bibliography and footnote references. Reviewed in: *Centaurus*, 2 (1953), 367-368; *Isis*, 43 (1952), 291-292.

COMPENDIA

122.   [Annual Reviews]. *Annual Reviews of Astronomy and Astro-physics*. Volume 1-. Palo Alto: Annual Reviews, 1963-.

Edited in turn by L. Goldberg, A.J. Deutsch, D. Layzer, J. Phillips and G. Burbidge, this comprehensive review treats all aspects of astronomy with chapters by noted specialists. Some chapters contain recent history. Extensive citations.

123.   Beer, Arthur, ed. *New Aspects in the History and Philosophy of Astronomy*. *Vistas in Astronomy*, 9 (1967), 1-317.

Transcripts of 32 papers presented at the first joint symposium of the International Astronomical Union and the Union Internationale d'Histoire et de Philosophie des Sciences, held at the University of Hamburg, 22 to 24 August 1964.
Contains items 35, 162, 164, 166, 168, 176, 196, 212, 219, 393, 399, 407, 414, 417, 425, 446, 522, 615, 650, 875.

124.   Beer, Arthur, and Peter Beer, eds. *The Origins, Achieve-ment and Influence of the Royal Observatory, Greenwich: 1675-1975*. *Vistas in Astronomy*, 20 (1976), 1-272.

43 papers presented at the Tercentenary Symposium held at Greenwich in July, 1975. Papers range from a review of astronomical progress in the past 300 years, the origins of the Greenwich Observatory, Greenwich Time and the establishment of the Prime Meridian to descriptions of other national observatories and the growth of astronomical

instrumentation. Reviewed in: *Annals of Science*, 35 (1978), 331–332.

125. Berendzen, R., ed. *Education in and History of Modern Astronomy. Annals of the New York Academy of Sciences*, 198 (1972). Pp. 275.

Collection of papers in five categories: (1) Education and Employment of Astronomers in the United States, 1920–1970; (2) International Issues in Astronomy Education; (3) University Level Astronomy Education for nonscience concentrators; (4) Noncollegiate Astronomy Education; (5) Personal Accounting of the Development of Modern Astronomy. Parts 1 and 3 contain some historical studies; parts 2 and 4 do not. Part 5 includes historical reminiscences by astronomers on the history of astronomical spectroscopy, radio astronomy, meteor and comet orbits, galactic structure, and rocket astronomy. Part 5 contains 117 citations. Reviewed in: *JHA*, 4 (1973), 208–209.
Contains items 182, 341, 491, 503, 936, 992, 1017, 1174, 1410.

126. Clerke, Agnes M., Alfred Fowler, and J. Ellard Gore. *The Concise Knowledge Astronomy*. London: Hutchinson, circa 1900. Pp. xvi + 581.

Provides a popular introduction to astronomy in four sections: History (Clerke); Geometrical astronomy and astronomical instruments (Fowler); the Solar System (Clerke); the Sidereal Heavens (Gore). The first section traces the general history of astronomy to Laplace, and then reviews "a century of progress" in astronomy. Later sections examine practical astronomy and contemporary issues, but draw upon late 19th-century studies. No direct citations.

127. [Copernicus, N.] *Avant, avec, après Copernic. La réprésenta-tion de l'univers et ses consequencès epistemologiques*. Paris: Blanchard, 1975. Pp. 440.

Proceedings of a 1973 symposium at Paris including generalist lectures by astronomers and historians of science on pre-Copernican astronomy, the life of Copernicus, and the influence of his work on science in the 16th to 18th centuries and on modern times. Reviewed in: *Centaurus* 23 (1980), 181.

128. Delauney, C.E. *Rapport sur les progrès de l'astronomie*. Paris: Imprimerie impériale, 1867. Pp. 38.

General review of 25 years of progress in astronomy, notably by French theoreticians and observers, arranged

by topic. Reviews advances in the theories of motion of
bodies in the Solar System, perturbation theory, discoveries of minor planets, the discovery of Neptune, lunar
theory, the motions of the Earth, eclipses, the solar
parallax, the constitution of the Sun (the work of Faye),
star catalogues, the solar motion, double stars, nebulae,
and variable stars. Numerous citations.

129. Dingle, Herbert, ed. *A Century of Science, 1851-1951*.
      New York, London: Hutchinson's Scientific and Technical
      Publ., 1951. Pp. ix + 338.

      Reviews progress in science since the Great Exposition
      of 1851. Includes chapters by W.H. McCrea and H. Spencer
      Jones on stellar evolution and cosmology. Reviewed in:
      Isis, 43 (1952), 377-378.
      Contains items 1144, 1210.

130. Eberhard, G., A. Kohlschütter, and H. Ludendorff, eds.
      Handbuch der Astrophysik. 7 volumes. Berlin: J. Springer,
      1928-1936. Pp. 5000+.

      Comprehensive and authoritative collection of 31 review
      papers by major astronomers covering all aspects of stellar
      astronomy and astrophysics, galactic structure and
      cosmology. Many papers include detailed commentary on the
      historical development of the subject area. Volume 7 is
      a condensed review of the previous six volumes.
      Contains items 466, 993, 1020, 1062, 1072, 1098, 1102,
      1108, 1152.

131. Ferguson, Allan, ed. *Natural Philosophy through the
      Eighteenth Century and Allied Topics*. London: Taylor
      and Francis, 1972. Pp. xi + 164.

      Collection of papers written in commemoration of the
      150th anniversary of the founding of the *Philosophical
      Magazine*. Includes a chapter by H. Spencer-Jones on
      astronomy. Reviewed in: *Annals of Science*, 30 (1973),
      458.

132. Flügge, S., ed. *Encyclopedia of Physics*. Volumes 50-54.
      Berlin: Springer, 1958-1962. Pp. 2700+.

      47 technical review papers on all aspects of astro-
      physics with extensive bibliographies and some brief
      historical commentary. Includes English-German subject
      index. Known as *Handbuch der Physik*.
      Contains items 493, 494, 1004, 1096, 1150.

133. Hartmann, J., ed. *Astronomie*. Leipzig: B.G. Teubner, 1921. Pp. viii + 639.

   Collection of twelve essays by eleven authors reviewing cultural aspects of astronomy and major developments in astronomy during all periods. Brief citations.

134. [International Astronomical Union]. *Transactions of the International Astronomical Union*. Volume 1-. London: Imperial College, 1922-. Published in later years by Cambridge University Press, Academic Press, and D. Reidel.

   Records the activities, resolutions, committee reports and major addresses at each General Assembly of the IAU. Since 1968 and the 13th General Assembly, the most significant results of the meetings; formal discourses, both joint and invited; and special meetings are recorded in companion volumes entitled "Highlights of Astronomy" edited in turn by prominent IAU members. The IAU also supports the publication of proceedings from its many (85+ to date) symposia, beginning with the first in 1955.

135. [International Union for History and Philosophy of Science]. "Histoire de la Physique y compris l'Astronomie. XIXe et XXe Siècles." *XIIe Congrès International d'Histoire des Sciences, Tome V*. Paris: Blanchard, 1971. Pp. 127.

   General brief review of the progress of astronomy and physics. Part of a major series sanctioned by the IUHPS at its 12th General Assembly in Paris in 1968, and based upon lectures presented there.

136. Kopal, Z., ed. *Advances in Astronomy and Astrophysics*. Volume 1-. New York: Academic Press, 1962-. Pp. 300 per volume average.

   Extensively documented review articles on all aspects of astronomy designed to emphasize recent developments but generally providing background on developments since 1950.

137. Kuiper, Gerard P., and Barbara M. Middlehurst, eds. *The Solar System*. 4 volumes. Chicago: University of Chicago Press, 1953-1963. Pp. 2800+.

   Comprehensive technical review series with volumes on solar physics; the Earth; planets and satellites; the Moon, meteorites and comets.

138. Kuiper, Gerard P., and Barbara M. Middlehurst, eds. *Stars and Stellar Systems*. 8 volumes. Chicago: University of Chicago Press, 1960–1975. Pp. 5000+.

Originally conceived in 1955 as a nine-volume review series, each with individual editors. General volume topics include: instrumentation and techniques; data acquisition; stellar atmospheres; nebulae and interstellar matter; stellar structure; galactic structure; galaxies and cosmology.

139. Lang, K., and O. Gingerich, eds. *Source Book in Astronomy and Astrophysics 1900–1975*. Cambridge, Mass.: Harvard University Press, 1979. Pp. 922.

An expanded and heavily annotated update of Harlow Shapley's *Sourcebook* (item 145), this large-scale book includes reproductions and excerpts of 132 important primary scientific papers that outline the development of modern astronomy and astrophysics. Numerous editorial citations. Reviewed in: *Sky and Telescope*, 60 (July 1980), 53–55; *JHA*, 11 (1980), 215–216; *Isis*, 72 (1981), 119–120.

140. Lovell, Bernard, ed. *The Royal Institution Library of Science: Astronomy*. 2 volumes. New York: American Elsevier, 1970. Pp. xvi + 416; x + 397.

Reprint of 70 transcriptions of the astronomical lectures given as the Friday Evening Discourses in Physical Sciences held at the Royal Institution between 1851 and 1939. Volume 1 contains lectures through 1894. In general, lecture topics represent research interests of lecturers. Reviewed in: *JHA*, 3 (1972), 70.
Contains items 290, 434, 442, 468, 662, 745, 802, 827, 832, 833, 837, 870, 882, 962, 963, 966, 989, 999, 1006, 1012, 1013, 1018, 1023, 1034, 1046, 1083, 1103, 1130, 1139, 1172, 1177, 1199, 1209, 1211.

141. Mayall, Margaret W., ed. *Centennial Symposia*. Cambridge, Mass.: Harvard College Observatory, 1948. Pp. viii + 385.

24 papers from four symposia reviewing recent astronomy and astrophysics. General topics include interstellar matter; electronic and computational devices of interest to astronomers; eclipsing binaries; and solar-terrestrial relationships. Symposia held at meetings of the American Astronomical Society in Cambridge in 1946, in honor of the centenary of active telescopic astronomy at Harvard. Brief footnote citations.
Contains item 1081.

*   Mikaylov, A.A., ed. *Forty Years of Astronomy in the U.S.S.R., 1917-1957.*

Cited herein as item 215.

142. Munitz, Milton K., ed. *Theories of the Universe from Babylonian Myth to Modern Science.* Glencoe, Ill.: Free Press, 1957. Pp. x + 457. Reprinted, New York: Free Press, 1965.

Anthology of 33 works, both primary and secondary, on various aspects of cosmology. Most are selected, annotated abstracts or excerpts. Bibliography of book titles and introductory notes. Reviewed in: *Isis*, 50 (1959), 160-161.

143. Page, Thornton, and Lou Williams Page, eds. *Sky and Telescope Library of Astronomy.* 8 volumes. New York: Macmillan, 1966-1969. Pp. 2800 (approx.).

Selections from articles that appeared in *Sky and Telescope* and its predecessors, *The Sky* and *The Telescope*, dating from the 1930s. Surveys all aspects of the recent history of astronomy and current issues and problems. Volumes include: *Wanderers in the Sky: The Motion of Planets and Space Probes* (rev.: *PASP*, 78 (1966), 96; *Science*, 149 (1965), 626); *Neighbors of the Earth* (rev.: *Sky and Telescope*, 31 (March 1966), 163); *The Origin of The Solar System* (rev.: *Sky and Telescope*, 32 (Sept. 1966), 155; *Science*, 152 (1966), 1496); *Telescopes* (rev.: *Sky and Telescope*, 33 (Apr. 1967), 239; *PASP*, 79 (1967), 274); *Starlight* (rev.: *Sky and Telescope*, 35 (Apr. 1968), 243); *The Evolution of the Stars* (rev.: *Sky and Telescope*, 35 (July 1968), 40); *Stars and Clouds of the Milky Way* (rev.: *Sky and Telescope*, 38 (June 1969), 379; *PASP*, 61 (1969), 914); *Beyond the Milky Way* (rev.: *PASP*, 81 (1969), 914).

144. Ronan, Colin A. *Illustrated Sources in History: Astronomy.* New York: Barnes & Noble, 1973. Pp. 112.

Thematic compilation of extracts from articles and books written during all periods and covering many areas, edited and introduced by Ronan. Reviewed in: *JHA*, 5 (1974), 65; *Isis*, 65 (1974), 529.

145. Shapley, Harlow, ed. *Source Book in Astronomy, 1900-1950.* Cambridge, Mass.: Harvard University Press, 1960. Pp. xv + 423.

Continuation of item 146. Includes 69 excerpts from
major reviews, commentaries, and primary sources by 70
authors. Reviewed in: *Isis*, 53 (1962), 430–431.

146.   Shapley, Harlow, and Helen E. Howarth, eds. *A Sourcebook
       in Astronomy*. New York: McGraw-Hill, 1929. Pp. xvi +
       412.

       Collection of excerpts from famous astronomical writings.
       Includes 65 authors and centers upon 19th century and
       the origins of astrophysics, but provides useful material
       in earlier periods. No annotation or summarizing remarks.
       Reviewed in: *Isis*, 13 (1929), 130–134.

147.   Stroobant, Paul. *Les progrès récents de l'astronomie*.
       9 volumes. Brussels: Royal Observatory, 1908-1922.
       Pp. 1700+.

       A general review of astronomical events and progress
       during the period of roughly 1907-1920. Covers all aspects
       of astronomy with numerous citations emphasizing observa-
       tional studies of planetary, solar, and cometary phenomena.
       No general index.

148.   Valentiner, Wilhelm, ed. *Handwörterbuch der Astronomie*.
       4 volumes. Breslau: Eduard Trewendt, 1897-1902. Pp.
       839; 644; 1100; 432.

       Astronomical part of the late 19th-century *Encyclopaedia
       of Science*. Includes detailed technical chapters on all
       areas of astronomy, including history. Extensive cita-
       tions.

                                ARTICLES

149.   Alter, Dinsmore. "A Century of Change." *The Griffith
       Observer*, 16 (1952), 90–93; 96.

       Compares the topics treated in Herschel's *Outlines of
       Astronomy* to topics in contemporary textbooks, and centers
       on the growth of spectroscopy. No direct citations.

150.   Bok, Bart J. "Report on Astronomy." *Popular Astronomy*,
       47 (1939), 356-372. *Harvard Reprint* No. 175. Pp. 28.

       General review of recent publications in several areas:
       popularization, textbooks, periodicals, educational

projects; societies and journals available to professionals and amateurs; review texts on the recent progress of astronomy. Only the reprint contains an 11-page bibliography.

151. Bondi, H. "Fact and Inference in Theory and Observation." *Vistas in Astronomy*, 1 (1955), 155-161.

Critically examines relative reliabilities claimed for both theoretical and observational work in astronomy. Concludes that contrary to general opinion, "observational papers are no less liable than theoretical papers to reach erroneous conclusions." 33 citations.

152. Eddington, A.S. "A Century of Astronomy." *Nature*, 109 (1922), 815-817.

From the presidential address delivered before the Royal Astronomical Society at its centenary celebrations, May 30, 1922. Identifies major events of the past century and provides an analysis of papers in the *Monthly Notices*.

*   Eddington, A.S. "Royal Astronomical Society Centenary Celebration."

Cited herein as item 361.

153. Finocchiaro, Maurice A. "A Curious History of Astronomy: Leopardi's *Storia Dell'Astronomia*." *Isis*, 65 (1974), 517-519.

Notes the obscurity of this early general history as due to fact that its author was known primarily as a poet. Describes the work, which includes historical chapters to the year 1811. 13 citations.

154. Gingerich, Owen. "The Development of Astronomical Theory and Practice from the 17th to the 20th Century." *Vistas in Astronomy*, 20 (1976), 1-9.

Presented as overview during the 1975 centenary celebrations of the Royal Observatory, Greenwich. Examines increased output of articles from about 13 per year in the late 17th century, to 56 per year a century later, to many thousands per year at present as an introduction to the many significant qualitative changes, including the emergence of gravitational theory and practice, galactic astronomy and astrophysics. Centers upon changes within the context of the growth of the Royal Greenwich Observatory. 20 citations.

155.   Herrmann, D.B., and J. Hamel. "Zur Fruhentwicklung der
       Astrophysik das Internationale Forscherkollektiv 1865-
       1899." *NTM-Schriftenreihe für Geschichte der Naturwissen-*
       schaften, Technik und Medizin, 12 (1975), 25-30.

       Examines the relative percentages of techniques used in
       early astrophysics and finds, from an examination of papers
       in the *Astronomischen Nachrichten*, that almost 50 per
       cent utilized spectroscopic techniques, 13 per cent
       examined colors, etc. Provides statistical analysis of
       distribution of authors in professional age as a function
       of time. 14 citations.

156.   McCrea, W.H. "Astronomer's Luck." *Quarterly Journal of*
       the Royal Astronomical Society, 13 (1972), 506-519.

       Thoughtful essay examines the apparently fortuitous
       conditions found in astronomy that aided advance and
       argues that "luck may play a considerable part in the
       picture we have of the astronomical universe and in the
       way we formulate our science at any epoch." Examples
       include Polaris, the angular sizes of Moon and Sun, the
       proximity of Sirius B, the Crab Nebula, and the bright-
       ness of 3C273. Also poses the question of possible se-
       lectivity in our knowledge due to situations where the
       "lucky example just is NOT there...." Reviews "accidental
       discoveries" and the "luck of history."

157.   Meadows, A.J. *The High Firmament*. Leicester: Leicester
       U. Press, 1969. Pp. x + 207.

       Quotations from English literature reacting to, or
       commenting upon, astronomical themes dating from medieval
       times through the 19th century. Descriptive historical
       narrative organized chronologically by intellectual period.
       Includes later commentary on the plurality of worlds.
       Extensive citations. Reviewed in: *JHA*, 1 (1970), 80;
       Isis, 61 (1970), 121-122; *Annals of Science*, 25 (1969),
       361-362.

158.   Rossiter, Margaret W. "'Women's Work' in Science, 1880-
       1910." *Isis*, 71 (1980), 381-398.

       Examines the segregated incorporation of women into
       scientific employment in America after a period, circa
       1870-1880, when women began to enter higher education.
       Identifies the type of scientific activities delegated
       to women and includes a section on women in astronomy.
       38 citations.

*   Rousseau, G.S. "Poiesis and Urania: The Relation of
        Poetry and Astronomy in the English Enlightenment."

        Cited herein as item 585.

159.   Shenynin, O.B. "Mathematical Treatment of Astronomical
        Observations." *Archive for History of Exact Sciences*,
        11 (1973), 97-126.

        Broad survey covering many periods examining the treat-
        ment of astronomical observational data "from the point
        of view of the classical theory of errors and of mathe-
        matical statistics." Bulk of discussion deals with work
        up to the mid 18th century. 86 citations.

*   Sokolovskaja, Zinaida. "L'importance des instruments
        dans l'evolution des connaissances astronomiques."

        Cited herein as item 406.

*   Stratton, F.J.M. "International Co-operation in Astronomy."

        Cited herein as item 377.

160.   Struve, Otto. "Fifty Years of Progress in Astronomy."
        Popular Astronomy, 51 (1943), 469-481.

        General survey covering many aspects but highlighting
        the development of observatories, the changing structure
        of research and institutions, and the changing structure
        of international contacts and cooperation. 3 citations.

*   Young, Charles A. "American Astronomy: Its History,
        Present State, Needs and Prospects."

        Cited herein as item 248.

161.   Young, Charles A. "Pending Problems in Astronomy."
        Science, 4 (1884), 192-203.

        Address by retiring president of the AAAS reviewing
        those astronomical problems "which seem to be most
        pressing, and most urgently require solution as a condi-
        tion of advance; and those which appear in themselves
        most interesting, or likely to be fruitful, from a
        philosophic point of view." Among the many discussed,
        those of note include the Earth's rotation and figure,
        especially the question of the variation of the Earth's
        rotation; needs for improvements to the theory of the
        motion of the moon; tidal evolution of the Earth-Moon
        system; the advance of Mercury's perihelion; Jupiter as
        a link between stars and planets; the constitution of the

Sun--whether it is gaseous, solid, or contains a crust;
solar-terrestrial magnetic relationships; the close,
continuous scrutiny of solar radiation; the origin and
maintenance of solar heat (Young prefers Helmholtz's
gravitational source); the need for better measures of
the distances to the stars. Young notes that his address
was confined primarily to solar and planetary astronomy.

HISTORIOGRAPHY

*   Beer, Arthur, ed. *New Aspects in the History and Philosophy
        of Astronomy*.

        Cited herein as item 123.

162.   Beer, Arthur. "Astronomical Dating of Works of Art." *New
        Aspects in the History and Philosophy of Astronomy*
        (item 123), 177-223.

        Surveys methods and tools used in the dating of paint-
        ings, frescos, horoscopes, and passages in literature.
        Reviews major dating projects. Commentary added by Willy
        Hartner, pp. 225-228. 42 citations.

163.   Berendzen, Richard. "On the Exponential Growth of Science."
        History of Science, xi (1973), 283-85.

        Discusses D.B. Herrmann's identification (see item
        294) of exponential growth in the number of observatories
        established during the 19th century. 9 citations. See
        also item 170.

164.   Biermann, Kurt-R. "Attempt at a Classification of Un-
        published Sources in the More Recent History of
        Astronomy in German-speaking Countries." *New Aspects
        in the History and Philosophy of Astronomy* (item 123),
        237-243.

        Constitutes a general survey of sources contemporary
        to the life of Alexander von Humboldt. 6 citations.

*   Doublet, Édouard L. *Histoire de l'astronomie*.

        Cited herein as item 88.

165.   Edge, David. "The Sociology of Innovation in Modern
        Astronomy." *Quarterly Journal of the Royal Astronomical
        Society*, 18 (1977), 326-339.

Examines sociological factors influencing the development of astronomy since 1945, centering upon radio astronomy. Notes high frequency of "serendipitous discoveries."
19 citations.

*   Evans, David S. "Historical Notes on Astronomy in South
Africa."

Cited herein as item 196.

166. Gingerich, Owen. "Applications of High-Speed Computers
to the History of Astronomy." *New Aspects in the History
and Philosophy of Astronomy* (item 123), 229-236.

Discusses how modern computers can be employed to
generate planetary and lunar tables useful for historical
work, and also how they might provide aid in the interpretation of historical themes in gravitational theory.
12 citations.

167. Gingerich, Owen. "Was Ptolemy a Fraud?" *Quarterly Journal
of the Royal Astronomical Society*, 21 (1980), 253-266.

Contemporary review of question of Ptolemy's reliability
in his claims of performing direct observations, contested
by a number of people, most recently by R.R. Newton. Of
interest here for a review of earlier analyses of Ptolemy's
work by Tobias Mayer and Delambre, though Gingerich concentrates on R.R. Newton and argues for the retention of
Ptolemy as "the greatest astronomer of antiquity." 24
citations. See also B. Goldstein's valuable review:
Science, 199 (24 Feb. 1978), 872, and Newton's rebuttal
to Gingerich: *Quarterly Journal of the Royal Astronomical
Society*, 21 (1980), 388-399.

167a. Grünbaum, Adolf. "*Ad Hoc* Auxiliary Hypotheses and Falsi-
ficationism." *The British Journal for the Philosophy
of Science*, 27 (1976), 329-362.

A philosophical analysis of theory modification utilizing
the discovery of Neptune as a case study wherein the
existence of Neptune was introduced as an auxiliary hypothesis to "immunize" Newtonian planetary theory from falsification. Compares the successful Neptune case to Leverrier's
later *ad hoc* introduction of Vulcan to account for the
motion of Mercury. Examines conditions necessary to
justify auxiliary hypotheses and argues that in the case
studies cited, only a "non-pejorative caveat" is required.
36 citations.

168.   Hammer, Franz. "Problems and Difficulties in Editing
       Kepler's Collected Works." *New Aspects in the History
       and Philosophy of Astronomy* (item 123), 229–236.

       Reports on experiences associated with the production
       of the first 15 volumes initiated in 1937 by the Bavarian
       Academy of Sciences, now entitled "Johannes Kepler
       Gesammelte Werke."

169.   Hoskin, M.A., and O. Gingerich. "On Writing the History
       of Modern Astronomy." *Journal for the History of
       Astronomy*, 11 (1980), 145–146.

       Identifies stylistic problems facing writers of the
       history of recent astronomy. Supports the contention
       that "The primary duty of the historian of astronomy is
       to illuminate his science as a creative human activity
       of the astronomical community of the time."

170.   Jaschek, Carlos. "Data Growth in Astronomy." *Quarterly
       Journal of the Royal Astronomical Society*, 19 (1978),
       269–276.

       Shows that data growth in selected topical areas in
       astronomical research does not follow a single pattern
       and is not necessarily exponential, in contrast to
       Jaschek's earlier work and that of D.B. Herrmann (items
       294 and 823). Selected topical areas, some extending
       back into the mid-19th century, include: number of known
       asteroids with known period; number of determined radial
       velocities; number of trigonometric parallaxes; visual
       binaries; stars catalogued on major photometric systems;
       number of stars classified for spectral type; and number
       of variable stars known. Notes a surprising 28 per cent
       redundancy where objects catalogued are not new, or have
       been previously measured, "which shows that a real
       'bibliographic inaccessibility' problem exists." 34
       citations. See also item 163.

*   Knight, David. *Sources for the History of Science 1660–
       1914.*

       Cited herein as item 30.

*   Kulikovsky, Piotr G. "Sources for the History of Astronomy
       in the Scientific Archives of the U.S.S.R."

       Cited herein as item 35.

* Lankford, John. "A Note on T.J.J. See's Observations of
   Craters on Mercury."

   Cited herein as item 652.

171. Linsley, Earle G. "An Approach to the History of
   Astronomy." *Popular Astronomy*, 48 (1940), 253-256.

   Suggests various modes for teaching the history of
   astronomy including biographies, institutional histories
   and topical histories. No direct citations.

* McKenna, Susan M.P. "Astronomy in Ireland from 1750."

   Cited herein as item 212.

172. Olmsted, John W. "The 'Application' of Telescopes to
   Astronomical Instruments, 1667-1669; A Study in
   Historical Method." *Isis*, 40 (1949), 213-224.

   Traces historical accounts of the application of
   telescopic sights to graduated arcs for measuring angles
   arguing that this "was the final step in the conversion
   of the telescope from an instrument of celestial discovery
   into an instrument of precision." Main intention of the
   essay is, however, to illustrate that earlier accounts
   of developments during this period were misled by incomplete
   access to relevant historical data, chiefly the critical
   edition of Huygens' works that began only in 1888, and
   that writers on this period after 1895 failed to use this
   important source, and thus retained an inaccurate picture
   of the application of telescopic sights. 66 citations.

173. Reingold, Nathan. "A Good Place to Study Astronomy."
   Quarterly Journal of Current Acquisitions, 20 (1963),
   211-217.

   Reviews Library of Congress sources for the historical
   study of astronomy.

* Rufus, W. Carl. "Proposed Periods in the History of American Astronomy."

   Cited herein as item 227.

174. Sarton, George. "Remarks Concerning the History of
   Twentieth Century Science." *Isis*, 26 (1936), 53-62.

   Based upon discussions with W.S. Adams of Mount Wilson
   and with John C. Merriam of the Carnegie Institution
   circa 1935. Assesses difficulties faced by historians

of 20th-century science in light of the acceleration
of progress and the explosion in documentation. Discusses
problems of "contemporary blindness" and argues that they
are specious because nothing in science is ever considered
to be final. "Science is gradually built up by the method
of successive approximations; the history of science is
necessarily built up the same way." Suggests useful modes
of doing contemporary history, including chronological
abstracts and "chronicles" as well as narratives to
facilitate the work of later historians.

175.   Stephenson, F.R., and D.H. Clark. *Applications of Early*
       Astronomical Records. Oxford: Oxford University Press,
       1978. Pp. x + 114.

       Examines the modern use of old observations of eclipses,
       aurorae, and sunspots using primary sources. A continua-
       tion of their work in the applied history of science (see
       item 1090). Extensive citations and bibliography. Reviewed
       by J. Eddy in: *JHA*, 11 (1980), 202-204.

*   Sticker, Bernhard. "Historical Scientific Instruments
       as Cultural Landmarks."

       Cited herein as item 407.

*   Whiteside, D.T. "The Expanding World of Newtonian Research."

       Cited herein as item 735.

176.   Zinner, Ernst. "Some Reflections on Research in the History
       of Astronomy." *New Aspects in the History and Philosophy*
       of Astronomy (item 123), 171-175.

       Demonstrates that original manuscript sources for
       astronomical data are usually more reliable than printed
       sources. Reviews his work on the compilation of manuscript
       data, since 1919.

NATIONAL AND INSTITUTIONAL HISTORIES

REGIONAL HISTORIES

*   Abalakin, Victor K. "The Development of Theoretical
       Astronomy in the U.S.S.R."

       Cited herein as item 741.

177.   Ambartsumyan, V.A., ed. *Razvitie astronomii v SSSR, 1917-
       1967*. Moscow: Nauka, 1967. Pp. 475.

       Reviews (in Russian) the development of astronomy in
       the USSR from 1917. Extensive bibliography.

178.   Ansari, S.M. Razaullah. "On the Early Development of
       Western Astronomy in India and the Role of the Royal
       Greenwich Observatory." *Archives Internationale d'His-
       toire des Sciences*, 27 (1977), 237-262.

       Provides a general review of Indian astronomy from the
       16th through 18th centuries concentrating on the develop-
       ment of observatories in the 19th century at Madras,
       Calcutta, Lucknow and elsewhere. Emphasizes the influence
       of Maskelyne, Airy, Christie and Lockyer, and their insti-
       tutions--the Royal Society and the Royal Greenwich
       Observatory. 153 citations.

179.   Ansari, S.M. Razaullah. "The Establishment of Observa-
       tories and the Socioeconomic Conditions of Scientific
       Work in Nineteenth Century India." *Indian Journal for
       the History of Science*, 13 (1978), 62-70.

       Traces history of the establishment by the East India
       Company of the Madras Observatory from 1792, and the Royal
       Observatory at Lucknow, created by N. Haydar, King of
       Oudh, in 1832. Notes astronomical work accomplished at
       these observatories, problems encountered by their staffs
       in maintaining support, and comments on aspects of the

study of the history of Indian astronomy. Points out that, contrary to many contemporary and earlier studies, Jai Singh did employ telescopic equipment as well as purely naked-eye sights. 29 citations.

*   Bailey, Solon I. "The Cambridge Astronomical Society of 1854."

Cited herein as item 353.

180.   Ball, A.W. Lintern. "England's Astronomical Education?" *Quarterly Journal of the Royal Astronomical Society*, 13 (1972), 486-505.

Provides a broad historical introduction to an assessment of the present condition. Notes significant place of popular astronomy education in the 16th through the 19th centuries but fears that it has been on the decline since the mid-19th century.

181.   Bedini, Silvio A. *Thinkers and Tinkers*. New York: Scribner's, 1975. Pp. xix + 520.

Documents astronomy in early America from Hariott to Thomas Jefferson emphasizing instrumentation and work within the general milieu of early American science. Includes episodes such as the Mason-Dixon survey and the 1769 transit of Venus. Provides 85 pages of bibliography, citations, and glossary. Reviewed in: *Annals of Science*, 34 (1977), 212-214; *Isis*, 68 (1977), 324-325.

182.   Berendzen, R., and Mary Treinen Moslen. "Manpower and Employment in American Astronomy." *Education in and History of Modern Astronomy* (item 125), 46-65.

Exposition of manpower and employment data between 1920 and 1970 with emphasis on problems of the 1970s. 28 citations.

*   Biermann, Kurt-R. "Attempt at a Classification of Unpublished Sources in the More Recent History of Astronomy in German-speaking Countries."

Cited herein as item 164.

183.   Bigourdan, Guillaume. *Histoire de l'astronomie d'observation et des observatoires en France*. 2 volumes bound as one. Paris: Gauthier-Villars, 1918; 1930. Pp. 184.

The first part outlines the origin and establishment of the Paris Observatory, introduced by an accounting of the principal French astronomers since Fabri de Peiresc in the early 17th century. Later chapters outline the growth of astronomy in France, organized by province. Numerous citations.

184.   Blair, G. Bruce. "Amateur Astronomy in America." *Astronomical Society of the Pacific Leaflet* 210 (Aug. 1946). Pp. 7.

Traces major periods of amateur astronomical activity from the 1830s.

*   Brasch, Frederick E. "John Winthrop (1714–1779), America's First Astronomer and the Science of His Period."

Cited herein as item 1251.

*   Browne, Charles Albert. "Scientific Notes and Letters of John Winthrop, Jr. (1600–1676), First Governor of Connecticut."

Cited herein as item 1253.

185.   Brush, Stephen G. "The Rise of Astronomy in America." *American Studies*, 20 (1979), 41–67.

Surveys major events, discoveries and trends in astronomy worldwide during the period 1800–1950 and attempts to identify those contributions that could be called American, as a prelude to an analysis of the phenomenon of American astronomy. Shows that in 1800, American contributions were practically nil, but overtook German contributions by 1900 and became predominant by 1930. Reviews the contributions and backgrounds of major American astronomers within three periods: "The Founding Parents"; "The transition to big astronomy"; "The last quarter-century." Includes a section on the role of women in American astronomy. Hints that the rise to preeminence of American astronomy and astronomers acted as a stimulus to other areas of American science but that, while Americans do not possess unique talents or methods, "they seem to have been able to stimulate and combine better than other cultures the essential factors for success." 58 citations, many of which contain several references to primary and secondary sources.

186.   [Canadian Astronomy]. "Astronomy in Canada." *Journal of the Royal Astronomical Society of Canada*, 61 (1967), 211–338.

Concentrates on contemporary aspects of Canadian
astronomy, but one of the nine papers deals with the origins
and growth of the Royal Astronomical Society of Canada.

187. Cannon, Susan Faye. *Science in Culture: The Early
      Victorian Period*. New York: Science History Publications,
      1978. Pp. xii + 296.

      Explores the social context of British science during
      the period including material on leading astronomical
      figures: John Herschel, George B. Airy, and the important
      astronomical commentator William Whewell. Reviewed in:
      Isis, 70 (1979), 593-595.

*   Cannon, Walter F. "John Herschel and the Idea of Science."

      Cited herein as item 1259.

*   Cawood, John. "Terrestrial Magnetism and the Development
      of International Collaboration in the Early Nineteenth
      Century."

      Cited herein as item 608.

188. Chapin, Seymour L. "The Astronomical Activities of
      Nicolas Claude Fabri de Peiresc." *Isis*, 48 (1957),
      13-28.

      Description of the work of this 17th-century amateur
      in Paris centering on his observational and practical
      studies. Notes Peiresc's influence on later French
      astronomers. 85 citations.

189. Chenakal, Valentin L. "The Astronomical Instruments of
      John Rowley in Eighteenth-Century Russia." *Journal for
      the History of Astronomy*, 3 (1972), 119-135.

      Discusses the instruments imported into Russia, how
      they were procured and their general description, chiefly
      orreries, quadrants, and geodetic instruments of John
      Rowley (1674-1728). 42 citations.

*   Collinder, Per. "Astronomical Books and Papers Printed
      in Sweden between 1881 and 1898: Bibliography and
      Historical Notes."

      Cited herein as item 10.

190. Collinder, Per. *Swedish Astronomers 1477-1900*. Stockholm:
      Almquist and Wiksell, 1970. Pp. 73.

General, brief survey of Swedish astronomy. Highlights
the growth of observatories and scientific institutions
supporting astronomy. Bibliography. Reviewed in: *Isis*,
63 (1972), 259.

191. D'Elia, Pasquale M. *Galileo in China*. Rufus Suter and
      Matthew Sciascia, translators. Cambridge, Mass.:
      Harvard University Press, 1960. Pp. xv + 115.

      Presents background to the reception and transmission
      of Galileo's astronomy in China after Father Emmanuel
      Diaz's 1615 publication of *Problems in Astronomy* in
      Chinese. Originally published in Italian in 1946 and
      reprinted in 1947. Reviewed in: *Isis*, 53 (1962), 409-410;
      Isis, 41 (1950), 220-222.

192. Daniels, George H. "The Process of Professionalism in
      American Science. The Emergent Period, 1820-1860."
      Isis, 58 (1967), 151-166.

      Examines the transition of American science into a
      professional community within an institutional framework
      through a study of relevant "social conditions" in mid-
      19th-century America. Includes B.A. Gould's establishment
      of the *Astronomical Journal* in 1849, and Gould's generally
      controversial but prominent role in American science.
      57 citations.

193. Diukov, I.A. "Twenty-Five Years of Soviet Astronomy."
      Popular Astronomy, 52 (1944), 383-388.

      Centers upon work at Pulkovo and other major centers
      of activity in fundamental astrometry and positional
      astronomy. No direct citations.

194. Dizer, M., ed. *International Symposium on the Observa-
      tories in Islam, 19-23 September, 1977*. Istanbul:
      Milli Egitim Basimevi, 1980. Pp. 272.

      22 papers presented in honor of 400th anniversary of
      the founding of the Istanbul Observatory. Includes papers
      on medieval and early-modern astronomy in Islam high-
      lighting observatories, instruments, European influences
      and contacts, and manuscript sources. Numerous citations.

195. Dreyer, J.L.E. "Indian Astronomical Instruments." *Nature*,
      103 (1919), 166-168.

      Brief exposition centering upon Jai Singh's observa-
      tories. Written as a review of "The Astronomical Observa-

tories of Jai Singh," by G.R. Kaye (*Archaeological Survey of India, New Imperial Series.* Volume xl. Calcutta, 1918. Pp. viii + 151 + 26 plates). 3 citations.

*   Dvoichenko-Markoff, Eufrosina. "The Pulkovo Observatory and Some American Astronomers in the Mid-Nineteenth Century."

Cited herein as item 280.

*   Edge, David O., and Michael J. Mulkay. *Astronomy Transformed: The Emergence of Radio Astronomy in Britain.*

Cited herein as item 486.

196.   Evans, David S. "Historical Notes on Astronomy in South Africa." *New Aspects in the History and Philosophy of Astronomy* (item 123), 265-282.

Traces the history of astronomy in South Africa beginning with the European settlement at the Cape noting that Lacaille's visit in 1751-3 began serious astronomical work. Identifies sources for the study of the history of astronomy in South Africa. 13 citations.

197.   Goldberg, Leo, and Lois Edwards, eds. *Astronomy in China: A Trip Report of the American Astronomy Delegation.* Washington, D.C.: National Academy of Sciences Committee on Scholarly Communication with the People's Republic of China, Report No. 7, 1979. Pp. ix + 109.

Includes (in Chapter 2) a short history of astronomy in China.

198.   Gorman, Mel. "Gassendi in America." *Isis*, 55 (1964), 409-417.

Traces the influence of the astronomical writings of Pierre Gassendi (1592-1655) in 17th- and early 18th-century America, primarily his 1645 text evaluating the work of Galileo and Kepler. Concludes that Gassendi's influence was widespread, "not dwindling completely until the commencement of the nineteenth century." 53 citations.

199.   Greene, John C. "Some Aspects of American Astronomy 1750-1815." *Isis*, 45 (1954), 339-358.

After brief mention of main characters during the period, including John Winthrop, David Rittenhouse, and Nathaniel Bowditch, this study concentrates on areas of research during the period: transits of Mercury and Venus; solar and lunar eclipses; observations of comets and meteors; geodetic boundary surveys. Examines American thought concerning the structure and stability of the universe mainly through citations from early American textbooks. Concludes that astronomy was an active pursuit in America during the period and that it was in up-to-date contact with European advances. 33 citations.

200. Gross, Walter E. "The American Philosophical Society and the Rise of Astronomy in the United States in the Middle of the Nineteenth Century." *Annals of Science*, 31 (1974), 407-427.

Reviews the growth of astronomy in America at mid-century and analyzes why the American Philosophical Society, at the time the leading organization of its kind in science, was not able to take a leading role in the further development of astronomy. This role was assumed by groups closer to the federal government with the resources available to "establish observatories, hire scientists, and establish cooperation between different observers...." 109 citations.

201. Herrmann, D.B. "Das Astronomentreffen im Jahre 1798 auf dem Seeberg bei Gotha." *Archive for History of Exact Sciences*, 6 (1970), 326-344.

A general study of those astronomers and astronomical institutions and trends significant to the later development of German astronomy in the 19th century. 82 citations.

*   Herrmann, D.B. "B.A. Gould and His *Astronomical Journal*."

Cited herein as item 382.

202. Herrmann, D.B. "Zur Frühentwicklung der Astrophysik in Deutschland und in den USA." *NTM-Schriftenreihe für Geschichte der Naturwissenschaften, Technik und Medizin*, 10 (1973), 38-44.

Compares quantitatively the number of astrophysical papers written by German and American authors in four major journals (*Astronomischen Nachrichten*, *Astrophysical Journal*, *Astronomical Journal*, and the *Publikationen des Astrophysikalischen Observatoriums Potsdam*) during the period 1850s-1905. Concludes that the rapid rise of American authorship in new areas of research resulted from the relative freedom from tradition early American astrophysicists enjoyed, in comparison with the more established German school steeped in classical mathematical astronomy, with the exception of Potsdam staff. 14 citations.

203. Hetherington, Norriss S. "Cleveland Abbe and a View of
     Science in Mid-Nineteenth Century America." *Annals of
     Science*, 33 (1976), 31-49.

     Examines trends in various aspects cf astronomy and
     science in the life and training of Cleveland Abbe, a
     late 19th-century astronomer considered a typical person
     of his time. In choosing such a person for study, a
     detailed glimpse is provided of the state of American
     astronomy and the limitations for professional develop-
     ment within the discipline. Includes Abbe's studies to
     become an astronomer circa 1856-1871 during a time when
     graduate education was just being introduced in America
     when government and private support for scientific
     activities was stimulating a rapid growth in American
     observatories, and creating a need for trained scientists.
     120 citations.

204. Iba, Yasuaki. "Fragmentary Notes on Astronomy in Japan."
     Popular Astronomy, 42 (1934), 243-252; 45 (1937), 301-
     310; 46 (1938), 89-96, 141-148, 263-267.

     Outline of general history of astronomy in Japan,
     including Chinese origins and Western influences in the
     19th century. Includes records of observations of celestial
     events, observatories and personalities. Numerous cita-
     tions.

205. Inkster, Ian. "Robert Goodacre's Astronomy Lectures
     (1823-1825), and the Structure of Scientific Culture
     in Philadelphia." *Annals of Science*, 35 (1978), 353-
     363.

     Shows that Goodacre's lectures were accepted during a
     period of significant transition toward public awareness
     of science. Suggests "that the success of Goodacre re-
     flected the highly partial social bias of Philadelphia's
     scientific culture in that decade." 63 citations.

*   Jarrell, Richard A. "Astronomical Archives in Canada."

     Cited herein as item 26.

206. Johnson, Francis R. *Astronomical Thought in Renaissance
     England*. Baltimore: Johns Hopkins Press, 1937. Pp. xi +
     357.

     Centers on the years 1500-1645. Examines cosmological
     thought in Britain and the acceptance of Copernican doc-
     trine by Robert Recorde and John Dee; the importance of

Thomas Digges; Gilbert's arguments for the Earth's motion; the role of popular almanacs. Includes detailed bibliographical notations and an appendix of English astronomy books to 1640. Reviewed in: *Isis*, 28 (1938), 514-516.

* Kaye, George Rusby. *The Astronomical Observatories of Jai Singh*.

    Cited herein as item 413.

207. Kharadze, E.K. "Astronomy in the Georgian SSSR." *Bulletin of the Abastumani Astrophysical Observatory*, No. 25 (1960), 3-22.

    Reviews (in Russian) recent developments in local astronomy, primarily the establishment of the Abastumani Observatory as a modern astronomical facility. Kharadze is a contemporary figure in Russian astronomy and is the director of the observatory.

208. Kuiper, Gerard P. "German Astronomy during the War." *Popular Astronomy*, 54 (1946), 263-287.

    Reviews progress of German astronomy during World War II. Organized by major observatories. Limits discussions to scientific activities. Annotated with bibliography.

* Kulikovsky, Piotr G. "Sources for the History of Astronomy in the Scientific Archives of the U.S.S.R."

    Cited herein as item 35.

209. Lange, Erwin F. "The Founders of American Meteoritics." *Meteoritics*, 10 (1975), No. 3.

    Brief review of events that initiated the study of meteorites in America, notably several major falls in the early 19th century. 19 citations.

* Lavrova, N.B., ed. *Bibliography of Works on Astronomy Performed in the USSR from 1917 to 1957*.

    Cited herein as item 39.

210. Lockyer, J. Norman. "The Development of Astronomy in America." *Publications of the Astronomical Society of the Pacific*, 12 (1900), 109-117.

    Reviews 60 years of American astronomy to 1900 centering upon the creation of Lick and Yerkes Observatories, discoveries there and at Harvard, Princeton and Dartmouth,

and the development of professional training centers of
astronomy in America. Compares the training of astronomers
in "astronomical physics" in Great Britain and America,
and their comparative support for the astronomical enter-
prise.

*   Loomis, Elias. *The Recent Progress of Astronomy; Especially
in the United States.*

Cited herein as item 104.

*   Loomis, Elias. "Astronomical Observatories in the United
States."

Cited herein as item 310.

*   Lovell, Bernard. *The Origins and International Economics
of Space Exploration.*

Cited herein as item 506.

211.   McCluskey, Stephen C. "The Astronomy of the Hopi Indians."
Journal for the History of Astronomy, 8 (1977), 174–195.

Examines, through the Hopi, "the actual operation of
a system of prescientific astronomy through the comparison
of the recorded dates on which solar and lunar festivals
were held with the known position of the Sun and Moon at
that time." The Hopi remained uninfluenced by Western
thought until the 1870s and therefore detailed records
survive. 35 citations.

212.   McKenna, Susan M.P. "Astronomy in Ireland from 1750."
New Aspects in the History and Philosophy of Astronomy
(item 123), 283–296.

Provides accounts of the origins and development of
nine important Irish observatories centering upon staff,
instrumentation and accomplishments. 29 citations.

213.   McKenney, Anne P. "What Women Have Done for Astronomy in
the United States." *Popular Astronomy*, 12 (1904), 171–
182.

Reviews the contributions of 12 women, and of observatories at prominent women's colleges and observatories
with female staff members. No direct citations.

214.   Mendillo, Michael, D.H. DeVorkin, and Richard Berendzen.
"History of American Astronomy." *Astronomy*, 4 (1976),
20–65; 87–107.

Popular review of origins and development of astronomy
and astrophysics in America.

215. Mikaylov, A.A., ed. *Forty Years of Astronomy in the
U.S.S.R., 1917-1957* [1960]. Translated by Wright Patterson Air Force Base, Foreign Technology Division, Ohio.
In two parts: FTD-TT-63-1149/1+2+4; AD429 455; FTD-TT-
63-1149/1+2+4; AD429 520. Pp. 601 + 672.

Translation of 1960 Russian collection of 24 articles
reviewing all aspects of astronomy and astrophysics
written by Soviet astronomers. Extensive bibliography.

216. Miller, Howard S. "Astronomical Entrepreneurship in the
Gilded Age." *Astronomical Society of the Pacific Leaflet*
No. 479 (May 1969). Pp. 8.

Examines how American astronomers of the late 19th
and early 20th century gained support for the construction of major observatories including Lick, Yerkes, and
Mount Wilson. Shows briefly how James Lick was convinced
to build an observatory and how George Ellery Hale
approached Charles Yerkes and later the Carnegie Institution of Washington to build his observatories.

217. Miller, Howard S. *Dollars for Research. Science and Its
Patrons in Nineteenth-Century America*. Seattle: Univ.
of Washington Pr., 1970. Pp. xiv + 258.

Traces growth of financial and institutional support
for all aspects of American science. Includes the Smithsonian Institution, the acquisition of James Lick's
Trust for an astronomical observatory, George Ellery Hale's
many entrepreneurial successes, and E.C. Pickering's
attempt to establish an endowment for astronomy along
with his successes in securing funds for major Harvard
projects. Reviewed in: *Isis*, 62 (1971), 116-117.

*   Mitchell, Samuel E. "Astronomy During the Early Years
of the American Philosophical Society."

Cited herein as item 371.

218. Moore, Patrick, and Pete Collins. *The Astronomy of Southern
Africa*. London: Robert Hale, 1977. Pp. 160.

Descriptive anecdotal illustrated review beginning
with late 17th- and 18th-century expeditions for the
determination of longitude and the establishment of the
Royal Observatory at the Cape of Good Hope in 1820.

Identifies the proliferation of observatories and observing
stations in the 19th and 20th centuries. No citations.
Reviewed in: *JHA*, 10 (1979), 63-65.

*   Mourot, Suzanne, and D.J. Cross. "Astronomy Archives in
          Australia."

          Cited herein as item 44.

219.   Musto, David F. "A Survey of the American Observatory
          Movement, 1800-1850." *New Aspects in the History and
          Philosophy of Astronomy* (item 123), 87-92.

          Traces the successes and failures in the development
          of early American observatories noting sources of funds,
          intended research, and the role of the federal government.
          15 citations.

220.   Nakayama, Shigeru. *A History of Japanese Astronomy: Chinese
          Background and Western Impact*. Cambridge, Mass.: Harvard
          University Press, 1969. Pp. xiii + 329.

          Detailed discussion of Chinese phase of Japanese
          astronomy (6th to 16th century), including Japanese
          horoscopes and calendars, followed by an extensive exposi-
          tion of changes in Japanese astronomy after contact with
          European science in the late 16th century. Later chapters
          bring history into mid-19th century and provide a summary
          of the development in general of astronomy in the Far
          East. Extensive documentation and 44-page bibliography.
          Essay review in: *JHA*, 3 (1972), 139-145. Also reviewed
          in: *PASP*, 81 (1969), 917-918; *Annals of Science*, 26
          (1970), 269-270.

221.   Newcomb, Simon. "Aspects of American Astronomy." *Astro-
          physical Journal*, 6 (1897), 289-309.

          Address given at dedication of Yerkes Observatory,
          October 22, 1897. General commentary on reception of
          astronomy by the public, support for astronomy, and the
          early history of American astronomy and astronomers in-
          cluding John Winthrop, the observation of the 1769 transit
          of Venus, Nathaniel Bowditch's translation of Laplace's
          Mécanique Céleste, the origins of the Naval Observatory
          and Harvard College Observatory, the disruption of
          astronomy by the American Civil War, and the great observa-
          tories of the late 19th century. Discusses the reception
          of American astronomy in the world community.

*   Numbers, Ronald L. *Creation by Natural Law: Laplace's
        Nebular Hypothesis in American Thought.*

        Cited herein as item 579.

222.   Osterbrock, Donald E. "The California-Wisconsin Axis in
        American Astronomy." *Sky and Telescope*, 51 (1976),
        9-14; 91-97.

        Traces migration to Lick Observatory of staff from
        positions at the University of Wisconsin and at Yerkes
        Observatory and shows that a significant proportion of
        research interests at Lick originated in Wisconsin.
        Includes George Ellery Hale's transfer of activity from
        Yerkes to Southern California to found the Mount Wilson
        Solar Observatory.

222a.   Pilz, Kurt. *600 Jahre Astronomie in Nürnberg.* Nürnberg:
        Hans Carl, 1977. Pp. 376.

        Chronological series of some 150 short sketches of
        astronomical workers linked in their activities to
        Nuremberg. Includes artisans as well as astronomers
        beginning in the 15th century and emphasizes the Re-
        naissance. Extensive citations. Reviewed in: *JHA*, 11
        (1980), 208-210.

223.   Plotkin, Howard. "Edward C. Pickering and the Endowment
        of Scientific Research in America, 1877-1918." *Isis*,
        69 (1978), 44-57.

        Traces Pickering's quest for the establishment of a
        secure research fund from his 1877 address as vice-
        president of the AAAS through World War I, and the limited
        success he achieved for others when compared to the sub-
        stantial endowments he was able to foster for astronomical
        work at his home institution--Harvard College Observatory.
        64 citations.

224.   Reingold, Nathan, ed. *Science in Nineteenth Century
        America: A Documentary History.* New York: Hill and
        Wang, 1964. Pp. xii + 308.

        A collection of scientific correspondence including
        Newcomb and Michelson on the velocity of light; Michelson
        and Morley; Nathaniel Bowditch, Thomas Jefferson, A.M.
        Legendre and Laplace on Bowditch's translation of *Mécanique
        Céleste*; J.W. Draper and H. Draper; H.A. Rowland. Re-
        viewed in: *Isis*, 56 (1965), 246-247.

225. Reingold, Nathan. "Cleveland Abbe at Pulkovo: Theory and
     Practice in the Nineteenth Century Physical Sciences."
     Archives Internationale d'Histoire des Sciences, 17
     (1964), 133–147.

     Detailed account of Abbe's studies under Otto Struve
     at Pulkovo during the years 1864–1866, and the role of
     Abbe's contact in establishing liaison between American
     and Russian astronomy. Notes the influence research
     styles at Pulkovo had on the later development of Abbe's
     science in America. Numerous citations.

226. Ronan, Colin A. *Their Majesties' Astronomers: A Survey
     of Astronomy in Britain Between the Two Elizabeths*.
     London: The Bodley Head, 1967. Pp. xi + 240.

     General popular history of astronomy in Britain from
     the early Copernican Robert Recorde's work in the 1550s
     through to the present. Examines the works of John Dee,
     Thomas Digges, and others; the historical problem of
     navigation; the stormy relations between the Royal
     Society and the Royal Observatory at Greenwich (Newton
     and Flamsteed); the discovery of Neptune. Includes topical
     chapters on 20th-century astronomy. Published in the
     United States as: *Astronomers Royal*. New York: Doubleday,
     1969. 83 citations and brief reading list. Reviewed in:
     Annals of Science, 24 (1968), 341–342.

*   Rossiter, Margaret W. "'Women's Work' in Science, 1880–
     1910."

     Cited herein as item 158.

*   Royal Astronomical Society. "Sesquicentenary Commemora-
     tion Issue."

     Cited herein as item 374.

227. Rufus, W.C. "Proposed Periods in the History of American
     Astronomy." *Popular Astronomy*, 29 (1921), 393–404;
     468–475.

     Considers the purposes of the study of the history of
     astronomy to be, after the accumulation of facts and
     organization of materials, the generation of a systematic
     account of the origin and development of astronomical
     ideas. Argues that to facilitate this study in American
     astronomy, systematic contextual eras must be identified.
     Proposes: "Introductory Period" (1490–1600); "Colonial
     Period" (1600–1780); "Apparently Stationary Period"

(1780-1830); "Popular Period" (1830-1860); "New Astronomy
Period" (1860-1890); and a "Correlation Period" (1890-).
Provides discussion of chronology of events supporting
the reality of his periods.

228. Rufus, W.C. "Astronomical Observatories in the United
States Prior to 1848." *Scientific Monthly*, 19 (1924),
120-139.

Narrative review traces astronomical work in America
from the early 18th century, but notes Thomas Hariot's
observations from America circa 1585. Identifies observers
in Colonial America, early astronomy at Harvard and in
Philadelphia. 24 citations.

229. Rumrill, H.B. "Early American Astronomy." *Popular Astronomy*,
50 (1942), 408-419.

Describes 17th- and 18th-century observatories and
research activities centering on Rittenhouse, Winthrop
and major observational events such as the 1769 transit
of Venus. Numerous indirect citations.

230. Russell, Henry Norris. "America's Role in the Development
of Astronomy." *Proceedings of the American Philosophical
Society*, 91 (1947), 10-16.

Presented at APS Symposium on "America's Role in the
Growth of Science," October 18, 1946. Begins with Rittenhouse, the origins of the Harvard College Observatory,
the American Observatory movement, Newcomb's researches
on the theories of the motions of the Moon and planets,
meteor astronomy at Yale, the applications of physics
to astronomy.

* Russell, John L. "Cosmological Teaching in the Seventeenth-Century Scottish Universities."

Cited herein as item 586.

* Rybka, E. *Four Hundred Years of the Copernican Heritage*.

Cited herein as item 587.

231. Rybka, E., ed. *Historia Astronomii w Polsee*. Warsaw:
Ossolineum, 1975. Pp. 330.

Volume 1 of a general history of astronomy in Poland,
this work extends from prehistoric times to the 18th
century. Centers around Copernicus and the acceptance
of the heliocentric system at the Jagiellonian University

of Cracow. Written in Polish, a short summary in English
by A. Przybylski appears in the *Journal for the History
of Astronomy*, 10 (1979), 58-60.

232.   Sagan, Carl. "The Past and Future of American Astronomy."
        Physics Today, 27 (1974), 23-31.

        Anecdotal review of the growth of American astronomy
        within the context of the progress of world astronomy
        since 1900. Centers upon the formation of the *Astrophysical
        Journal* and the contents of its first issues; the opening
        of the Yerkes Observatory; salaries of astronomers and
        observatory budgets circa 1900; the rise of physics and
        its applications to astronomy; problems of nomenclature;
        lunar and solar observations. 64 citations.

233.   Sayili, Aydin. *The Observatory in Islam*. Publications of
        the Turkish Historical Society Series VII, No. 38.
        Ankara: T.T.K. Basimevi, 1960. Pp. xi + 472.

        Comprehensive study of the origins and development of
        the Islamic observatory from the 9th and 10th centuries
        through the 14th century. Includes detailed summary and
        commentary on Islamic astronomy, its influences on later
        European astronomy, and causes for the decline of sci-
        entific work in Islam. Numerous citations and extensive
        bibliography.

234.   See, T.J.J. "The Services of Nathaniel Bowditch to
        American Astronomy." *Popular Astronomy*, 2 (1895), 385-
        394.

        Brief sketch of Bowditch's life and primary work, the
        translation of Laplace's *Méchanique Céleste*. Argues that
        Bowditch was the founder of American mathematics and
        physical astronomy. No direct citations.

*   Shapley, Harlow S. "Astronomy and International Coopera-
        tion."

        Cited herein as item 375.

235.   Singh, Prahlad. *Stone Observatories in India*. Varanasi,
        India: Bhanata Manisha, 1978. Pp. iv + 216.

        Illustrated descriptions of non-optical observatories
        and instruments of late 18th-century India built by Jai
        Singh, Raja of Amber.

236. Sivin, Nathan. "Copernicus in China." *Studia Copernicana,
     vi.* Warsaw: Polish Academy of Science, 1973. Pp. 60.

     Examines problems of transmission, through the Jesuit
     mission, of Copernican science into China. Centers primarily
     in the late 16th and 17th centuries when Jesuits failed
     to properly transmit Copernican concepts, but continues
     to the mid-19th century when they finally attempted to
     convey Copernicanism properly, though ineffectually.
     Reviewed in: *JHA*, 5 (1974), 204–205.

*   Skabelund, Donald. "Cosmology on the American Frontier:
     Orson Pratt's Key to the Universe."

     Cited herein as item 590.

237. Slouka, H., et al. *Astronomy in Czechoslovakia from the
     Earliest Times to Today.* Prague: Osveta, 1952. Pp. 346.

     Brief introductory histories of Czech astronomy intro-
     duce (in Czech) about 300 pages of illustrations of old
     manuscripts, portraits and astronomical artifacts. Anno-
     tations to the illustrations are in Czech, Russian and
     English. Reviewed in: *Isis*, 45 (1954), 100–101.

238. TenBruggencate, P., ed. *Astronomy, Astrophysics and
     Cosmology.* Wiesbaden: Field Technical Information
     Agency, 1948. Pp. 441.

     Part of FIAT (Field Information Technical Agency)
     series reviewing German science during World War II.
     Extensive citations.

239. Thomson, Malcolm M. *The Beginning of the Long Dash:
     A History of Timekeeping in Canada.* Toronto: University
     of Toronto Press, 1978. Pp. x + 190.

     Traces the development of the use of radio time signals
     but includes introductory material on earlier telegraphic
     methods and modern microwave techniques. The "long dash"
     is taken from the signal used to denote the hour mark.
     Reviewed in: *Isis*, 71 (1980), 157–158.

*   Thoren, Victor E. "Kepler's Second Law in England."

     Cited herein as item 713.

*   Tsu, Wen Shion. "The Observations of Halley's Comet in
     Chinese History."

     Cited herein as item 661.

*   Tsu, Wen Shion. "A Statistical Survey of Solar Eclipses
         in Chinese History."

         Cited herein as item 630.

240.   Tumanian, B.E. *History of Armenian Astronomy*. 2 volumes.
         Armenia: Erevan, 1964; 1968.

         General description of history of astronomy in Armenia
         with extensive resumes in English. Reviewed in: *Scientia*,
         105 (1970), 668-670; 106 (1971), 328-329.

241.   Tumanian, B.E. *The Geocentric and Heliocentric Systems
         in Armenia*. Armenia: Erevan University Press, 1973.
         Pp. 135.

         Describes development of cosmological views in Armenia
         beginning in the 5th century B.C. and extending through
         the 19th century. Notes that even in the late 18th century,
         Copernican Heliocentrism was regarded with caution. In
         Armenian with English summary. Reviewed in: *JHA*, 6 (1975),
         67.

242.   Van Tassell, David D., and Michael Hall, eds. *Science
         and Society in the United States*. Homewood, Ill.:
         Dorsey Press, 1966. Pp. vi + 360.

         Includes various essays and brief bibliography on the
         history of astronomy in America.

*   Warner, Brian. "Astronomical Archives in Southern Africa."

         Cited herein as item 63.

243.   Warner, Deborah Jean. "Science Education for Women in
         Antebellum America." *Isis* 69 (1978), 58-67.

         Within a general survey, provides commentary on observa-
         tories, courses of instruction, and textbooks in astronomy
         commonly found at women's colleges. 39 citations.

244.   Warner, Deborah Jean. "Women Astronomers." *Natural History*,
         (May 1979), 12-26.

         Reviews the growth of professionalization of women
         astronomers in the United States since the late 19th
         century. Demonstrates that while women did gain profes-
         sional status, until recently it was secondary to the
         role of men in astronomy.

245. Warner, Deborah Jean. "Astronomy in Antebellum America."
      N. Reingold, ed. *The Sciences in the American Context:
      New Perspectives*. Washington, D.C.: Smithsonian Insti-
      tution Press, 1979. Pp. 55-75.

      Provides comprehensive review of early 19th-century
      astronomy concluding that the American appearance of John
      Herschel's *Treatise on Astronomy* stimulated astronomical
      activity in America.

*   Wright, Helen. *Explorer of the Universe: A Biography of
      George Ellery Hale*.

      Cited herein as item 1390.

246. Yamamoto, Issei. "Japanese Astronomy in Past and Present."
      Isis, 17 (1932), 425-427.

      Table of contents of a lecture delivered on July 24,
      1931, in Kyoto.

247. Yeomans, Donald K. "The Origins of North American
      Astronomy: Seventeenth Century." *Isis*, 68 (1977), 414-
      425.

      Examines the transmission of astronomical technique
      and style from England to America noting the first almanacs
      that appeared in Cambridge and the teaching of Ptolemaic
      astronomy at Harvard; the first telescopes brought in by
      John Winthrop, Jr.; and studies of late 17th-century
      comets. 47 citations.

248. Young, Charles A. "American Astronomy--Its History,
      Present State, Needs and Prospects." *Proceedings of
      the American Association for the Advancement of Science*,
      25 (1876), 35-48.

      Address of the vice president of the AAAS delivered
      at the Buffalo meeting, August 1876.

                    INSTITUTIONAL HISTORIES

*   Abbot, Charles G. *Adventures in the World of Science*.

      Cited herein as item 1223.

*   Bennett, J.A. "The Manuscript Archives of the Royal
      Astronomical Society."

      Cited herein as item 5.

*   Berendzen, R., ed. *Education in and History of Modern
          Astronomy*.

          Cited herein as item 125.

249.   Chapin, Seymour L. "The Academy of Science during the
          Eighteenth Century: An Astronomical Appraisal." *French
          Historical Studies*, 5 (1968), 371-404.

          Detailed examination of the role of the French Academy
          in the support for astronomy to the eve of the Revolution,
          with some commentary on conditions during the revolutionary
          period. Shows that while the Academy had been created to
          promote science and did provide initial support for French
          astronomy, by the end of the century, "it had acted as
          a brake rather than an accelerator upon the scientific
          movement." 98 citations.

*   Chapin, Seymour L. "In a Mirror Brightly: French Attempts
          to Build Reflecting Telescopes using Platinum."

          Cited herein as item 429.

*   Christie, W.H.M. "Universal Time."

          Cited herein as item 870.

*   Edge, David O. "The Sociology of Innovation in Modern
          Astronomy."

          Cited herein as item 165.

*   Edge, David O., and Michael J. Mulkay. *Astronomy Trans-
          formed: The Emergence of Radio Astronomy in Britain*.

          Cited herein as item 486.

*   Fox, Philip. *Adler Planetarium and Astronomical Museum
          of Chicago*.

          Cited herein as item 395.

250.   Hahn, Roger. *The Anatomy of a Scientific Institution.
          The Paris Academy of Sciences, 1666-1803*. Berkeley:
          University of California Press, 1971. Pp. xiv + 433.

          Within this general institutional history, background
          is provided for appreciating the support of progress in
          celestial mechanics in France through the Academy's
          establishment of "prizes" for the solution of leading
          problems in the theory of the solar system. Reviewed
          in: *Annals of Science*, 29 (1972), 313-316; *Isis*, 63
          (1972), 405-407. See also item 249.

251. Hall, R. Cargill. *Lunar Impact, A History of Project Ranger*. Washington, D.C.: National Aeronautics and Space Administration, 1977. Pp. xvii + 450.

Narrative account of the trials of the long ill-fated Ranger series which achieved a measure of success only with the seventh launch. Examines changing priorities of Ranger program from primary scientific research of the lunar surface to a data acquisition step in the Apollo Program, and the many internal problems faced by the Ranger program staff which came as a result of the early launch failures and changing priorities. Extensive citations.

* Herrmann, D.B. "Das Astronomentreffen im Jahre 1798 auf dem Seeberg bei Gotha."

Cited herein as item 201.

251a. Kargon, Robert H. "Temple to Science: Cooperative Research and the Birth of the California Institute of Technology." *Historical Studies in the Physical Sciences*, 8 (1977), 3-31.

Examines George Ellery Hale's efforts at mobilizing resources to transform a small college in Pasadena into a major scientific institution through the establishment of major laboratories for modern physics and chemistry. Follows Hale's success at bringing Robert A. Millikan and other major figures to the California campus. 105 citations.

* Kevles, Daniel J. "Hale and the Role of a Central Scientific Institution in the United States."

Cited herein as item 367.

* Lovell, Bernard. *The Origins and International Economics of Space Exploration*.

Cited herein as item 506.

252. Lovell, Bernard. "The Effects of Defence Science on the Advance of Astronomy." *Journal for the History of Astronomy*, 8 (1977), 151-173.

Examines period 1935 to 1957 "as an epoch when defence science not only influenced but actually revolutionized observational astronomy." Discusses the development of radar and its effect upon radio astronomy; the growth of rocketry to Sputnik; and the growth of scientific manpower and the general supply of astronomers in post-World War II Britain and America. 41 citations.

*   Miller, Howard S. "Astronomical Entrepreneurship in the Gilded Age."

     Cited herein as item 216.

*   Miller, Howard S. *Dollars for Research.*

     Cited herein as item 217.

*   Olmsted, John W. "The Scientific Expedition of Jean Richer to Cayenne (1672-1673)."

     Cited herein as item 899.

253.   Schreiber, John, S.J. "Jesuit Astronomy." *Popular Astronom* 12 (1904), 9-20; 94-112; 230-239; 303-310; 375-385.

     Translation, by William F. Rigge, S.J., of "De Jesuiten des 17. und 18. Jahrhunderts und ihr Verhältnis zur Astronomie," *Natur und Offenbarung*, 49 (1903). Reviews contributions during the first period, 1540-1773, and continues with a detailed topical exposition of Jesuit contributions in the modern period after restoration in 1814. Notes influence in the Orient and studies of the spectra of the sun and stars. Numerous citations.

254.   Schur, Wilhelm. *Beiträge zur Geschichte der Astronomie in Hannover.* Berlin: Weidmann, 1901. Pp. 61.

     Brief biographical sketches of 55 scientists associated with Hannovarian astronomy from the 16th through mid-19th centuries. Notes major bibliographical sources.

*   Struve, Otto, and Velta Zebergs. *Astronomy of the 20th Century.*

     Cited herein as item 113.

255.   Swenson, Loyd S., James M. Grimwood, and Charles C. Alexander. *This New Ocean: A History of Project Mercury.* Washington, D.C.: NASA, 1966. Pp. xv + 681.

     Describes the organizational program, technological progress and scientific contributions of the project. Reviewed in: *Isis*, 58 (1967), 441-443.

*   Wright, Helen. *Explorer of the Universe: A Biography of George Ellery Hale.*

     Cited herein as item 1390.

* Wright, Helen, J.N. Warnow, and Charles Weiner, eds. *The Legacy of George Ellery Hale.*

  Cited herein as item 1391.

OBSERVATORIES

256. Abbot, C.G. "The Astrophysical Observatory of the Smith-sonian Institution." *Astronomical Society of the Pacific Leaflet* No. 216 (Feb. 1947). Pp. 8.

   Describes fifty years of personal recollections of research centering upon continued bolometric observations of the Sun and the detection of some 16 solar cycles ranging in period from seven months to 23 years in length, and all "nearly integral fractions of 273 months." Reprinted in *Annual Reports of the Smithsonian Institution* (1948), 167-175.

257. Abetti, Giorgio. "Astronomy on the Heights." *Astronomical Society of the Pacific Leaflet* No. 237 (Dec. 1948). Pp. 7.

   Briefly reviews the early recognition of advantages of high altitude observing stations. Centers upon establishment and use of high altitude observatories in Italy and Europe in the 19th century, notably the Catania Observatory on Mt. Etna; Pic du Midi in the Pyrenees; the Federal Observatory of Zurich annex at Arosa.

258. Adams, Walter S. "Early Days at Mount Wilson." *Popular Astronomy*, 58 (1950), 64-82; 97-115.

   Reprint of paper from the *Publications of the Astronomical Society of the Pacific*, 59 (1947), based upon reminiscences prepared for the staff of the observatory by the retired director.

259. André, Charles, and G. Rayet. *L'astronomie practique et les observatoires en Europe et en Amérique.* Paris: Gauthier-Villars, 1874. 5 volumes.

   A series of reviews by various authors including André, Rayet, and Angot, of observatories in England, Italy, North and South America, and in the British Colonies.

260.   André, Charles, and A. Angot. *L'astronomie practique et*
        les observatoires en Europe et en Amérique. Paris:
        1874-1881. Pp. 850.

        In five parts, usually bound in three volumes. Describes
        major observatories active in the 19th century. Includes
        illustrations of buildings and equipment.

261.   Bailey, Solon I. "History of the Expedition." *Annals of*
        the Harvard College Observatory, 34 (1895), 1-48.

        History of the Harvard expedition to Peru from 1889 to
        1891 for the photometric observations of southern stars.

262.   Bailey, Solon I. *The History and Work of Harvard Observa-*
        tory, 1839 to 1927. New York: McGraw-Hill, 1931. Pp.
        xiii + 301.

        Historical review of astronomy at Harvard since the
        establishment of the College, with background on New
        England astronomy. Outlines growth of instrumentation,
        eclipse expeditions, the establishment of remote observing
        stations, staff membership and publications. Provides
        a topical review of scientific accomplishments and in-
        cludes extensive section on biographical sketches of
        staff, research associates, and benefactors. Numerous
        citations.

*   Ball, W. Valentine, ed. *Reminiscences and Letters of Sir*
        Robert Ball.

        Cited herein as item 1239.

263.   Bateman, Alfredo D. *El Observatorio Astronomico de*
        Bogotá, 1803-1953. Bogotá: Universidad Nacional, 1954.
        Pp. 189.

        Chronological review of the activities of astronomers
        at the university and at the observatory, in Spanish.
        52 citations.

264.   Beer, Arthur, ed. *Vistas in Astronomy.* Volume 1. London:
        Pergamon Press, 1955. Pp. xvi + 776.

        General review of many aspects of astronomy including
        chapters on its history and philosophy. Volume 1 includes
        review articles on celestial mechanics, dynamics, instru-
        mentation, radio astronomy and solar physics. Subsequent
        volumes contain significant materials on the history of
        astronomy. Reviewed in: *Isis,* 49 (1958), 446-449.

*   Beer, Arthur, and Peter Beer, eds. "The Origins, Achieve-
     ment and Influence of the Royal Observatory, Greenwich:
     1675-1975."

     Cited herein as item 124.

265. Bell, Trudy E. "Garblings and Perversions: The Dudley
     Observatory Controversy." *The Griffith Observer*, 38
     (1974), 2-9.

     Reviews a famous institutional controversy over the
     establishment of observatory policy between B.A. Gould
     and the Trustees of the Observatory circa 1858. No di-
     rect citations. See also items 267 and 326.

266. Bond, William Cranch. "History and Description of the
     Astronomical Observatory of Harvard College." *Annals
     of the Harvard College Observatory*, 1 pt. 1 (1856),
     1-191.

     The history of the establishment of the observatory
     from the early 19th century to the mid-1850s written by
     the first director of the observatory. Includes descrip-
     tion of instrumentation, observatory statutes, listings
     of contributors to the observatory, and yearly reports of
     the Visiting Committee from 1845 to 1855.

267. Boss, Benjamin. *History of the Dudley Observatory: 1852-
     1956*. Albany: Dudley Observatory, 1968. Pp. vi + 123.

     Brief outline including historical narrative; descrip-
     tion of major programs and the formation of the *General
     Catalogue* and the *Astronomical Journal*; and later re-
     search. Eight appendices list staff, bibliography, re-
     search areas, funding sources. Author was director from
     1912 to 1956, succeeding his father, Lewis Boss. See
     also: *Dudley Observatory: Controversy between Dr. Gould
     and the Trustees, 1858-1859*. Albany: Van Benthuysen,
     1858-1859. See also item 326.

*   Bruhns, C., ed. *Vierteljahrsschrift der Astronomischen
     Gesellschaft*.

     Cited herein as item 356.

268. Bushnell, David. *The Sacramento Peak Observatory, 1947-
     1962*. Washington, D.C.: Office of Aerospace Research,
     1962. Pp. v + 77.

Descriptive narrative of the history of the high altitude
solar observatory operated by the Air Force Cambridge
Research Laboratories. Reviews the origins, the development of staff, facilities and funding, and concentrates
on the scientific mission of the agency, centering on
the close reconnaisance of solar activity. Numerous citations.

269.   Cassini, J.D. *Mémoires pour servir à l'histoire des
       sciences et à celle de l'Observatoire royal de Paris*.
       Paris: Bluet, 1810.

A history of the Paris Observatory centering upon its
reorganization and reinstrumentation after 1785. See
also items 274 and 342.

270.   Cincinnati Observatory. *The Centenary of the Cincinnati
       Observatory, 1843-1943*. Cincinnati: University of
       Cincinnati, 1944. Pp. 63.

Series of lectures and papers on the founding of the
Observatory by O.M. Mitchel in 1843 and a description
of the Cincinnati telescope, at its installation the
largest in the United States.

271.   Collins, A. Frederick. *The Greatest Eye in the World*.
       New York: D. Appleton-Century, 1942. Pp. xviii + 266.

Popular and descriptive history of the telescope provid-
ing detailed illustrations of telescopes and accessories.
Includes chapters on the histories of major observatories:
Royal Greenwich, U.S. Naval, Harvard, Yale, Lowell, Lick,
Yerkes, Mount Wilson and Palomar. Light documentation.

*   Cotter, Charles H. "George Biddell Airy and His Mechanical
       Correction of the Magnetic Compass."

Cited herein as item 514.

272.   Cowan, Frances M. "An Account of the Founding of the
       Central Observatory of Poulkova, Russia." *Popular
       Astronomy*, 20 (1912), 415-423.

Reviews the early history of Pulkova and its parent,
the Russian Academy of Sciences, based upon F.G.W. Struve'
original 1845 account entitled: "Description de l'observatoire astronomique central de Poulkova" (St. Petersbourg:
l'Academie imperiale des sciences, 1845). No direct
citations.

273. Dadaev, Aleksandr N. *Pulkovo Observatory: An Essay on Its History and Scientific Activity* [1972]. Kevin Krisciunas, translator. Washington, D.C.: NASA Technical Memorandum 75083, 1978 (National Technical Information Service). Pp. vi + 232.

This translation of a 1972 work concentrates on the 20th century.

274. Denisse, J.F. *Trois siècles d'astronomie, 1667-1967.* Paris: Tournon, 1967. Pp. vi + 78.

Brief review of the foundation and development of the Paris Observatory providing glimpses of the observatory as run by the Cassinis; major geodetic and astronomical projects; celestial mechanics in the 18th century; Le Verrier; optics and photography. Includes an annotated bibliography of rare books and instruments housed at the Observatory, itemized within each topical section describing the many aspects of the observatory. 298 citations.

275. DeSitter, W. *Short History of the Observatory of the University at Leiden, 1633-1933.* Haarlem: J. Enschede, 1933. Pp. 48.

Brief illustrated review of founding of observatory and its development noting major historical figures, growth of staff and instrumentation. Abstract in: *Nature*, 132 (1933), 771-772.

276. Dewhirst, D.W. "Observatories and Instrument Makers in the Eighteenth Century." *Vistas in Astronomy*, 1 (1955), 139-143.

The growth of positional astronomy in 18th-century Britain and Europe is explored within the context of available instrumentation and the positional accuracies obtained. 6 citations.

277. Dick, Julius. "The 250th Anniversary of the Berlin Observatory." *Popular Astronomy*, 59 (1951), 524-535.

Translated from *Die Sterne* (1950). Constitutes a history of this observatory now located in Babelsberg. Examines contributions of its staff including J.E. Bode, J.F. Encke, J. Bernoulli, W.J. Foerster, K.H. Struve and Paul Guthnick, and concentrates on major instrumentation and research areas. No direct citations.

278.   Donelly, Marian Card. *A Short History of Observatories*.
        Eugene: University of Oregon, 1973. Pp. xvi + 163.

        Illustrated popular history from Galileo to the present.
        10-page bibliography and numerous citations.

279.   Dreyer, J.L.E. "Lord Rosse's Six-foot Reflector." *The
        Observatory*, 37 (1914), 399-402.

        Brief description of this major speculum reflector
        based upon direct knowledge. Numerous indirect citations.

280.   Dvoichenko-Markoff, Eufrosina. "The Pulkovo Observatory
        and Some American Astronomers in the Mid-Nineteenth
        Century." *Isis*, 43 (1952), 243-246.

        Traces contact between American and Russian astronomers
        since the 1761 and 1769 transit expeditions which brought
        them together. Notes reciprocal memberships in Russian
        and American scientific societies, highlighting the sig-
        nificance of F.G.W. Struve's election to the American
        Academy of Arts and Sciences in 1834 as one which brought
        knowledge of the Pulkovo Observatory to America, and
        established its influence in the design of the Harvard
        College Observatory. Examines S. Newcomb's later contacts
        with Pulkovo and Russian astronomy. 28 citations.

*   Edge, David O., and Michael J. Mulkay. *Astronomy Trans-
        formed: The Emergence of Radio Astronomy in Britain*.

        Cited herein as item 486.

281.   Evans, D.S., T.J. Deeming, B.H. Evans, and S. Goldfarb,
        eds. *Herschel at the Cape, Diaries and Correspondence,
        1834-1838*. Austin: University of Texas Press, 1969.
        Pp. xxxv + 398.

        Transcriptions of John Herschel's diaries from 1833
        to 1838 while he was working at the Cape of Good Hope
        completing general sweeps of the sky for nebulae and
        double stars initiated by his father. Includes letters,
        travel diaries, sketches, plans; all edited with detailed
        annotation and introductory remarks. Reviewed in: *Isis*,
        60 (1969), 581-583; *Annals of Science*, 26 (1970), 171-
        172.

282.   Fernie, Donald. *The Whisper and the Vision*. Toronto:
        Clarke, Irwin & Co., 1976. Pp. viii + 189.

        Popular anecdotal narrative of 18th- and 19th-century
        eclipse expeditions, geodetic expeditions, and observing

stations set up in remote parts of the world. Examines
the French and British expeditions to observe the transits
of Venus in 1761 and 1769, David Gill's measurement of
the solar parallax by the geocentric parallax of Mars,
and the history of the Cape Observatory and Harvard's
stations in South America. Indirect citations to secondary
sources.

283. Feuillebois, Genevieve. "Les Manuscrits de la Biblio-
      thèque de l'Observatoire de Paris." *Journal for the
      History of Astronomy*, 6 (1975), 72-74.

      Brief description of the history of the Observatory
      and the manuscripts held in its library including those
      of Laplace and LeVerrier. Notes organization of library
      holdings.

284. Forbes, Eric G. "Dr. Bradley's Astronomical Observations."
      Quarterly Journal of the Royal Astronomical Society,
      6 (1965), 321-328.

      Recounts Bradley's efforts at securing better instrumenta-
      tion for the Royal Greenwich Observatory, and the observa-
      tional programs he maintained. Concentrates upon the
      difficulties which resulted from the subsequent removal
      of Bradley's observations from Greenwich by the executors
      of Bradley's estate, and their eventual return to the
      public domain. Examines Bessel's reduction of Bradley's
      observed star places some years later. 18 citations.

285. Forbes, Eric G. "Index of the Board of Longitude Papers
      at the Royal Greenwich Observatory." *Journal for the
      History of Astronomy*, 1 (1970), 169-179; 2 (1971),
      58-70, 133-145.

      Description and short history of the Board of Longitude
      prefaces the Index.

286. Forbes, Eric G. "The Foundation of the First Göttingen
      Observatory: A Study in Politics and Personalities."
      Journal for the History of Astronomy, 5 (1974), 22-29.

      Examines the influential role of Johann Andreas Segner
      in creating and operating the Observatory and the personal
      conflict between Segner and Tobias Mayer in these develop-
      ments circa 1750-1754. 50 citations.

287. Forbes, Eric G., A.J. Meadows, and Derek Howse. *Greenwich
      Observatory: The Royal Observatory at Greenwich and
      Herstmonceaux 1675-1975*. 3 volumes. London: Taylor and
      Francis, 1975. Pp. xv + 204; xi + 135; xix + 178.

Detailed history of the institution founded to provide
new and better methods for determining longitude at sea,
but which, from its inception under Flamsteed, made signal
contributions to all aspects of positional astronomy. The
three volumes, authored respectively by the named historians, cover the entire history and all aspects of research at the Observatory, including the vicissitudes of
its removal, post-World War II, to Herstmonceaux in Sussex
and its expansion into modern astrophysics. Volume 1 is
largely a chronological development carrying through the
Astronomers Royal up to the appointment of G.B. Airy in
1835. Volume 2 examines Airy's tenure in detail and then
divides up the recent history of the Observatory by topic
reviewing positional astronomy, the time service and the
introduction of modern observational astrophysics after
the turn of the century. Volume 3 identifies buildings
and instrumentation. Extensive notes and bibliography.
Essay reviews in: *Annals of Science*, 34 (1977), 63-70;
JHA, 10 (1979), 122-126; *Isis*, 68 (1977), 296-299.

*   Forbes, G. *David Gill: Man and Astronomer*.

Cited herein as item 1286.

*   Frost, Edwin Brant. *An Astronomer's Life*.

Cited herein as item 1288.

288.   Furness, Caroline E. "The Longitude of the Vassar College
Observatory." *Publications of the Vassar College Observatory* No. 4 (1934). Pp. 1-23.

Provides a descriptive history of the Observatory from
Maria Mitchell's tenure (1865-1877) to the present, based
upon historical accounts of efforts to establish the
longitude of the Observatory.

289.   Gautier, Raoul, and Georges Tiercy. *L'Observatoire de
Genève*. Genève: A. Kundig, 1930. Pp. 172.

General history of astronomical activities in Geneva
beginning with the observatory of Jacques-André Mallet
in 1772; the establishment of the Geneva Observatory
in 1830, and the state of the Observatory in 1930. Brief
citations.

290.   Gill, David. "An Astronomer's Work in a Modern Observatory." *Royal Institution Library of Science: Astronomy*
(item 140), 362-377.

Lecture dated May 29, 1891. Describes the buildings and instruments of the Royal Observatory, Cape of Good Hope, and its astrometric programs, and then turns to descriptions of the astrophysical observatories of William Huggins and Potsdam.

291. Gill, David. *History and Description of the Royal Observatory at the Cape of Good Hope*. London: H.M. Stationery Office, 1913. Pp. 190 + 136.

Detailed review of instrumentation and techniques of observation with historical chapters covering the period from F. Fallows' directorship beginning in 1820 through those of Henderson, Maclear, Smyth, Gill and others, to that of Hough, commencing in 1907.

*   Greenstein, Jesse L. "The Seventieth Anniversary of Professor Joel Stebbins and of the Washburn Observatory."

Cited herein as item 1292.

292. Hale, George Ellery. "The Aim of the Yerkes Observatory." *Astrophysical Journal*, 6 (1897), 310-321.

Read at dedication ceremonies of the Yerkes Observatory, October 19, 1897. Describes his general plan of work for the Observatory including all areas of solar research, double star measures, parallaxes, radial velocities, analysis of stellar spectra and laboratory spectroscopy. Discusses need to establish competent support services in optics and instrumentation.

293. Hayes, E. Nelson. *Trackers of the Skies*. Cambridge, Mass.: Howard A. Doyle, 1968. Pp. xiii + 169.

Reviews the origin and development of the Smithsonian Astrophysical Observatory program to organize an optical tracking program in the mid-fifties for the U.S. satellite program. Reviewed in: *Technology & Culture*, 10 (1969), 620-622; *Isis*, 63 (1972), 127.

294. Herrmann, D.B. "An Exponential Law for the Establishment of Observatories in the Nineteenth Century." *Journal for the History of Astronomy*, 4 (1973), 57-58.

Derives an empirical relationship for the number of observatories in existence as a function of time through the 19th century and finds that the number in existence doubled every 34 years. Abstract of detailed analysis in: *Die Sterne*, 49 (1973), 48-52. See also items 163, 170, and 823.

295. Herrmann, D.B. "Zur Vorgeschichte des Astrophysikalischen
      Observatorium Potsdam (1865 bis 1874)." *Astronomische
      Nachrichten*, 296 (1975), 245-259.

      Examines the origin and influences of Wilhelm Foerster's
      1871 paper "Denkschrift betreffend die Einrichtung einer
      Sonnenwarte" on the foundation of the Potsdam Astrophysical
      Observatory, the first national institute created for the
      study of astrophysics. Discusses K.F. Zöllner's influence
      on the research programs at Potsdam, and the attempts to
      secure Kirchhoff as first director. 67 citations.

296. Herrmann, D.B. "Wilhelm Foerster und die Gründung des
      Astrophysikalischen Observatoriums Potsdam." *Sternzeiten*,
      2 (1977), 29-33.

      Examines to what extent Foerster contributed to the
      creation of Potsdam and his relationship with other major
      figures of the early history of the Observatory, notably
      Herman Vogel.

*   Herrmann, D.B. "Julius Scheiner und der Erste Lehrstuhl
      für die Astrophysik an der Universität Berlin."

      Cited herein as item 347.

297. Hirshfeld, Alan W. "The Leviathan of Parsonstown: An
      Astronomer's Dream." *The Griffith Observer*, 38 (1974),
      2-6; 10.

      Reviews the construction of large speculum metal re-
      flecting telescopes by William Parsons, the Third Earl
      of Rosse, and his subsequent observations of spiral
      nebulae. 5 citations.

298. Hodge, John E. "Charles Dillon Perrine and the Transforma-
      tion of the Argentine National Observatory." *Journal
      for the History of Astronomy*, 8 (1977), 12-25.

      Chronicles development of the Cordoba facility in 1870
      and its early history, and then focusses upon Perrine's
      appointment as director in 1909 and his many projects
      at Cordoba including the construction of a 60-inch re-
      flector during some twenty years of his tenure until his
      forced retirement in 1936. Analyzes Perrine's difficulties
      with the faculty of the University of Cordoba and with
      Argentine political figures. 52 citations.

299. Hoffleit, Dorrit. "The Library of Harvard College
      Observatory." *Harvard Library Bulletin*, 5 (1951), 102-
      111.

Contrasts the nature of this major library in 1950 from its character in 1850. As of 1951 it held 15,000 monographs and serials, 19,000 pamphlets and 6000 star maps. Reviews the origins of the library, its early supporters, and the history of its cataloguing efforts, and provides a short list of its holdings of books printed before 1600. Adds a brief comparison of the size of the Harvard collection to those at the Naval Observatory, Mount Wilson Observatory, and the Lick and Yerkes Observatories. 9 citations.

300. Hogg, A.R. "The Commonwealth Observatory." *Publications of the Astronomical Society of the Pacific Leaflet* No. 294 (Oct. 1953). Pp. 8.

Brief history of the origins (in 1905) and development of the Commonwealth Observatory, now called the Mt. Stromlo Observatory. Describes funding problems, delays, the opening of the "Commonwealth Solar Observatory" in 1924, and first instrumentation.

301. Holden, Edward S. *Handbook of the Lick Observatory*. San Francisco: Bancroft Co., 1888. Pp. 135.

Popular exposition including a short biographical sketch of the life of James Lick, the history of the building of the Lick Observatory, a description of its buildings and instruments, and a detailed accounting of "the principal observatories of the world."

302. Howse, Derek. *Francis Place and the Early History of the Greenwich Observatory*. New York: Science History Publications, 1975. Pp. 64.

Description of the design and building of the Greenwich Observatory through the copper plates of the 17th-century artist Francis Place illustrating many of the original instruments. Essay review in: *JHA*, 10 (1979), 122–126. Reviewed also in: *Annals of Science*, 34 (1977), 63–70; *Isis*, 68 (1977), 296–299.

*   Howse, Derek, et al. *An Inventory of the Navigation and Astronomy Collections in the National Maritime Museum, Greenwich*.

Cited herein as item 400.

303. Hoyt, William Graves. "Historical Note: Astronomy on the San Francisco Peaks." *Plateau: The Quarterly of the Museum of Northern Arizona*, 47 (1975), 113–117.

Reviews establishment and maintenance of a high altitude observing station by the Lowell Observatory to make spectroscopic observations in the ultra-violet and infrared, of specific interest in planetary studies. 10 citations.

* Hoyt, William Graves. *Lowell and Mars*.

Cited herein as item 954.

* Huggins, William, ed. *The Scientific Papers of Sir William Huggins*.

Cited herein as item 1316.

304. Jones, Bessie Zaban. *Lighthouse of the Skies. The Smithsonian Astrophysical Observatory: Background and History 1846-1955*. Washington, D.C.: Smithsonian Institution, 1965. Pp. xv + 339.

Traces the origins of the Smithsonian Institution and the long quest for the foundation there of an astronomical observatory through the efforts of Joseph Henry and Spencer F. Baird and its final establishment under S.P. Langley. Reviews the development of the Observatory's range of activities including Langley's early field stations for the measurement of solar radiation continued by Charles G. Abbot; Langley's experiments in aerodynamics; eclipse expeditions; Robert Goddard's experiments in rocketry supported by Abbot; and events leading up to the transfer of the Observatory to Cambridge in 1955-1956. Includes many sidelights on the related activities of the primary figures, especially Henry's involvement with the establishment of the Lick Observatory. Appendices include a list of staff members, 1890-1955, and major projects of the Observatory 1895-1955 compiled by C.G. Abbot. Extensive notes and citations.

305. Jones, Bessie Zaban, and Lyle Gifford Boyd. *The Harvard College Observatory: The First Four Directorships, 1839-1919*. Cambridge, Mass.: Harvard University Press, 1971. Pp. xiv + 495.

Traces the origins of the Observatory from the mid-17th century, and the official founding in 1839. Emphasizes the life and career of the fourth director, E.C. Pickering, and the many contributions he and his Observatory made during his long tenure, from 1876 to 1919, to the progress of stellar astrophysics including the establishment of vast projects to determine the magnitudes

and spectral classes of the stars. Traces the role of
women on the staff of the Observatory. Extensive citations. Reviewed in: *Isis*, 64 (1973), 560-63; *JHA*, 8 (1977),
214-215.

*   Kaye, G. Rusby. *The Astronomical Observatories of Jai
      Singh*.

      Cited herein as item 413.

*   Killick, Victor W. "California's Early Astro-Geodetic
      Observatories."

      Cited herein as item 895.

*   King, H.C. *The History of the Telescope*.

      Cited herein as item 99.

*   Knight, William H. "Some Telescopes in the United States."

      Cited herein as item 440.

306. Knox-Shaw, H. "The Radcliffe Observatory at Pretoria."
      Astronomical Society of the Pacific Leaflet No. 272
      (Dec. 1951). Pp. 8.

      Brief description of the founding of this Observatory
      by the closing of the Radcliffe Observatory at Oxford
      and its eventual transfer to and expansion in South
      Africa to take advantage of better climate and untapped
      southern skies. Reviews the construction of the 74-inch
      Radcliffe reflector.

307. Knox-Shaw, H. "The Radcliffe Observatory." *Vistas in
      Astronomy*, 1 (1955), 144-149.

      Short history of the Radcliffe Observatory of the
      University of Oxford, built in 1771 for the use of the
      Savilian Professor of Astronomy, Thomas Hornsby. Reviews
      the succession of Radcliffe Observers, and the removal
      of the Observatory, in 1935-1937, to Pretoria, South
      Africa. No direct citations.

308. Krisciunas, K. "A Short History of Pulkovo Observatory."
      Vistas in Astronomy, 22 (1978), 27-37.

      Follows developments since its founding in 1839 by
      F.G.W. Struve, and examines its instrumentation, staff,
      and research in fundamental positional astronomy and
      geodesy. Describes Pulkovo's difficulties during and

after the Revolution, its restoration, and hints at the magnificent historical holdings of its famous but largely destroyed library. 9 citations.

*   Kuiper, Gerard P. "German Astronomy during the War."

Cited herein as item 208.

309. Laurie, P.S. "The Board of Visitors of the Royal Observatory." *Quarterly Journal of the Royal Astronomical Society*, 7 (1966), 169–185; 8 (1967), 334–353.

Reviews the succession of "Royal Warrants" for the appointment of Visitors to the Royal Observatory as a body authorized to examine the operation of the facility and make recommendations to the Admiralty concerning the observed state of the facility and proposed future programs and staff changes. Examines changes in the nature of the Board as the Royal Astronomical Society gained in prominence necessitating a reorganization of the Board's Charter to include nominations from the RAS as well as from the Royal Society.

*   Laurie, P.S., and D.W. Waters. "James Bradley's New Observatory and Instruments."

Cited herein as item 441.

*   Lick Observatory Archives. *Preliminary Finding Aid to the Archives of the Lick Observatory*.

Cited herein as item 41.

310. Loomis, Elias. "Astronomical Observatories in the United States." *Harper's New Monthly Magazine*, 13 (June 1856), 25–52.

General itemized review extending his discussion in item 104.

311. Lou, Kao. *The Past and Future of the Peking Central Observatory*. Peking: The Central Observatory, 1930. Pp. 27 + 21 (Chinese and English texts).

Brief description of early astronomical instruments with commentary on present activities.

*   Lovell, Bernard. *The Story of Jodrell Bank*.

Cited herein as item 490.

*   MacPike, E.F. *Hevelius, Flamsteed, Halley: Three Contem-
       porary Astronomers and Their Mutual Relations.*

       Cited herein as item 1338.

312.   Mailly, E. *Tableau de l'astronomie dans l'hemisphère
       austral et dans l'Inde.* Bruxelles: F. Hayez, 1872.
       Pp. 232.

       Reviews 18th- and early 19th-century expeditions to
       the south and east for astronomical observations and
       describes observatories at Madras, Lucknow, Santiago,
       and in Australia. Some citations.

313.   Maunder, E. Walter. *The Royal Observatory Greenwich.*
       London: Religious Tract Society, 1900. Pp. 320.

       Well-illustrated popular account centering upon the
       established departments of the Observatory circa 1900.
       The first four chapters provide a historical introduction
       from Flamsteed to Christie. No direct citations.

314.   Mayall, N.U. "Mexico Dedicates a New Observatory."
       Publications of the Astronomical Society of the Pacific,
       54 (1942), 117–123.

       Describes the founding of the Mexican National Astro-
       physical Observatory in Tonanzintla, its instrumentation,
       and the international ceremonies at its dedication,
       which included a public audience estimated at 10,000
       persons.

315.   McCrea, W.H. *The Royal Greenwich Observatory.* London:
       H.M. Stationery Office, 1975. Pp. 80.

       Introduces the history of the Observatory through an
       illustrated chronological survey of the Astronomers Royal
       and a description of present-day instrumentation.

316.   McCrea, W.H. "The Royal Greenwich Observatory, 1675–1975."
       Quarterly Journal of the Royal Astronomical Society,
       17 (1976), 4–24.

       Provides a brief history of the Observatory. Centers
       upon traditionally "fringe" activities in its past, in-
       cluding its relations with the Board of Longitude, its
       interests in geomagnetism and its participation in inter-
       national projects. Attempts to determine the origins of
       these originally peripheral interests, and argues that
       "the sort of activities that have been 'fringe' in the
       past are now in process of becoming central for the fore-
       seeable future." 5 citations.

317. Melmore, Sidney. "Nathaniel Pigott's Observatory 1781–
      1793." *Annals of Science*, 9 (1953), 281–286.

      Brief description based upon extracts from drafts of
      letters by Pigott, and from journals and observing books
      kept by Pigott and later by his son Edward. Pigott, an
      associate of John Goodricke, made extensive observations
      of meridian transits, zenith distances, and observed the
      transit of Mercury on 3 May 1786. 10 citations.

318. Mikhelson, N.N. "The Construction of Astronomical Instru-
      ments at Pulkovo Observatory." *Izvestiya Glavnoi Astro-
      nomicheskoi Observatorii v Pulkove*, 24 (1964), 12–29.

      Describes (in Russian) construction and history of use
      of major instrumentation at Pulkovo since the mid-19th
      century. 135 citations.

319. Milham, Willis I. *Early American Observatories*. Williams-
      town, Mass.: Williams College, 1938. Pp. 158.

      Examines the question of the identity of the first
      permanent astronomical observatory in the United States.
      Considers eleven early observatories including those of
      David Rittenhouse circa 1769–1796, and installations at
      Yale, Harvard, North Carolina, Wesleyan (Middletown, Ct.),
      Williams College and elsewhere. Makes no definite con-
      clusion except for noting that the Hopkins Observatory
      of Williams College and the Observatory of Western Reserve
      College are the oldest still in existence. See also:
      Popular Astronomy, 45 (1937), 465–474; 523–540. Includes
      numerous citations and selected bibliography.

320. Moore, J.H. "Fifty Years of Research at the Lick Observa-
      tory." *Publications of the Astronomical Society of the
      Pacific*, 50 (1938), 189–203.

      General survey of principal research programs completed
      or in progress including S.W. Burnham's first catalogues
      of double stars and the extension of this work by R.G.
      Aitken; R.H. Tucker's early positional work; E.E. Barnard's
      wide-field photographic surveys; the construction of
      spectroscopic equipment and their use by Keeler, Campbell,
      Curtis, Moore and Wright for determining stellar radial
      velocities and the nature of the chief nebular line.
      Reviews the many expeditions, including temporary eclipse
      sites and the semi-permanent D.O. Mills Expedition to
      Chile to complete the radial velocity survey for southern
      hemisphere stars. Includes mention of later work, notably
      that of Trumpler on clusters and the detection of absorp-
      tion in space.

321. Moore, Patrick. *The Astronomy of Birr Castle*. London: Mitchell Beazley, 1971. Pp. xii + 81.

Portrayal of the history of the lords of Birr Castle centering upon the Third and Fourth Earls of Rosse who in the 19th century produced the largest speculum reflector (6-foot) ever in existence which led to the discovery of spiral nebulae. Reviewed in: *JHA*, 3 (1972), 146.

322. Müller, Peter. *Sternwarten-Architektur und Geschichte der astronomischen Observatorien*. Frankfurt: Lang, 1975. Pp. 260.

A collection of data on observatories from all times and locales identified by their architecture. Bibliographical references. Reviewed in: *JHA*, 7 (1976), 207-208.

* Musto, David F. "A Survey of the American Observatory Movement, 1800-1850."

Cited herein as item 219.

* Needham, Joseph. "The Peking Observatory in A.D. 1280 and the Development of the Equatorial Mounting."

Cited herein as item 415.

323. Neubauer, F.J. "A Short History of the Lick Observatory." *Popular Astronomy*, 58 (1950), 201-222; 318-334; 369-388.

General review of the origins, construction and early staff of the Observatory with a discussion of its general institutional research programs and instruments to the death of James Keeler in 1900.

324. Newcomb, Simon. "Astronomical Observatories." *The Observatory*, 5 (1882), 247-253.

Reviews the history of the development of observatories through an identification of funding sources. No direct citations.

325. Nourse, Joseph Everett. "Memoir of the Founding and Progress of the United States Naval Observatory." *Washington Observations, 1871*. Washington, D.C.: Government Printing Office, 1873. Appendix IV. Pp. 52.

Detailed account of the governmental deliberations involved in setting up the Observatory, from 1810 through the mid-19th century. Includes extensive citations to official state papers and Congressional Reports.

326.   Olson, Richard G. "The Gould Controversy at Dudley Ob-
        servatory. Public and Professional Values in Conflict."
        Annals of Science, 27 (1971), 265-276.

        Reviews the origins of the Dudley Observatory in
        Albany, New York, which was in part made possible through
        the public speeches of O.M. Mitchell calling for popular
        support of astronomy in Cincinnati and later in Albany.
        As a result of Mitchell's orations, Dudley was founded,
        and its acting director, B.A. Gould, was installed. But
        Gould's vision of Dudley was that of a pure astronomical
        observatory that would not have to cater to popular
        interests. The Trustees of the Dudley Observatory thought
        otherwise, hence the controversy. 23 citations.

*   Osterbrock, Donald E. "The California-Wisconsin Axis in
        American Astronomy."

        Cited herein as item 222.

327.   Page, Thornton. *Observatories of the World*. Cambridge:
        Smithsonian Astrophysical Observatory, 1967. Pp. vii +
        41.

        Provides listings by countries of the instrumentation
        in place at major optical and radio observatories after
        a series of short notes on the various programs observa-
        tories follow, the types of instruments available for
        contemporary research, and a brief historical note on
        the development of observatories. 7 citations.

328.   Pickering, David B. "The Astronomical Fraternity of the
        World." *Popular Astronomy*, 35 (1927), 157-169.

        Beginning of a series of illustrated articles based
        upon a tour of European observatories in 1926.

329.   Plotkin, Howard. "Astronomers versus the Navy: The Revolt
        of American Astronomers over the Management of the
        United States Naval Observatory, 1877-1902." *Proceedings
        of the American Philosophical Society*, 122 (1978), 385-
        399.

        Shows that in its early years, the Naval Observatory
        did not operate under a clear mandate from the U.S.
        Congress, its mission was vague, and though well supported,
        it was not a vital institution responding to the needs
        of modern astronomy. As a result, many American astronomers
        began to petition the government to put the USNO under
        civilian control resulting "in a full scale revolt over

the management of the USNO during the last quarter of
the nineteenth century." 135 citations.

330. Polish Astronomical Society, ed. *Astronomical Observatories
in Poland*. Warsaw: Polish Scientific Publishers, 1973.
Pp. 61.

Description of major observatories with commentary on
their recent history.

331. Potsdam Observatory. *Die Königlichen Observatorien für
Astrophysik, Meterologie und Geodasie bei Potsdam*.
Berlin: Mayer & Müller, 1890. Pp. 159.

Detailed summary of the first years of the formation
of this national facility, the first in the world for
pure astrophysics. Early staff included Auwers, Foerster
and Kirchhoff as co-directors in 1877. H.C. Vogel, J.
Wilsing, J. Scheiner and many others soon joined the
staff, which then also became a major international center
for advanced training in astrophysics.

* Price, Derek J. "The Early Observatory Instruments of
Trinity College, Cambridge."

Cited herein as item 404.

332. Pulkova Observatory. "Poulkovo Observatory and Nikolayev
Observatory." *Izvestiya Glavnoi Astronomicheskoi
Observatorii v Pulkove*, 22 (1965), 3-73.

Eight papers (in Russian) reviewing the past 50 years'
work of these observatories specializing in astrometry
and time service.

* Reingold, Nathan. "Cleveland Abbe at Pulkova: Theory and
Practice in the Nineteenth Century Physical Sciences."

Cited herein as item 225.

* Royal Astronomical Society. *General Index*.

Cited herein as item 50.

* Rufus, W.C. "Astronomical Observatories in the United
States Prior to 1848."

Cited herein as item 228.

333. Schweiger-Lerchenfeld, A. *Atlas der Himmelskunde*. Wien:
A. Hartleben, 1898. Pp. 259 + 62 plates.

Lavishly illustrated popular history in folio centering
upon late 19th-century observatories, their instrumentation, and techniques of observation.

334.   Searle, Arthur. "Historical Account of the Observatory
       from October 1855, to October 1876." *Annals of the
       Harvard College Observatory*, 8 (1876), 1-65.

Extends the history of the Observatory from Bond's
account (item 266) to 1876 and the beginning of E.C.
Pickering's tenure as director. Includes brief biographical notes on the Bonds, Joseph Winlock, and P.S. Coolidge;
an itemization of the property and funds held by the
Observatory; equipment on hand; and the work of the Observatory.

*   Shane, Mary Lea. "The Archives of Lick Observatory."

Cited herein as item 56.

*   Sidgreaves, Walter. "Spectroscopic Studies of Astrophysical
        Problems at Sonyhurst College Observatory."

Cited herein as item 1083.

335.   Skinner, A.N. "The United States Naval Observatory."
       Science, 9 (Jan. 6, 1899), 1-16.

Reviews scope and mission of the Observatory within
the context of the question of a need for a national
observatory. See various statements by many contemporary
astronomers and officials: *Science*, 9 (March 31, 1899),
465-467. See also item 329.

336.   Sterns, Mabel. *Directory of Astronomical Observatories
       in the United States*. Ann Arbor: J.W. Edwards, 1947.
       Pp. 162.

Listing includes name, location, ownership and description of hundreds of public, private, amateur and professional observatories.

336a. Stratton, F.J.M. "The History of the Cambridge Observa-
       tories." *Annals of the Solar Physics Observatory,
       Cambridge*, 1 (1949). Pp. iv + 26 + 8 plates.

Describes early observatories dating from 1739, the
founding of the University Observatory in the 1820s, the
origin of the Plumian Professorship and Observatory
Syndicate, as well as the many historical astronomical
instruments in use there. Identifies the origins of the

Solar Physics Observatory, moved from London to Cambridge circa 1911, and provides listings of staff members of the combined observatories. Some citations.

337. Stroobant, Paul, ed. *Les observatoires astronomiques et les astronomes.* Brussels: Observatoire Royal de Belgique, 1907; 1931; 1936; 1959. Pp. 316 + 314 + 106 + 452.

Four surveys compiling data on observatories worldwide including exact location, staff, brief history, equipment. The 1931, 1936 and 1959 surveys were conducted under the auspices of the IAU. After World War II, survey reorganized under direction of F. Rigaux.

* Szanser, Adam J. "F.G.W. Struve (1793-1864). Astronomer at the Pulkovo Observatory."

Cited herein as item 1373.

* Szanser, Adam J. "Johannes Hevelius (1611-1687). Astronomer of Polish Kings."

Cited herein as item 1374.

338. Tatham, W.G., and K.A. Harwood, "Astronomers and other Scientists on St. Helena." *Annals of Science*, 31 (1974), 489-510.

Reviews the history of astronomical work undertaken on this island from Edmond Halley's establishment of an observatory there from 1677 through March 1678; David Gill's visits in 1877 and the visit of David Todd and Cleveland Abbe in 1890. Describes other expeditions for magnetic observations, the transits of Venus, and contemporary efforts to locate the site of Halley's observatory. 33 citations.

* Warner, Brian. "Cape of Good Hope Royal Observatory Papers in the Archives of the Royal Greenwich Observatory."

Cited herein as item 64.

339. Warner, Brian. *Astronomers at the Royal Observatory Cape of Good Hope.* Cape Town: A.A. Balkema, 1979. Pp. xii + 132.

Brief study, centering upon 19th-century developments, of the growth of astronomy and the establishment of observatories, but especially the establishment and history of the Royal Observatory at the Cape of Good Hope. Re-

views the contributions of Fearon Fallows, Thomas Henderson, Thomas Maclear, E.J. Stone, David Gill, and 20th-
century directors of the Observatory. No direct citations;
brief bibliography. Reviewed in: *JHA*, 11 (1980), 214-215.

*   Watson, R.D., and J.M. Watson. "The Great Melbourne
        Telescope."

        Cited herein as item 461.

340.   Weber, Gustavus Adolphus. *The Naval Observatory; Its
        History, Activities, and Organization*. Baltimore: Johns
        Hopkins Press, 1926. Pp. xii + 101.

        Part of the service monograph series by Weber conducted
        for the Institute for Government Research. Includes
        chapters on the early efforts to found an astronomical
        observatory, the growth of the Nautical Almanac Office
        from 1849 to 1894, and a history of the Naval Observatory
        from 1866 to 1926. Identifies activities of the Observa-
        tory; its organization, administrative and scientific
        structure; and provides appendices on its financial and
        legal structure. Includes detailed bibliography.

341.   Whitford, A.E. "Astronomy and Astronomers at the Mountain
        Observatories." *Education in and History of Modern
        Astronomy* (item 125), 202-210.

        Review of early interests and activity in locating
        observatories upon remote mountain sites. Centers upon
        work at the Lick and Hale Observatories. 28 citations.

*   Wilson, Margaret. *Ninth Astronomer Royal, The Life of
        Frank Watson Dyson*.

        Cited herein as item 1387.

342.   Wolf, C. *Histoire de l'observatoire de Paris de sa
        fondation à 1793*. Paris: Gauthier-Villars, 1902. Pp.
        xii + 392.

        Comprehensive study detailing the laborious planning
        of the Royal Observatory in the late 17th century with
        an exhaustive description of the early Observatory, its
        instrumentation and intended function. Includes commentary
        on the development of the Observatory, its periods of
        inactivity, and its major restoration by Cassini IV.
        Provides chronological listing of staff and diagrams of
        the floor plans of the buildings. Numerous citations and
        annotation. See also items 269 and 274.

343. Woodbury, David O. *The Glass Giant of Palomar*. New York: Dodd, Mead, 1939. Pp. xiii + 368.

   Popular narrative of the making of the 200-inch Palomar telescope from its inception by George Ellery Hale in the 1920s, efforts to secure funds for a large telescope, design studies, and the long period of final design and construction of the telescope. Written while the telescope was still under construction and when the approaching war threatened to delay the project. Provides a brief popular review of the history of telescope making. No direct citations and no index.

* Woolley, Richard. "James Bradley, Third Astronomer Royal."

   Cited herein as item 1388.

344. Wright, Helen. *The Great Palomar Telescope*. London: Faber and Faber, 1953. Pp. 176.

   Traces the history of the telescope as used in astronomy from Galileo emphasizing the growth of observatories in America and the progress of spectroscopic techniques. The origin, funding, design and construction of the 200-inch Palomar telescope is then presented centering on George Ellery Hale's arguments for the need of a great telescope, and his organizational and entrepreneurial efforts that led to its construction. No direct citations and no index. Reviewed in: *Annals of Science*, 10 (1954), 80-81.

* Wright, Helen. *Explorer of the Universe: A Biography of George Ellery Hale*.

   Cited herein as item 1390.

* Wright, Helen, J.N. Warnow, and Charles Weiner, eds. *The Legacy of George Ellery Hale*.

   Cited herein as item 1391.

345. [Yerkes Observatory Staff]. "The Yerkes Observatory, 1897-1947." *Science*, 106 (1947), 195-220.

   Series of articles by Robert Hutchins, Walter S. Adams, G.P. Kuiper, W.W. Morgan, O. Struve, G. van Biesbroeck, S. Chandrasekhar and G. Herzberg commemorating the 50th anniversary of the founding of the Observatory and of the American Astronomical Society, both the results of the efforts of George Ellery Hale. Topics include remi-

niscences of the early years of the Observatory, studies
of the sun and stars at the Observatory including astrometry and spectroscopy and theoretical astrophysics.

UNIVERSITIES AND COLLEGES

346. Brown, Sanborn C., and Leonard M. Rieser. *Natural
     Philosophy at Dartmouth. From Surveyors' Chains to
     the Pressure of Light.* New Hampshire: University Press
     of New England, 1974. Pp. xii + 127.

     Provides biographical sketches of ten professors of
     Natural Philosophy at Dartmouth including the late 19th-
     century pioneer in astronomical spectroscopy, Charles A.
     Young. Reviewed in: *Isis,* 67 (1976), 289-290.

*   Cohen, I. Bernard. *Some Early Tools of American Science.*

     Cited herein as item 392.

*   Furness, Caroline E. "The Longitude of the Vassar College
     Observatory."

     Cited herein as item 288.

347. Herrmann, D.B. "Julius Scheiner und der Erste Lehrstuhl
     für die Astrophysik an der Universität Berlin." *NTM-
     Schriftenreihe für Geschichte der Naturwissenschaften,
     Technik und Medizin,* 14 (1977), 33-42.

     Traces the trends in the growth and interests of the
     Potsdam Astrophysical Observatory staff that eventually
     aided in the establishment of a professorship in astro-
     physics at the University of Berlin first held by Julius
     Scheiner, an influential spectroscopist and author of
     major astrophysics texts (see item 930). Traces Scheiner's
     professional research interests in Bonn and Potsdam, and
     surveys his course offerings at Berlin between 1895 and
     1911, which included photography, photometry, spectral
     analysis of the stars, the temperatures of the Sun and
     stars, and the theory of spectra. 19 citations.

*   Hodge, John E. "Charles Dillon Perrine and the Transforma-
     tion of the Argentine National Observatory."

     Cited herein as item 298.

348. Musto, David F. "Yale Astronomy in the Nineteenth Century."
     Ventures, 8 (1968), 7-18.

     Demonstrates that the lack of institutional support for
     basic science at Yale during the 19th century caused Yale
     astronomers to seek out and maintain commercially viable
     technical services. These services included telegraphy
     of time signals and the calibration of thermometers and
     clocks. No direct citations.

349. Plotkin, Howard. "Henry Tappan, Franz Brünnow, and the
     Founding of the Ann Arbor School of Astronomers, 1852-
     1863." *Annals of Science*, 37 (1980), 287-302.

     Traces the origins and development of professional
     training in astronomy at the University of Michigan; the
     adaptation of and dependence upon the Prussian system
     of education through one of its foremost expatriate ex-
     ponents, Franz Brünnow. Follows Brünnow's application of
     the German method at Michigan, and identifies notable
     members of the "Ann Arbor school of astronomers" during
     the period 1854-1905 when it was one of the most active
     centers for the professional training of astronomers in
     the United States. 63 citations.

349a. Rochester, G.D. "The History of Astronomy in the University
     of Durham from 1835 to 1939." *Quarterly Journal of the
     Royal Astronomical Society*, 21 (1980), 369-378.

     Reviews the progression of researchers and teachers,
     and the growth of instrumentation. Highlights the career
     of R.A. Sampson who made notable contributions to the
     theory of the motions of the Jovian satellites and to
     the study of the solar interior. 29 notes and citations.

350. Schmidt-Schönbeck, Charlotte. *300 Jahre Physik und
     Astronomie an der Kieler Universität*. Kiel: F. Hirt,
     1965. Pp. 261.

     Includes a general survey from the 17th century through
     mid-20th century outlining development of astronomy
     teaching and research. Reviews the establishment of the
     observatory and its research programs. Identifies major
     astronomical figures at Kiel including Elis Strömgren
     and other celestial mechanicians, and the pioneer astro-
     physicist Albrecht Unsöld's establishment of astrophysics
     after World War II as director of Kiel's Institute of
     Theoretical Physics. Reviewed in: *British Journal for
     the History of Science*, 3 (1967), 400-401.

SOCIETIES (NATIONAL AND INTERNATIONAL)

351. Adams, Walter S. "The History of the International
      Astronomical Union." *Publications of the Astronomical
      Society of the Pacific*, 61 (1949), 5-12.

      Brief overview beginning with its origins as the "Inter-
      national Union for Cooperation in Solar Research" in
      1904. Describes evolution of the Union, its breakdown
      and reorganization into the IAU after World War I, and
      its present structure and purpose.

352. Aitken, Robert G. "The Origin of the Astronomical Society
      of the Pacific." *The Griffith Observer*, 14 (1950),
      86-93.

      Provides a brief historical sketch showing the simul-
      taneous and related development of the Society and of
      the Lick Observatory, with Director E.S. Holden as the
      connecting link. 3 citations. For related material see:
      Louis Berman. "The 80th Anniversary of the Astronomical
      Society of the Pacific." *Astronomical Society of the
      Pacific Leaflet* No. 476 (Feb. 1969). Pp. 8.

353. Bailey, Solon I. "The Cambridge Astronomical Society of
      1854." *Popular Astronomy*, 36 (1928), 226-229.

      Brief account of an early astronomical society in the
      United States, considered to be oldest by Bailey, possibly
      predated only by a Cincinnati public group. Cambridge
      society meetings held in 1854 were led by Benjamin
      Peirce. Meetings lasted until 1857.

354. Bell, Trudy E. "Towers Reaching to the Skies: The First
      American Astronomical Society." *The Griffith Observer*,
      42 (April 1978), 2-10.

      Describes formation and short life of this amateur
      society based in Brooklyn, New York, that existed between
      1883 and 1890.

355. Berendzen, Richard. "Origins of the American Astronomical
      Society." *Physics Today*, 27 (1974), 32-39.

      Provides brief overview of historical context for
      understanding the origin of the AAS; the foundation of
      the Lick Observatory and the establishment there of the
      Astronomical Society of the Pacific; the foundation of
      Yerkes by G.E. Hale and his drive to establish the AAS

at the time of the dedication of Yerkes in 1895-97. Notes that there were unfavorable reactions to Hale's aggressiveness which, among other things, delayed the foundation of the AAS until late 1899. 28 citations.

* Biermann, Kurt-R. "Alexander von Humboldt als Initiator und Organisator Internationaler Zusammenarbeit auf Geophysikalischen Gebeit."

Cited herein as item 604.

356. Bruhns, C., ed. *Vierteljahrsschrift der astronomischen Gesellschaft.* 77 volumes. Leipzig: W. Engelmann, 1866-1942.

Annual review of the activities of members (both institutional and individual) of the Astronomischen Gesellschaft including observatory reports on staff activity, extensive biographical memoirs of prominent members, abstracts of important scientific papers, book reviews, and activities of the international European society. Some original papers on astronomy and the history of astronomy appeared from time to time. Later editors included P. ten Bruggencate, H. Siedentopf, P. Guthnik, R. Prager, R. Lehmann-Filhes and G. Müller.

357. Campbell, W.W. "The International Research Council and the International Astronomical Union." *Publications of the Astronomical Society of the Pacific*, 31 (1919), 249-256.

Reviews the formative meetings of the IAU, the role of the International Research Council, and problems with the governments of Germany and Austria-Hungary in the wake of World War I. Sympathizes with the French decision to boycott the Central Powers.

358. Campbell, W.W., and Joel Stebbins. "Report on the Organization of the International Astronomical Union." *Proceedings of the National Academy of Sciences*, 6 (1920), 349-396.

Report of the chairman and secretary of the American delegation to the formative meeting of "representatives of Astronomy in the allied and associated nations" in Brussels in July 1919 that created the IAU. Reviews formation of American section, their initial meetings prior to Brussels, the formation of subcommittees and rosters of membership, the proceedings of the Brussels meeting, the development of the IAU Constitution, IAU

Commissions and Committees, suggestions on notation and
nomenclature, duties of astronomical information bureaus,
almanacs and ephemerides.

*   Christie, W.H.M., ed. *The Observatory*.

Cited herein as item 380.

359.   Dreyer, J.L.E., and H.H. Turner, eds. *History of the
       Royal Astronomical Society, 1820-1920*. London: Royal
       Astronomical Society, 1923. Pp. vii + 258.

Collection of seven chronological chapters by six authors
written in honor of the centenary of the Society. Provides
biographical sketches of major figures in the Society,
general activities, membership and structure, and major
projects and expeditions. Constitutes a history of British
astronomy in the 19th and early 20th centuries. Extensive
citations.

360.   Dvoichenko-Markoff, Eufrosina. "Benjamin Franklin, the
       American Philosophical Society, and the Russian Academy
       of Science." *Proceedings of the American Philosophical
       Society*, 91 (1947), 250-257.

Traces the relations between the two societies within
the context of American-Russian cultural relations in
the 18th century. Examines Franklin's influence upon
M.V. Lomonosov, noted early Russian chemist and astronomical commentator. 42 citations.

361.   Eddington, A.S. "Royal Astronomical Society Centenary
       Celebration." *Monthly Notices of the Royal Astronomical
       Society*, 82 (1922), 431-443.

Records the minutes and transcription of discussions
at the RAS centenary including a synopsis of a review by
J.L.E. Dreyer of the first 100 years of the RAS and a
brief overview by A.S. Eddington of progress in astronomy
in the past 100 years. Analyzes the distribution of papers
in the *Monthly Notices* by topic and year. See also:
Monthly Notices, 82 (1922), 33.

362.   Forbes, Eric Gray, ed. *Human Implications of Scientific
       Advance*. Edinburgh: Edinburgh University Press, 1978.
       Pp. 596.

Proceedings of the 15th International Congress of the
History of Science in Edinburgh, 10-15 August 1977.
Comprises papers from ten symposia including one on

cosmology since Newton.
Contains items 424, 553, 604, 1196, 1208, 1215.

*   Gill, David. "The Application of Photography in Astronomy."

Cited herein as item 468.

363.   Hansen, Julie Vinter. "The International Astronomical
News Service." *Vistas in Astronomy*, 1 (1955), 16-21.

Traces the development of the Service from sporadic
19th-century attempts; improvements brought about by the
telegraph; Foerster's arguments in 1879 to establish a
central bureau under the auspices of the Astronomischen
Gesellschaft and the establishment of the International
Astronomical Union Service in 1920. 10 citations.

364.   Herrmann, D.B. "K.F. Zöllner in seinen Beziehungen zu
O.W. Struve und Russland." *Die Sterne*, 53 (1977), 226-
236.

Provides brief analysis of Zöllner's influence and
relationship with Struve, notably in the organization of
the international "Astronomischen Gesellschaft" in 1863
and their efforts to coordinate large-scale research
programs in positional astronomy. Records the various
meetings of Zöllner and Struve at meetings of the AG
between 1865 and 1875, and their correspondence. 42 citations.

*   [International Astronomical Union]. *Transactions of the
International Astronomical Union*. Volume 1- . London:
Imperial College, 1922- . Published in later years by
Cambridge University Press, Academic Press, and D.
Reidel.

Cited herein as item 134.

365.   Kevles, Daniel J. "George Ellery Hale, the First World
War, and the Advancement of Science in America." *Isis*,
59 (1968), 427-437.

Traces Hale's efforts, through the National Academy
of Sciences, to promote international scientific cooperation; to establish the National Research Council as a
service to governmental needs and as a strengthening
agent for the NAS itself; and to advance pure research
in America. 54 citations.

366. Kevles, Daniel J. "'Into Hostile Political Camps': The
      Reorganization of International Science in World War
      I." *Isis*, 62 (1971), 47-60.

      Examines the role of George Ellery Hale as a central
      figure in the International Research Council debate on
      the exclusion of the Central Powers from international
      scientific society in the post-war era. Notes the pleas
      of Europeans, notably those of Dutch astronomer J.C.
      Kapteyn, to abstain from the ultimate boycott which lasted
      for over a decade after Versailles. 52 citations.

367. Kevles, Daniel J. "Hale and the Role of a Central Scien-
      tific Institution in the United States." *The Legacy of
      George Ellery Hale* (item 1391), 273-282.

      Hale's legendary and highly successful efforts in
      organizing cooperative research networks within an insti-
      tutional framework are reviewed. The origins and growth
      of the National Academy of Sciences and Hale's role in
      the activities of the Academy are at the center of this
      discussion. 21 citations.

368. Mayall, R. Newton. "The Story of the AAVSO." *Review of
      Popular Astronomy*, 55 (1961), (Sept.-Oct.), 4-9; (Nov.-
      Dec.), 8-12.

      Reviews the origins and activities of the American
      Association of Variable Star Observers during its first
      50 years. Includes some additional commentary on the
      history of variable star work. No direct citations.

369. McCrea, W.H. "Einstein: Relations with the Royal Astro-
      nomical Society." *Quarterly Journal of the Royal
      Astronomical Society*, 20 (1979), 251-260.

      Reviews justifications for RAS support for Einstein's
      work during and immediately after World War I. Notes
      that "The Society took the first steps to gain personal
      recognition for Einstein in the world that had so recent-
      ly been at war with Germany." Examines the origins of
      the Society's support through its co-sponsorship of the
      Joint Permanent Eclipse Committee that organized the
      1919 expeditions to test General Relativity; the Society's
      printing of de Sitter's work assimilating general
      relativity; Eddington's "Council Notes" on Einstein's
      work; and the embarrassing delay in awarding Einstein
      the RAS Gold Medal in 1926, "a case of third time lucky--
      in a truly ironical manner." See also item 761a.

370. Minnaert, M. "International Cooperation in Astronomy." *Vistas in Astronomy*, 1 (1955), 5-16.

    Examines the many ways astronomers worldwide have maintained contact through collaborative work and identifies four major areas of international cooperation: (1) coordinated observations of celestial phenomena; (2) accumulation of data on a large scale; (3) development of international centers and programs for research; (4) standardization of nomenclature. Reviews the origins of the International Astronomical Union. No direct citations.

371. Mitchell, Samuel E. "Astronomy During the Early Years of the American Philosophical Society." *Proceedings of the American Philosophical Society*, 86 (1943), 13-21.

    Reviews mid- to late 18th-century astronomical activity in America centering on T. Jefferson, D. Rittenhouse, and support for astronomy by the American Philosophical Society. 5 citations with detailed reference to the minutes and memoranda of the Society.

*   Newcomb, Simon. *Side-Lights on Astronomy and Kindred Fields of Popular Science*.

    Cited herein as item 1403.

372. Plotkin, Howard. "Edward Charles Pickering's Diary of a Trip to Pasadena to Attend Meeting of Solar Union, August 1910." *Southern California Quarterly*, 60 (1978), 29-44.

    Reprints unpublished diary by Harvard College Observatory director that recalls the scientific and social events surrounding an important international meeting of astronomers when the scope of the International Union for Cooperation in Solar Research was expanded to include stellar astrophysics, and where Pickering played a major role in the expansion.

373. Redman, R.O. "The JEPC." *Quarterly Journal of the Royal Astronomical Society*, 12 (1971), 39-44.

    Briefly recounts the formation and history of the Joint Permanent Eclipse Committee of the Royal Society and the Royal Astronomical Society, founded in 1894 and dissolved in 1971.

*   [Royal Astronomical Society]. *Catalogue of the Library of the Royal Astronomical Society to June 1884*.

    Cited herein as item 51.

374.   [Royal Astronomical Society]. "Sesquicentenary Commemora-
        tion Issue." *Quarterly Journal of the Royal Astronomical
        Society*, 11 (1970), 379-459.

        In addition to a general review by G.J. Whitrow, articles
        include an examination of Caroline Herschel's telescope
        and "Sidelights on Astronomy during the Society's History"
        by H. Dingle.

*   [Royal Astronomical Society]. *General Index*.

        Cited herein as item 50.

375.   Shapley, Harlow S. "Astronomy and International Coopera-
        tion." *Proceedings of the American Philosophical
        Society*, 91 (1947), 73-74.

        Presented at APS Symposium on America's role in the
        growth of science, October 21, 1946. Summarizes various
        ongoing astronomical projects that require international
        cooperation including: continuing refinement of the
        astronomical unit (the distance between the Earth and
        Sun); preparation of tables for nautical almanacs;
        restoration of libraries destroyed during World War II;
        problems in galactic structure.

376.   Stebbins, Joel. "The American Astronomical Society, 1897-
        1947." *Popular Astronomy*, 55 (1947), 404-413.

        Review of the first half-century of the Society read
        at the anniversary meeting of the Society held at the
        Yerkes Observatory, September 6, 1947. Reviews origin
        and structure of Society.

377.   Stratton, F.J.M. "International Co-operation in Astronomy."
        Monthly Notices of the Royal Astronomical Society, 94
        (1934), 361-372.

        Reviews the many areas of international astronomical
        cooperation since the early 19th century including star
        charts, an organized scheme for the circulation of astro-
        nomical news via the mail and then telegrams, the founda-
        tion of the Astronomischen Gesellschaft at Heidelberg
        in 1863, the standardization of stellar magnitude systems,
        the Washington Conference of 1884 establishing a Prime
        Meridian, the organization of the Carte du Ciel in 1887,
        the Eros Campaign of 1901, and the formation of the In-
        ternational Solar Union in 1904 and its extension to the
        International Astronomical Union in the period 1919-1922.
        Examines the structure of the Union, and the many special-

ist cooperative ventures it has fostered, and which have been incorporated into its structure. No direct citations.

378. Whitrow, G.J. "Some Prominent Personalities and Events in the Early History of the Royal Astronomical Society." *Quarterly Journal of the Royal Astronomical Society,* 11 (1970), 89-104.

Reviews the origin of the Society in 1820 through the efforts of John Herschel, William Pearson, and "12 other gentlemen." Provides brief sketches of Herschel and Pearson, as well as Francis Baily, another prominent member of the original committee and an early Newton biographer and star catalogue organizer. 29 citations.

* Wilkins, G.A. "The System of Astronomical Constants, Part i."

Cited herein as item 890.

* [Yerkes Observatory Staff]. "The Yerkes Observatory, 1897-1947."

Cited herein as item 345.

JOURNALS

379. Armitage, Angus. "Baron von Zach and His Astronomical Correspondence." *Popular Astronomy,* 57 (1949), 326-333.

Reviews the early history of the development of a purely astronomical journal in the late 18th century, separate from almanacs or ephemerides, which was largely the result of the efforts of the Hungarian astronomer Franz Xavier von Zach. Zach's "Correspondence" eventually developed into a series that ran to 1826 in various forms. 12 citations.

* Bruhns, C., ed. *Vierteljahrsschrift der astronomischen Gesellschaft.*

Cited herein as item 356.

380. Christie, W.H.M., ed. *The Observatory.* Volume 1- . London: Taylor and Francis, 1878- .

A monthly review of astronomical activities centering upon the Royal Astronomical Society (London). Includes

minutes of RAS meetings, major addresses, discussions,
news notes, letters, and announcements of celestial
phenomena. In later years short papers appear. Considered
to be the RAS "house organ" whereas the later runs of the
Monthly Notices contain purely scientific papers. Christie
was the founding editor.

*   Daniels, George H. "The Process of Professionalism in
American Science. The Emergent Period, 1820-1860."

        Cited herein as item 192.

*   Ferguson, Allan, ed. *Natural Philosophy through the
Eighteenth Century and Allied Topics.*

        Cited herein as item 131.

380a. [The Griffith Observer]. "Index of Articles in the
Griffith Observer, February 1937-December 1948."
The Griffith Observer, 12 (1948), 145-147.

       Index to the first ten years of this popular review
periodical. A substantial fraction of the short papers
during this period, as well as in subsequent years, deal
with historical topics.

381. Heinemann, K. "Astronomische Bibliographie: zum 50.
Jubiläumsband des 'Astronomischen Jahresberichts.'"
Die Sterne, 29, 1 (1953), 33-39.

       Describes the founding of the *Astronomischer Jahres-
bericht* in the 1890s. See item 69.

382. Herrmann, D.B. "B.A. Gould and His *Astronomical Journal.*"
Journal for the History of Astronomy, 2 (1971), 98-108.

       Describes the state of American astronomy during the
years (circa 1849) that Gould created the *Astronomical
Journal.* Highlights the German influence upon Gould
and the first years of the journal's existence until
its temporary 14-year lapse in publication, primarily
due to the American Civil War. 52 citations.

383. Herrmann, D.B. *Die Entstehung der astronomische Fach-
zeitschriften in Deutschland (1798-1821). Publications
of Archenhold Observatory* No. 5. Berlin-Treptow: 1972.
Pp. x + 150.

       Outlines development first of scientific journals
generally and then focusses upon origins and development
of late 18th- and early 19th-century journals specializ-

ing in astronomy culminating in Schumacher's *Astronomischen Nachrichten* in 1821. Attempts to evaluate role of journals in the development of astronomy. Reviewed in: *JHA*, 4 (1973), 63-64.

* Herrmann, D.B., and J. Hamel. "Zur Frühentwicklung der Astrophysik das Internationale Forscherkollektiv 1865-1899."

   Cited herein as item 155.

384. Kahrstedt, A. "175 Jahre Berliner Astronomisches Jahrbuch." *Die Sterne*, 25 (1949), 111-116.

   Reviews the origins of this yearly ephemeris beginning with early 18th-century attempts to create an "Astronomischen Kalenders" by various German scientific organizations. Examines Johann Elert Bode's organization of the "Astronomisches Jahrbuch oder Ephemeriden für das Jahr 1776...," Johann Franz Encke's succession in 1830, and later developments. Includes discussion of content changes in the yearbook.

385. McCormmach, Russell. "Ormsby MacKnight Mitchel's *Sidereal Messenger*, 1846-1848." *Proceedings of the American Philosophical Society*, 110 (1966), 35-47.

   Detailed review of the origins and development of the *Sidereal Messenger* as a monthly journal of astronomy created during the first emergence in America of intense popular interest in astronomy and observatory building. 114 citations.

* [Popular Astronomy]. *A General Index*.

   Cited herein as item 48.

* [Royal Astronomical Society]. *General Index*.

   Cited herein as item 50.

386. Sadler, D.H. "The Bicentenary of the Nautical Almanac." *Quarterly Journal of the Royal Astronomical Society*, 8 (1967), 161-171.

   Traces history of *The Nautical Almanac and Astronomical Ephemeris* developed for the determination of longitude at sea since its inception by Nevil Maskelyne, fifth Astronomer Royal, in 1765. 5 citations.

387. [Sky Publishing Corporation]. *Sky and Telescope*. Cambridge,
     Mass.: Sky Publishing Corporation. Published monthly
     since 1941 and the combination of *The Sky* (1936-1941)
     and *The Telescope* (1933-1941).

     General popular review covering all aspects of astronomy.
     Short historical and review papers frequently appear.
     Of special interest are the multitude of historical
     notes by Joseph Ashbrook. Many articles have been in-
     cluded in item 143. No cumulative index.

388. [Deleted]

*   Wright, Helen. *Explorer of the Universe: A Biography of
     George Ellery Hale*.

     Cited herein as item 1390.

INSTRUMENTATION

GENERAL

389. Ambronn, L. *Handbuch der astronomischen Instrumentenkunde.*
2 volumes. Berlin: Julius Springer, 1899. Pp. 1276.

Treats all aspects of instrumentation in astronomy
including design, construction and use. Highly detailed
and illustrated with over 1000 woodcuts. Reviewed in:
PASP, 12 (1900), 117-121.

* Bedini, Silvio A. *Thinkers and Tinkers.*

Cited herein as item 181.

* Beer, Arthur, ed. *New Aspects in the History and Philosophy
of Astronomy.*

Cited herein as item 123.

390. Bell, Louis. *The Telescope.* New York: McGraw-Hill, 1922.
Pp. viii + 287.

Popular introduction to the construction and use of
telescopes with some historical material. Includes many
illustrations of use of contemporary instrumentation.

391. Bowen, Ira S. "Astronomical Instrumentation in the
Twentieth Century." *The Legacy of George Ellery Hale*
(item 1391), 239-249.

Discusses the application of new observational techniques
to astronomical measurements including spectroscopy,
photometry and interferometry, and Hale's role in
generating these new techniques. Special attention is
paid to the development of infra-red, photoelectric, and
interferometric techniques. Concludes that in the present
century a significant shift in the development of instrumentation from positional to physical astronomy has

taken place and that the most sensitive and sophisticated receivers in the physical sciences are to be found in astronomical observatories. 24 citations.

*   Chauvenet, W. *Manual of Spherical and Practical Astronomy*.

Cited herein as item 820.

392.   Cohen, I. Bernard. *Some Early Tools of American Science: An Account of the Early Scientific Instruments and Mineralogical and Biological Collections in Harvard University*. Cambridge, Mass.: Harvard University Press, 1950. Pp. xxi + 201.

Descriptions of 18th- and early 19th-century equipment at Harvard including their history and use in teaching and research. Includes several reflecting and refracting telescopes and mechanical planetaria, especially one made in 1787 by Joseph Pope, a Boston clockmaker. Reviewed in: *Isis*, 41 (1950), 233-234.

393.   Czenakal, V.L. "The Astronomical Instruments of the Seventeenth and Eighteenth Centuries in the Museums of the U.S.S.R." *New Aspects in the History and Philosophy of Astronomy* (item 123), 53-77.

Well-illustrated introduction to the holdings, including telescopes, astrolabes, armillary spheres, globes, quadrants and sundials. 18 citations.

*   Czenakal, V.L. "The Astronomical Instruments of John Rowley in Eighteenth-Century Russia."

Cited herein as item 189.

394.   Daumas, Maurice. *Les Instruments Scientifiques aux XVIIe et XVIIIe Siècles*. Paris: Presses Universitaires de France, 1953. Pp. 417.

In three parts, including: "the manufacture and trade of instruments in the 17th century; factors modifying the evolution of the instrument trade"; and the trade in the 18th century. Identifies the invention of the telescope and its development mainly through improvements in optical glass and grinding techniques; social, economic, technical and practical factors, as well as the scientific problems that both stimulated and were aided by progress in instrumentation. Covers later improvements including the invention of achromatic optics and the evolution of the telescope and micrometrical instrumentation. Extensive

notes and bibliography. Translated into English in 1972.
Reviewed in: *Annals of Science*, 10 (1954), 78; *Isis*, 44
(1953), 391–392.

*   Dewhirst, D.W. "Observatories and Instrument Makers in
    the Eighteenth Century."

    Cited herein as item 276.

*   Dreyer, J.L.E. "Indian Astronomical Instruments."

    Cited herein as item 195.

395.   Fox, Philip. *Adler Planetarium and Astronomical Museum
    of Chicago*. Chicago: Lakeside Press, 1933. Pp. 61.

    Illustrated review of the major instrument collection
held at the Adler. Instruments include orreries, globes,
armillae, astrolabes, nocturnals, quadrants and octants,
as well as sundials, early clocks, and old optical telescopes and geodetic instruments.

396.   Gavine, David. "James Stewart MacKenzie (1719–1800) and
    the Bute MSS." *Journal for the History of Astronomy*,
    5 (1974), 208–214.

    Preliminary description of MacKenzie's papers and in-
struments, including cardboard sundials, telescopes by
various makers, and detailed descriptions of telescopes
and instruments of the period. 15 citations.

397.   Hagar, Charles F. "Through the Eyes of Zeiss." *The
    Griffith Observer*, 25 (1961), 62–70.

    Descriptive history of the Zeiss Works centering upon
the work of Carl Zeiss, Ernst Abbe, and Walther Bauers-
feld. 3 citations.

*   Hellman, C. Doris. "George Graham: Maker of Horological
    and Astronomical Instruments."

    Cited herein as item 1299.

*   Hellman, C. Doris. "John Bird (1709–1776). Mathematical
    Instrument Maker in the Strand."

    Cited herein as item 1300.

*   Henderson, E. *Life of James Ferguson in a Brief Auto-
    biographical Account, and Further Extended Memoir*.

    Cited herein as item 1301.

398. Horský, Zdeněk, and Otilie Škopová. *Astronomy Gnomonics.*
      A Catalogue of Instruments of the 15th to the 19th
      Centuries in the Collections of the National Technical
      Museum, Prague. Prague: National Technical Museum,
      1968. Pp. 202.

      Descriptive partially illustrated catalogue of astro-
      labes, telescopes, sundials, miscellaneous astronomical
      and geodetic instruments. Telescopes date to 1850.
      Reviewed in: *JHA*, 1 (1970), 81; *Isis*, 62 (1971), 530.

399. Hoskin, Michael A. "Apparatus and Ideas in Mid-nineteenth-
      century Cosmology." *New Aspects in the History and*
      Philosophy of Astronomy (item 123), 79–85.

      Examines the developments in astronomical instrumenta-
      tion from Herschel's telescope to the introduction of
      spectroscopic techniques and their influence upon cosmo-
      logical speculation, notably the existence of true nebulae.
      Concludes that many times, incomplete evidence and also
      "an insistent desire to arrive at a simple theory" misled
      astronomers to "adopt an extreme position that not only
      solved but also created problems." 32 citations.

400. Howse, Derek, et al. *An Inventory of the Navigation and*
      Astronomy Collections in the National Maritime Museum,
      Greenwich. 3 volumes. London: National Maritime Museum,
      1970. 35 sections.

      Alphabetical organization by instrument type with a
      list of instruments used at the Royal Observatory, 1676–
      1950, and notations on other collections. Extensive
      photographic illustrations.

*   King, H.C. *The History of the Telescope.*

      Cited herein as item 99.

401. Konkoly, Nicolaus von. *Praktische Anleitung zur Anstellung*
      Astronomischer Beobachtungen mit besonderer Rucksicht
      auf die Astrophysik. Nebst einer modernen Instrument-
      kunde. Braunschweig, 1883. Pp. xxii + 912.

      General exposition of astronomical and astrophysical
      observational techniques and appropriate instrumentation
      in use during the late 19th century. Over 300 woodcuts
      illustrate instruments, providing details of their con-
      struction and use.

*   Livingston, Dorothy Michelson. *The Master of Light: A Biography of Albert A. Michelson.*

     Cited herein as item 1331.

*   Lovell, Bernard. "The Effects of Defence Science on the Advance of Astronomy."

     Cited herein as item 252.

402.   Maddison, Francis. "Early Astronomical and Mathematical Instruments. A Brief Survey of Sources and Modern Studies." *History of Science*, 2 (1963), 17–50.

     Identifies collections in museums and libraries and general studies classified by period and nationality. Includes extensive information on pre-telescopic instrumentation. 39 detailed references and 16-page bibliography. Continued by G.L'E. Turner in: *History of Science*, 8 (1969), 53–93.

403.   Pipping, Gunnar. *The Chamber of Physics: Instruments in the History of Sciences Collections of the Royal Swedish Academy of Sciences, Stockholm.* Stockholm: Almquist and Wiksell, 1977. Pp. 250.

     Contains descriptions of astronomical, optical and related scientific instruments including a study of instrument making in Sweden. Examines the venerable Stockholm Observatory, begun circa 1750, and its historical instrumentation. Noted are a quadrant and telescope by John Bird, various reflectors by James Short and William Herschel, and refractors by J. Dolland and Son and by Utzschneider and Fraunhofer. Reviewed in: *Isis*, 70 (1979), 442; *JHA*, 10 (1979), 130–131.

404.   Price, Derek J. "The Early Observatory Instruments of Trinity College, Cambridge." *Annals of Science*, 8 (1952), 1–12.

     Describes instruments stored in the library of Trinity not noted in Gunther's *Early Science in Cambridge* (Oxford, 1937). Provides a brief inventory and analysis of these newly discovered artifacts, which are identified as "part of the original equipment of the first official astronomical observatory built for the use of the Plumian Professor, Roger Cotes, in the early years of the eighteenth century." 18 citations.

405. Repsold, J.A. *Zur Geschichte der astronomischen Mess-*
      werkzeuge. 2 volumes. Leipzig: W. Engelmann, 1908.
      Pp. 122 + 161.

      General, well-illustrated technical history of all
      types of astronomical instrumentation from 1450 to ap-
      proximately 1900. Volume 1 organized either by major
      astronomer, or maker, or type of instrument; Volume 2
      organized by maker. The Repsold firm is well represented
      by the author, but other 19th-century firms in Germany,
      Britain and the United States are given detailed attention.
      Numerous citations.

*   Ronchi, L.R., and G. Abetti. "Psycho-physiological Effects
      in Visual Astronomical Observations--the Planet Mars."

      Cited herein as item 658.

*   Schweiger-Lerchenfeld, A. *Atlas der Himmelskunde.*

      Cited herein as item 333.

*   Smart, William M. *Text-Book on Spherical Astronomy.*

      Cited herein as item 686.

406. Sokolovskaja, Zinaida. "L'importance des instruments dans
      l'evolution des connaissances astronomiques." *XIIe*
      Congrès International d'Histoire des Sciences, Tome
      XA (Paris, 1968). Paris: Blanchard, 1971. Pp. 93-99.

      Identifies five periods of development of modern
      astronomy based upon the character of instrumentation.
      The periods are: visual observations (to 1600); visual
      use of telescopic devices (to 1850); telescopic devices
      used with photography and spectroscopy (to 1950); optical
      and radio telescopes and electronic devices (to 1960);
      space-borne astronomical instrumentation (from 1960).
      10 citations.

407. Sticker, Bernhard. "Historical Scientific Instruments
      as Cultural Landmarks." *New Aspects in the History and*
      Philosophy of Astronomy (item 123), 102-108.

      Calls for the cooperation of observatories and museums
      in the identification and preservation of astronomical
      instruments, noting that the bulk of early optical in-
      strumentation is now lost.

*   Todd, David P. *Stars and Telescopes.*

      Cited herein as item 1408.

408. Turner, G. L'E. *Van Marum's Scientific Instruments in
      Teyler's Museum, Part 1, Essays on Van Marum and the
      Museum; Part 2, Descriptive Catalogue.* Leyden: Nordhoff,
      1973. Pp. 401.

      Analyzes scientific instruments circa late 18th century
      including telescopes by Herschel, Dolland, van Deijl and
      Ramsden, and astronomical models by George Adams. Exten-
      sive bibliography.

*   Van Dyck, Walther. *Georg von Reichenbach.*

      Cited herein as item 1379.

*   Volkmann, Harald. *Carl Zeiss und Ernst Abbe: Ihr Leben
      und ihr Werk.*

      Cited herein as item 1380.

409. Warner, Deborah Jean. "Lewis Morris Rutherfurd: Pioneer
      Astronomical Photographer and Spectroscopist." *Technology
      and Culture*, 12 (1971), 190–216.

      Identifies the significance of Rutherfurd's astronomical
      instrumentation for the advancement of astronomical
      photography and spectroscopy. Itemizes known artifacts
      of Rutherfurd's creation including telescopes, diffrac-
      tion gratings, micrometers, multi-prism astronomical
      spectroscopes and assesses their quality and scientific
      use. 76 citations.

410. Wynter, Harriet, and Anthony Turner. *Scientific Instru-
      ments.* New York: Scribner's, 1975. Pp. 239.

      Survey from mid-16th through first half of 19th cen-
      tury. Reviews astronomy to the extent of an appreciation
      of the role of instrumentation. Limits discussion to
      smaller, usually portable, equipment. Reviewed in: *JHA*,
      7 (1976), 209–210.

411. Zinner, Ernst. *Deutsche und niederlandische astronomische
      Instrumente des 11-18 Jahrhunderts.* Munich: Beck, 1956.
      Pp. x + 680.

      Provides descriptive data arranged by instrument type
      with cross-referenced alphabetical listing by instrument
      maker. Includes a section on unidentified trademarks.
      Examines astrolabes, sundials, quadrants, and early
      optical instruments. Reviewed in: *Isis*, 49 (1958), 87–88.

PRE-TELESCOPIC INSTRUMENTATION

412. Edgecomb, D.W. "Notes on the Invention of the Telescope."
      Popular Astronomy, 7 (1899), 184-193.

      Reviews early development of eyeglasses and lenses, and
      precursors to Galileo. 6 citations.

412a. Gibbs, Sharon L., Janice A. Henderson, and Derek J. de
      Solla Price. *A Computerized Checklist of Astrolabes*.
      New Haven: Department of History of Science, Yale
      University, 1973. Pp. 118.

      Extension of *An International Checklist of Astrolabes*
      prepared by D. Price in 1955: *Archives Internationale
      d'Histoire des Sciences*, No. 32; No. 33 (1955). Covers
      all periods and includes information sorted by maker,
      date, dimensions, and collection. For illustrations and
      descriptions of astrolabes, see: R.T. Gunther, *The Astro-
      labes of the World* (Oxford, 1932); and H. Michel, *Traité
      de l'Astrolabe* (Paris, 1947).

413. Kaye, George Rusby. *The Astronomical Observatories of
      Jai Singh*. Calcutta: Archaeological Survey of India,
      New Imperial Series No. 40, 1918. Pp. viii + 153.

      Describes the enormous astronomical buildings and metal
      astronomical sighting instruments built by Jai Singh in
      five cities in India in the early 18th century. Discusses
      his astronomical observations and influences upon his
      work from the Mideast and Europe, but especially from
      Mohammedan traditions. One detailed chapter attempts to
      provide background for appreciating Jai Singh's accomplish-
      ments and is a brief survey of one thousand years of
      Greek, Arabic and European astronomy. Reviewed in: *Isis*,
      2 (1914), 421-423.

414. Kirchvogel, Paul A. "Wilhelm IV, Tycho Brahe, and
      Eberhard Baldewein--The Missing Instruments of the
      Kassel Observatory." *New Aspects in the History and
      Philosophy of Astronomy* (item 123), 109-121.

      Illustrated description of the instruments found in a
      double portrait of Wilhelm IV and his wife showing their
      similarity to ones used by Tycho in closely succeeding
      years. Numerous citations.

*   Maddison, Francis. "Early Astronomical and Mathematical
      Instruments. A Brief Survey of Sources and Modern

Studies."

Cited herein as item 402.

415. Needham, Joseph. "The Peking Observatory in A.D. 1280 and the Development of the Equatorial Mounting." *Vistas in Astronomy*, 1 (1955), 67-81.

Describes how the equatorial mounting was developed from the Arabic and European torquetum by Kuo Shou-Ching after contact with the Persian astronomer Jamal Al-Din after the year 1267. Traces the influence of Kuo Shou-Ching's equatorial devices upon later European instrumentation, notably that of Tycho Brahe. 51 footnotes and 57 bibliographical citations.

416. Price, Derek J. "A Collection of Armillary Spheres and Other Antique Scientific Instruments." *Annals of Science*, 10 (1954), 172-187.

Includes a short catalogue of instruments held at various sites in Britain. 13 citations.

417. Przypkowski, Tadeusz. "The Art of Sundials in Poland from the Thirteenth to the Nineteenth Century." *New Aspects in the History and Philosophy of Astronomy* (item 123), 13-23.

Well-illustrated review, in French, covering theory and construction of sundials. 23 citations.

* Singh, Prahlad. *Stone Observatories in India.*

Cited herein as item 235.

OPTICAL INSTRUMENTATION AND TECHNIQUE

418. Allen, Phyllis. "Problems Connected with the Development of the Telescope (1609-1687)." *Isis*, 34 (1943), 302-311.

Reviews Galileo's construction of a telescope; Kepler's suggestion of convex lenses for eyepieces and Scheiner's independent execution of this design improvement; problems of creating erect, color-free images; various methods devised by Hooke and others to overcome chromatic aberration and Newton's belief that the problem was incurable with lenses; Newton's and Gregory's construction of reflecting telescopes in 1668 followed by Cassegrain's design in 1672. Methods of observing the Sun with transparent reflecting mirrors are also discussed as well as Huygen's suggestion of using a paraboloidal surface for better definition. Other problems included the production of durable reflecting surfaces and light scattering in long focus "aerial" telescopes. 124 citations.

419. Ariotti, Piero E. "Bonaventura Cavalieri, Marin Mersenne, and the Reflecting Telescope." *Isis*, 66 (1975), 303-321.

Argues that the conception of the reflecting telescope was not a consequence of Newton's conclusion that lenses could not provide color-free images. Traces the studies and attempts of Cavalieri to produce a successful reflecting design and the causes of his failures. 45 citations.

420. Bedini, Silvio A. "An Early Optical Lens-Grinding Lathe." *Technology and Culture*, 8 (1967), 74-77.

Description of a lathe based upon a woodcut from an early 17th-century work on optics by Count Carlo Antonio Manzini (1607-1677) who devoted much of his life to astronomy. 2 citations.

421. Bedini, Silvio A. "The Aerial Telescope." *Technology and Culture*, 8 (1967), 395-401.

Reviews late 17th-century experimentation with long-focus refracting telescopes, identifies some early telescope makers and examines the problems they faced in telescope manufacturing and design. Claims that Giovanni Domenico Cassini, at the Royal Observatory at Paris between 1670 and 1690, provided "the most substantial contribution to the development of the aerial and the air telescope...." 12 citations.

422. Bennett, J.A. "On the Power of Penetrating into Space: The Telescopes of William Herschel." *Journal for the History of Astronomy*, 7 (1976), 75-108.

Provides a general overview of the character of William Herschel's many telescopes beginning with small specula in the 1770s and ending with his 20-foot and 40-foot instruments and later revisions and variations. Describes the operation of each telescope, changes in optical and mechanical design, and identifies the primary goal of obtaining light gathering power. Many detailed illustrations. 193 citations.

423.   Bennett, J.A. "William Herschel's Large Twenty-foot
        Telescope." *Quarterly Journal of the Royal Astronomical
        Society*, 17 (1976), 303-305.

        Brief review of telescope completed in 1783.

424.   Bennett, J.A. "The Giant Reflector, 1770-1870." *Human
        Implications of Scientific Advance* (item 362), 553-558.

        Reviews the origins and role of the reflectors of the
        Herschels, Lassell, and the Earls of Rosse in advancing
        the study of faint extended objects. 8 citations.

425.   Birkenmajer, Alexander. "Alexius Sylvius Polonus (1593-
        ca. 1653), a Little Known Maker of Astronomical Instru-
        ments." *New Aspects in the History and Philosophy of
        Astronomy* (item 123), 11-12.

        Briefly notes telescopes made by Sylvius for the observa-
        tion of sunspots. 2 citations.

426.   Brashear, J.A. "Glass for Optical Instruments with Especial
        Reference to Telescope Objectives." *Popular Astronomy*,
        1 (1894), 221-224; 241-243; 291-295; 447-449; 2 (1895),
        9-12; 57-59.

        Covers basic aspects of production of large lens blanks
        including photographic record of history of 40-inch Yerkes
        blanks. Author was famous optician and builder of major
        astronomical telescopes. No direct citations.

427.   Bryden, D.J. *James Short and His Telescopes*. London:
        H.M. Stationery Office, 1968. Pp. 34.

        Description of a 1968 exhibit at the Royal Scottish
        Museum of the work of James Short, a noted astronomical
        instrument maker of the late 18th century. Includes brief
        review of Short's life and the development of astronomical
        telescopes during his lifetime. Reviewed in: *Annals of
        Science*, 25 (1969), 92.

*   Buttmann, Günther. *The Shadow of the Telescope: A Biography
        of John Herschel*.

        Cited herein as item 1257.

428.   Carl, Philipp. *Die Principien der astronomischen Instru-
        mentenkunde*. Leipzig: Voight and Günther, 1863. Pp.
        x + 166 + 185 illustrations.

Detailed illustrated exposition of the theory and construction of astronomical instrumentation from micrometers and vernier scales on transit and meridian circles to a general account of equatorial mountings. Includes listing of instrument makers by country and 6-page bibliography.

429. Chapin, Seymour L. "In a Mirror Brightly: French Attempts to Build Reflecting Telescopes Using Platinum." *Journal for the History of Astronomy*, 3 (1972), 87-104.

Discusses late 18th-century attempts at the Paris Observatory to construct reflecting telescopes using a platinum alloy in contrast to the more common copper/tin/arsenic alloys of the English. Abortive plans for a large French telescope comparable to William Herschel's are described. Relationship of French science to French politics during period is explored. 98 citations.

*   Curtis, H.D. "The Comet-Seeker Hoax."

Cited herein as item 646.

430. Dreyer, J.L.E. "On the Invention of the Sextant." *Astronomische Nachrichten*, 155 (1886), 33-36.

Clarifies the independent origins of the reflecting sextant invented in the 1730s in Philadelphia by Thomas Godfrey and by John Hadley, instrument maker in the Strand, London. Brief indirect citations.

431. Eddy, John A. "The Schaeberle 40-ft. Eclipse Camera of Lick Observatory." *Journal for the History of Astronomy*, 2 (1971), 1-22.

Description of the many eclipse expeditions conducted by the staff of the Lick Observatory during the 40-year active life of a long-focus portable solar camera with a 5-inch diameter lens. 33 citations.

*   Flamsteed, John. *The Gresham Lectures of John Flamsteed*.

Cited herein as item 540.

*   Forbes, Eric G. "Dr. Bradley's Astronomical Observations."

Cited herein as item 284.

432. Forbes, Eric G. "Tobias Mayer's New Astrolabe (1759). Its Principles and Construction." *Annals of Science*, 27 (1971), 109-116.

Describes the development of this measuring instrument, which utilized micrometric adjustments, and which was used for the measurement of azimuth angles. Forbes provides a translation of Mayer's description of the instrument and identifies it as a primitive form of theodolite. 17 citations.

433. Glaze, Francis W. "The Optical Glass Industry, Past and Present." *Annual Report of the Smithsonian Institution* (1948), 217-225.

Reviews progress since the 17th century centering on astronomical needs, production in the World Wars, changing techniques, fabrication of the 200-inch Palomar blank, and work at the National Bureau of Standards. Notes the advance in glass production in the United States in recent years, thus making it possible to procure acceptable blanks from domestic sources. 8 citations. Reprinted from *Sky and Telescope*, 6 (1947), No. 3; No. 4.

434. Grubb, Howard. "The Development of the Astronomical Telescope." *Royal Institution Library of Science: Astronomy* (item 140), 398-416.

Lecture dated May 25, 1894. Points out the many design changes required for large telescopes to allow for the "introduction of the new photographic method" into astronomy. Changes include improved equatorial mounts, driving clocks and remote controls allowing the functions of the observatory to be operated from the telescopic eyepiece over long periods of time while the observer is engaged in the exposure of the photographic plate. Reviews his own suggestions for a rising floor at Lick Observatory, and other improvements there and at Yerkes Observatory.

* Hale, George Ellery. *The Study of Stellar Evolution.*

Cited herein as item 1141.

435. Hastings, C.S. "The History of the Telescope." *Sidereal Messenger*, 10 (1891), 335-354.

A concise review, partly based upon Poggendorff's *Geschichte der Physik* but emphasizing late 19th-century opticians and problems in optical design. Reviews methods of lens production.

436. Herrmann, D.B. "Karl Friedrich Zöllner und die 'Potsdamer
      Durchmusterung,' Versuch einer Rekonstruktion." *Die
      Sterne*, 50 (1974), 170-180.

      Reviews the development of techniques of photometric
      analysis of the brightnesses of stars centering upon
      Zöllner's development of accurate photometric instrumenta-
      tion capable of precise calibration, and his influence
      upon Foerster, von Bülow and Vogel that eventually resulted
      in the production, in the 1880s and 1890s, of the compre-
      hensive "Photometrischen Durchmusterung des Nördlichen
      Himmels," by G. Müller and P. Kempf. 31 citations.

437. Herschel, John F.W. *The Telescope*. Edinburgh: Adam and
      Charles Black, 1861. Pp. vii + 190.

      Reprint, with some alterations, of *Encyclopaedia
      Britannica* article. Provides general review of telescopes;
      their design and construction; with comparisons of the
      value of reflectors and refractors. Brief bibliography.

438. King, H.C. "The Invention and Early Development of the
      Achromatic Telescope." *Popular Astronomy*, 56 (1948),
      75-88.

      Traces the slow development of refracting telescopes
      during the 18th century until the successful use of
      crown and flint glass to make achromatic lenses by John
      Dolland at mid-century. Follows Peter Dolland's work
      after his father's death, and similar work in France by
      Clairaut. 37 citations.

439. King, H.C. "The Optical Work of Charles Tulley." *Popular
      Astronomy*, 57 (1949), 74-79.

      Brief assessment of the optical work of this prominent
      early 19th-century British telescope maker. 9 citations.

440. Knight, William H. "Some Telescopes in the United States."
      Sidereal Messenger, 10 (1891), 393-399.

      Lists the equipment of some 90 amateur and professional
      observatories including location, owner, description of
      instruments, maker and date, staff and specialities.

441. Laurie, P.S., and D.W. Waters, "James Bradley's New
      Observatory and Instruments." *Quarterly Journal of the
      Royal Astronomical Society*, 4 (1963), 55-61.

Describes the Royal Observatory buildings in Bradley's
time, circa 1765, and Bradley's astronomical instruments:
a 12 1/2-foot Zenith sector; 8-foot brass quadrant;
equatorial sector; 8 1/2-foot transit; and level.

442. Lockyer, William J.S. "The Growth of the Telescope."
     Royal Institution Library of Science: Astronomy (item
     140), 258-270.

     Lecture delivered April 20, 1923. Briefly reviews
     history of the astronomical telescope. Provides graphical
     illustrations identifying 100 years of growth of instru-
     mentation (1820-1923) showing their geographical distri-
     bution.

*   Loomis, Elias. *The Recent Progress of Astronomy: Especially
       in the United States*.

       Cited herein as item 104.

443. Mayall, N.U. "Bernhard Schmidt and His Coma-Free Reflector."
     Publications of the Astronomical Society of the Pacific,
     58 (1946), 282-290.

     Translation of Schmidt's 1932 paper announcing the general
     form of his famous optical design and a translation of
     an obituary notice of Schmidt by R. Schorr in 1936.

444. McColley, Grant. "Josephus Blancanus and the Adoption of
     Our Word 'Telescope.'" *Isis*, 28 (1938), 364-365.

     Brief note arguing that the repeated use of various
     forms of the term "telescope" in Blancanus's 1620 astro-
     nomical work *Sphaera Mundi* aided in the adoption of the
     term. See also item 451. 3 citations.

445. Meadows, A.J. "Observational Defects in Eighteenth-Century
     British Telescopes." *Annals of Science*, 26 (1970),
     305-317.

     Examines the problems of obtaining suitable lens glass,
     and discusses the usual imperfections found in available
     glass both before and after the introduction of achromatic
     systems. Includes a review of efforts to improve speculum
     metal reflectors to get around the constant problem of
     chromatic abberation, and notes the developments of
     various refinements in mountings, micrometers, and heli-
     ometers. 28 citations.

446.   Michkovitch, Voislav V. "A Historical Study on the Pris-
       matic Astrolabe." *New Aspects in the History and
       Philosophy of Astronomy* (item 123), 93-95.

       Brief review of the origins of this instrument in the
       early 19th century. 11 citations.

447.   Multhauf, Robert, ed. "Holcomb, Fitz and Peate: Three
       19th-Century American Telescope Makers." *United States
       National Museum Bulletin*, 228 (1962), 155-184.

       Three biographical accounts by Amasa Holcomb, J. Fitz
       Howell, F.W. Preston and W.J. McGrath, Jr., with a
       coordinating introduction by Robert Multhauf. Provides
       well-illustrated accounting of the role of these telescope
       makers in the period just prior to that of the Clarks
       and Brashear. Includes a short list of artifacts of their
       work in the Smithsonian Institution. Numerous citations.

*   Nassau, J.J. "Ambrose Swasey, Builder of Machines, Tele-
       scopes, and Men."

       Cited herein as item 1345.

448.   Nielsen, Axel V. "Ole Romer and His Meridian Circle."
       Vistas in Astronomy, 10 (1968), 105-112.

       Roemer's meridian circle, made famous as the one used
       to determine the speed of light, is described together
       with a discussion of instrumental errors and general
       methodology employed by Roemer. 16 citations.

*   Olmsted, John W. "The 'Application' of Telescopes to
       Astronomical Instruments, 1667-1669; A Study in Historica
       Method."

       Cited herein as item 172.

449.   Pease, F.G. "Astronomical Telescopes." *Publications of
       the Astronomical Society of the Pacific*, 40 (1928),
       11-23.

       Popular lecture reviewing the growth of the telescope.
       Emphasizes modern instrumentation at the Mount Wilson
       Observatory including the stellar interferometer of
       Michelson and Pease, the solar telescopes, and present
       plans for a "very large" telescope which called for a
       25-foot mirror--the early model for the Palomar 200-inch.
       4 citations.

450. Pendray, G. Edward. *Men, Mirrors and Stars*. New York:
         Funk and Wagnalls, 1935; 1939. Pp. x + 341.

         Popular introduction to the history of telescopic
      astronomy highlighting the growth of technique in the
      production of telescopes and the "battle of the telescopes"
      or the relative usefulness of reflectors and refractors
      at various times in the past. Includes chapters on the
      history of glassmaking and the growth of telescopic
      equipment in America. No direct citations.

451. Rosen, Edward. *The Naming of the Telescope*. New York:
         Henry Schuman, 1947. Pp. xvi + 110.

         Brief study of the earliest users and producers of
      telescopes. Concludes that Frederick Cesi coined the
      term "telescope" in 1611, though he did not prefer it,
      and that Cesi himself did not lay claim to the term but
      gave priority to the poet and theologian John Demisiani,
      who argued that scientific instruments should take ancient
      Greek names. See also item 444. 249 citations. Reviewed
      in: *Annals of Science*, 6 (1949), 209-210; *Isis*, 41 (1950),
      219-220.

*   Scaife, W. Lucien, ed. *John A. Brashear*.

         Cited herein as item 1363.

452. Shane, C.D. "Astronomical Telescopes since 1890." *The
         Legacy of George Ellery Hale* (item 1391), 209-219.

         Identifies significant advances in the optical and
      mechanical design and construction of large astronomical
      telescopes. 12 citations.

453. Simms, William. *The Achromatic Telescope and Its Various
         Mountings, Especially the Equatorial*. London, 1852.
         Reprinted, Kent: P.M.E. Erwood, 1980. Pp. 74 + 16.

         Popular descriptive review of the theory, construction
      and use of refracting telescopes. Includes many line
      drawings and a contemporary catalogue of the instruments
      manufactured and marketed by Troughton and Simms. No
      citations or index.

*   Stetson, Harlan True. "Elihu Thomson: His Interest in
         Astronomy."

         Cited herein as item 1372.

454. Tucker, R.H. "Transit Circles Today." *Quarterly Journal of the Royal Astronomical Society*, 10 (1969), 223-232.

Provides brief historical note on the transit circle as an introduction to an examination of modern technical improvements. Programs of the present time are identified. 25 citations.

455. Turner, A.J. "Some Comments by Caroline Herschel on the Use of the 40-ft Telescope." *Journal for the History of Astronomy*, 8 (1977), 196-198.

Transcription of a letter from Caroline Herschel to her nephew J.F.W. Herschel in 1827 describing the difficulties in building and using the telescope. Hints that telescope was more extensively used than published record indicates. 10 citations.

456. Turner, G. L'E. "James Short and the Reflecting Telescope." *XIIe Congrès International d'Histoire des Sciences, Tome XA (Paris, 1968)*. Paris: Blanchard, 1971. Pp. 101-106.

Traces reflectors constructed by Short between 1735 and his death in 1768 and attempts to determine how many were made. 99 are identified but more are expected to exist. 12 citations.

457. Turner, G. L'E. "The Number Code on Reflecting Telescopes by Nairne and Blunt." *Journal for the History of Astronomy*, 10 (1979), 177-184.

Explains key to number code engraved on reflecting telescopes constructed by Nairne and by Nairne and Blunt, English instrument makers of the late 18th and early 19th centuries. Reconstructs, from code, the number of telescopes originally produced and thereby shows them to have disappeared more frequently than larger telescopes surviving today manufactured by Short and others. Notes that of "Short's total production of 1370 reflectors, over 9 percent survive today, whereas of Nairne's production of about 850, only 1.7 percent are so far known to exist." Suggests that most of Nairne's instruments were for private non-professional use. 8 citations.

458. Van Helden, Albert. "The Telescope in the Seventeenth Century." *Isis*, 65 (1974), 38-58.

Examines and assesses the symbiotic role of the telescope and the development of astronomy in the 17th century.

Notes that from 1610 to 1660, "the improvements made in the telescope were due almost solely to improvements in lens grinding techniques," as opposed to advances in theoretical optics. Concludes that the solar system, as examined by the telescope, was the one "that Newton quantified, not the cosmos of Copernicus." 108 citations.

459. Van Helden, Albert. "The Historical Problem of the Invention of the Telescope." *History of Science*, 13 (1975), 251-263.

Abstract discussion of material presented in item 460. 77 citations.

460. Van Helden, Albert. "The Invention of the Telescope." *Transactions of the American Philosophical Society*, 67, pt. 4, 1977. Pp. 67.

Rejects persistent claims to the invention of the telescope by Roger Bacon, Thomas Digges and Della Porta, and that its invention was a simple matter. Does not provide a definite origin to the telescope but credits Jacob Metius as well as Hans Lipperhey in its invention and practical demonstration circa 1608. Translates relevant documents and provides extensive bibliography. Reviewed in: *Annals of Science*, 36 (1979), 418-419; *JHA*, 10 (1979), 57-58; *Isis*, 70 (1979), 601-602.

*   Warner, Deborah Jean. *Alvan Clark & Sons: Artists in Optics*.

Cited herein as item 1381.

461. Watson, R.D., and J.M. Watson. "The Great Melbourne Telescope." *The Australian Physicist*, 14 (1977), 182-185.

Describes the procurement and installation of a 48-inch reflector in 1869, at the time the second largest telescope in the world, and its effect upon Australian science. Establishment of the Melbourne instrument arose from an 1852 petition by the Royal Society to provide a major facility in the southern hemisphere for the continued study of nebulae. Though the mounting by Grubb proved adequate for long and continued use, the severe restrictions caused by the use of speculum metal for the Cassegrain primary mirror made this telescope almost immediately obsolete, as it was the last major instrument to use speculum metal instead of glass with a silver overlay. 5 citations.

*   Woodbury, David O. *The Glass Giant of Palomar*.

        Cited herein as item 343.

*   Wright, Helen. *The Great Palomar Telescope*.

        Cited herein as item 344.

PHOTOGRAPHIC INSTRUMENTATION AND TECHNIQUE

462.   Barnard, E.E. "The Development of Photography in Astronomy
        Popular Astronomy, 6 (1898), 425-455.

        Reviews the first 50 years of the application of
        photography to astronomy, as the vice-presidential
        address before Section A of the American Association for
        the Advancement of Science, August 22, 1898. No direct
        citations.

463.   Baum, W.A. "Photosensitive Detectors." *Annual Review of
        Astronomy and Astrophysics*, 2 (1964), 165-184.

        Describes all forms of image detectors including the
        eye, photographic plates, photoelectric surfaces. 62
        citations.

464.   Bell, Trudy E. "History of Astrophotography." *Astronomy*,
        4 (1976), 66-79.

        Brief popular review centering upon the late 19th
        century. No direct citations.

465.   DeVaucouleurs, Gerard. *Astronomical Photography from the
        Daguerreotype to the Electron Camera*. R. Wright, tr.
        London: Faber and Faber, 1961. Pp. 94.

        Popular exposition of the applications of photography
        to astronomy. Few citations.

466.   Eberhard, G. "Photographische Photometrie." *Handbuch der
        Astrophysik* (item 130), 431-518.

        General review of all aspects of photographic measure-
        ment of brightness. Contained within part 2 of Volume 2
        of the *Handbuch*. Includes a historical introduction
        noting the pioneering work of Harvard College Observatory,
        Karl Schwarzschild and J. Hartmann and emphasizes the
        early confusion over the proper manner in which to convert

photographic images into an astronomical magnitude system. Numerous citations and bibliography.

467.   Gernsheim, Helmut, and Alison Gernsheim. *The History of Photography from the Earliest Use of the Camera Obscura in the Eighteenth Century up to 1914*. London: Geoffrey Cumberlege, Oxford University Press, 1955. Pp. xxviii + 395.

Includes material on the early scientific development of photography, noting John Herschel's contributions, etc., but does not include a discussion of the scientific applications of photography. Extensive citations and bibliography. Reviewed in: *Isis*, 49 (1958), 449-451.

468.   Gill, David. "The Application of Photography in Astronomy." *Royal Institution Library of Science: Astronomy* (item 140), 314-328.

Lecture dated June 3, 1887. Reports on the proceedings of the Astrographic Chart Conference, held in Paris in April 1887 to organize an international campaign to photograph the entire sky for the determination of stellar magnitudes and positions. Reviews history of photographic determinations of the positions of stars from Warren de la Rue's suggestions in 1861, L.M. Rutherfurd's photographs of star clusters in the 1860s, B.A. Gould's program in Argentina, and later work leading to the Conference. Highlights Paul and Prosper Henry's development of dry-plate wide-field astrophotography in the 1870s.

469.   Herrmann, D.B., and D. Hoffmann. "Astrofotometrie und Lichttechnik in der 2. Hälfte des 19. Jahrhunderts." *NTM-Schriftenreihe für Geschichte der Naturwissenschaften, Technik und Medizin*, 13 (1976), 94-104.

The development of techniques of measuring brightness is traced through the first half of the 19th century with discussions of earlier systems. Centers upon the developing relationship between astronomical photometry and the experimental techniques of laboratory photometry or "technischen fotometrie" in Germany that provided reliable photometric standards. 56 citations.

470.   Hoffleit, Dorrit. *Some Firsts in Astronomical Photography*. Cambridge, Mass.: Harvard College Observatory, 1950. Pp. 39.

Describes the growth of astronomical photography from Arago's announcement of Daguerre's discovery in 1839 to

the publication of Antonia Maury's detailed classification of the photographic spectra of bright stars in
1897. Numerous illustrations, citations, and chronology
of "firsts" in the 19th century.

471. Mees, C.E. Kenneth. "Astronomical Photography Looks to
      the Red." *The Telescope*, 1 (1934), 102–113.

      Reviews use of dyes for color sensitization from Vogel's
      discovery in 1873 that dyes added to emulsions sensitized
      them in the regions which the dyes themselves absorbed.
      Cyanine dyes are discussed as prepared at the turn of
      the century and as developed at Kodak and elsewhere to
      extend sensitivity to the infra-red for terrestrial and
      astronomical use, especially for the spectrographic
      detection of water vapor lines in terrestrial, solar and
      planetary spectra.

472. Mees, C.E. Kenneth. "Recent Progress in Astronomical
      Photography." *Annual Report of the Smithsonian*. Washing-
      ton, D.C.: Smithsonian, 1954. Pp. 205–218.

      James Arthur lecture given under Smithsonian auspices
      on May 21, 1953. Provides brief history of his contacts
      with F.H. Seares and G.E. Hale, and development of emul-
      sions useful in astronomy through the 103 series and
      new infra-red emulsions. Describes how the 103 series
      has been utilized in recent astronomy.

473. Norman, Daniel. "The Development of Astronomical Photog-
      raphy." *Osiris*, 5 (1938), 560–594.

      Detailed illustrated narrative of progress from Daguerre
      and J.W. Draper's pioneer attempts circa 1840, through
      the 1930s. Examines both technical advances and advances
      in astronomical knowledge. Identifies priorities for
      many discoveries, arguing that while John Draper and
      Henri Becquerel discovered, simultaneously and indepen-
      dently, the ultra-violet lines, Draper indisputably was
      the discoverer of the new infra-red lines. Highlights
      the work of Harvard College Observatory, the author's
      affiliation. 136 citations and a chronological outline
      of major events and trends.

474. Russel, H.C. "Progress of Astronomical Photography."
      Popular Astronomy, 2 (1894), 101–105; 170–176; (1895),
      310–316; 457–463.

      Reviews all aspects of astronomical photography during
      the 19th century beginning with daguerreotypes of the

Sun and Moon in the 1840s and including the rise of positional photography, photographic spectroscopy, and eclipse photography. Numerous citations.

474a. Stone, Ormond. "Photographers versus Old Fashioned Astronomers." *Sidereal Messenger*, 6 (1887), 1-4.

Suggests that "photography will not necessarily displace every other method of observation" but will be used in conjunction with earlier visual techniques which are still useful. Comments on earlier statements regarding the value of photography in astronomy.

475. Turner, H.H. "Some Reflections Suggested by the Application of Photography to Astronomical Research." *Popular Astronomy*, 13 (1905), 72-82; 122-129.

Address delivered to the Section of Astrophysics at the Congress of Arts and Sciences at St. Louis, September 21, 1904. Reviews the role of photography in astronomy and classifies this role into three sections: power, facility, and accuracy. No direct citations.

476. Warner, Deborah Jean. "George Willis Ritchey and the Development of Celestial Photography." *American Scientist*, 54 (1966), 64-93.

After a brief biographical sketch, traces Ritchey's career in the design and construction of major astronomical equipment, notably G.E. Hale's early reflectors, and identifies his role in the application of photography to astronomy through the development of suitable reflecting systems. 69 citations.

* Warner, Deborah Jean. "Lewis Morris Rutherfurd: Pioneer Astronomical Photographer and Spectroscopist."

Cited herein as item 409.

477. Weaver, Harold F. "The Development of Astronomical Photometry." *Popular Astronomy*, 54 (1946), 211-230; 287-299; 339-351; 389-404; 451-464; 504-526.

A comprehensive review covering all periods and techniques. Divided into four periods: eye estimates with or without telescopes; use of a visual comparator and establishment of standard scale of brightness; application of photography; application of "physical photometers" capable of broad band measurement. The first period is only briefly reviewed. Emphasis upon later periods. 218 citations.

*   Weaver, Harold F. "Photoelectric Photometry."

       Cited herein as item 493.

SPECTROSCOPIC INSTRUMENTATION AND TECHNIQUE

478.   Adams, C.W. "William Allen Miller and William Hallowes
       Miller." *Isis*, 34 (1943), 337-339.

       Clarifies roles of the two Millers in mid-19th-century
       spectroscopy. W.A. Miller was the associate of William
       Huggins while W.H. Miller produced important spectroscopic
       studies testing the coincidence of the Fraunhofer D double
       seen in the solar spectrum with the bright line doublet
       produced by a laboratory "spirit" lamp—a sodium source.
       This 1854 experiment by Miller is offered as an important
       anticipation of Kirchhoff's work in 1860. 20 citations.

479.   Bowen, I.S. "Spectrographs." *Stars and Stellar Systems*,
       Volume 2. Chicago: University of Chicago Press, 1962.
       Pp. 34-62.

       Includes a short history of the development of spectro-
       graphs within a general review of their design, construc-
       tion and use. 12 citations.

480.   Cornell, E.S. "Early Studies in Radiant Heat." *Annals of
       Science*, 1 (1936), 217-225.

       Concludes that the existence of invisible radiant heat
       rays was generally acknowledged prior to 1800. 19 cita-
       tions. See also items 497 and 507.

*   DeKosky, Robert K. "Spectroscopy and the Elements in the
       Late Nineteenth Century: The Work of Sir William Crookes

       Cited herein as item 912.

*   Dunham, Theodore. "Methods in Stellar Spectroscopy."

       Cited herein as item 1110.

*   Forbes, Eric Gray. "A History of the Solar Red Shift
       Problem."

       Cited herein as item 1003.

481. Hamor, W.A. "David Alter and the Discovery of Spectro-
      chemical Analysis." *Isis*, 22 (1934-5), 507-510.

      Brief review of mid-19th-century development of spectro-
      scopic technique centering on obscure work of Alter, an
      American trained in medicine and science, who examined
      spectra vaporized in arc and spark sources.

*   Hetherington, Norriss S. "Adriaan van Maanen's Measure-
      ments of Solar Spectra for a General Magnetic Field."

      Cited herein as item 1007.

*   Roscoe, Henry E. *Spectrum Analysis*.

      Cited herein as item 928.

*   Russel, H.C. "Progress of Astronomical Photography."

      Cited herein as item 474.

*   Scheiner, Julius. *Die Spectralanalyse der Gestirne*.

      Cited herein as item 930.

*   Schellen, H. *Spectrum Analysis*.

      Cited herein as item 931.

*   Siegel, Daniel M. "Balfour Stewart and Gustav Robert
      Kirchhoff: Two Independent Approaches to 'Kirchhoff's
      Radiation Law.'"

      Cited herein as item 932.

482. Smyth, Piazzi. "Practical Spectroscopy in 1880." *The
      Observatory*, 3 (1880), 491-500; 523-529; 555-564.

      Reviews the rapid rise of spectroscopic work but
      laments the lack of calibration, consistency, and re-
      peatability in spectroscopic studies of line structure
      in the laboratory and at the telescope. Examines tech-
      niques for determining wavelength scales harking back
      to Baden Powell's calibrations of 1835. Notes perplexing
      inconsistencies in the absence of nitrogen in the solar
      spectrum but its alleged presence in nebulae and in the
      Earth's atmosphere.

483. Sutton, M.A. "Sir John Herschel and the Development of
      Spectroscopy in Britain." *British Journal for the
      History of Science*, 7 (1974), 42-60.

Examines John Herschel's contributions during the period 1819–61, chiefly the influence of his arguments circa 1833 that new discoveries of line spectra had important implications for theories of the structure of matter, "in particular to their apparent conflict with the law of continuity." 107 citations.

484. Sutton, M.A. "Spectroscopy and the Chemists: A Neglected Opportunity." *Ambix*, 23 (1976), 16–26.

Examines causes for the "delayed acceptance of spectrum analysis by chemists" by showing that they were not so much hampered by a lack of theoretical understanding as were physicists, but had available to them already other useful and productive means for chemical analysis, and so were not pressured to apply new techniques. Useful analysis of pre-Kirchhoff studies of the solar spectrum by Brewster, Talbot, Herschel and others. 77 citations.

485. Thiele, Joachim. "Zur Wirkungsgeschichte des Dopplerprinzip im Neunzehnten Jahrhundert." *Annals of Science*, 27 (1971), 393–407.

Reviews origins of the Doppler Principle, its acoustical verification, and then its optical verification, first suggested by Fizeau in 1848 and carried out in astronomical observations by William Huggins in England in the 1860s. 72 citations.

*   Warner, Deborah Jean. "Lewis Morris Rutherfurd: Pioneer Astronomical Photographer and Spectroscopist."

Cited herein as item 409.

RADIO INSTRUMENTATION AND TECHNIQUE

*   Edge, David O. "The Sociology of Innovation in Modern Astronomy."

Cited herein as item 165.

486. Edge, David O., and Michael J. Mulkay. *Astronomy Transformed: The Emergence of Radio Astronomy in Britain.* New York: Wiley-Interscience, 1976. Pp. xvi + 482.

Concentrates on two major case studies in the development of radio astronomy: Jodrell Bank in Manchester and

the Cambridge group. Contrasts the social structures of
the two groups where the former allows for independent
research interests of subgroups and the latter requires
participation within the one major group. Examines the
effects of differing technical strategies at the two
institutions and argues that there have been "scientific
[i.e., cultural], technical and social constraints that
have allowed two different styles of leadership to operate
at the two centres." Heavily documented. Reviewed in:
Technology and Culture, 19 (1978), 580–583; *Isis*, 70
(1979), 636–637; *JHA*, 9 (1978), 142–150.

487. Hey, J.S. "Solar Radio Eclipse Observations." *Vistas
      in Astronomy*, 1 (1955), 521–531.

      Reviews the results of observations at eclipses dating
      from 1945. 28 citations.

488. Hey, J.S. *The Evolution of Radio Astronomy*. New York:
      Science History Publ., 1973. Pp. ix + 214.

      General history of radio astronomy to 1971 by a pioneer
      in the field. Includes extensive bibliography of published
      sources and a glossary of terms. Reviewed in: *JHA*, 5
      (1974), 64.

489. Kraus, John. *Big Ear*. Ohio: Cygnus Books, 1976. Pp. v +
      228.

      Personal recollections of the origins and growth of
      radio astronomy centering upon his work at the Ohio State
      University. Includes bibliography of Kraus's works, and
      a short reference list.

490. Lovell, Bernard. *The Story of Jodrell Bank*. New York:
      Harper & Row, 1968. Pp. xvi + 265.

      Description of the origins and development of the
      world's largest fully steerable radio telescope near
      Manchester, England, by its founder and director. Con-
      stitutes a detailed reminiscence based upon diaries and
      personal papers. Reviewed in: *Isis*, 60 (1969), 264–265.

*   McKinley, D.W. *Meteor Science and Engineering*.

      Cited herein as item 983.

491. Westerhout, Gart. "The Early History of Radio Astronomy."
      Education in and History of Modern Astronomy (item 125),
      211–218.

Brief review of late 19th-century attempts to detect
radio emission from celestial sources; first discoveries
in the 1930s; and post-World War II developments and
discoveries to 1955. 12 citations.

*   Ze-zong, Xi, and Bo Shu-ren. "Ancient Novae and Supernovae
Recorded in the Annals of China, Korea, and Japan and
Their Significance in Radio Astronomy."

Cited herein as item 1106.

PHOTOELECTRIC INSTRUMENTATION AND TECHNIQUE

*   Baum, W.A. "Photosensitive Detectors."

Cited herein as item 463.

*   Curtis, H.D. "The Comet-Seeker Hoax."

Cited herein as item 646.

*   DeVaucouleurs, Gerard. *Astronomical Photography from the
Daguerreotype to the Electron Camera*.

Cited herein as item 465.

*   Greenstein, Jesse L. "The Seventieth Anniversary of
Professor Joel Stebbins and of the Washburn Observatory."

Cited herein as item 1292.

492.   Huffer, C.M. "The Development of Photo-electric Photometry."
Vistas in Astronomy, 1 (1955), 491-498.

Reviews the stages of development of photoelectric instrumentation and technique from the first selenium cell
in the 1890s; the introduction of different cathodes in
improved cells through the early 20th century, the first
vacuum tube amplifiers that replaced the electrometers
of Lindemann and others, and the introduction of the
photomultiplier tube after World War II. 7 citations.

*   Weaver, Harold F. "The Development of Astronomical
Photometry."

Cited herein as item 477.

493. Weaver, Harold F. "Photoelectric Photometry." *Encyclo-paedia of Physics*, 54 (1962), 130-179.

    Comprehensive review of all aspects of photometric, photographic and photoelectric techniques and their progress since the late 19th century. Includes the establishment and refinement of magnitude scales, color indices, and spectral classification by photometry. 134 citations.

494. Whitford, A.E. "Photoelectric Techniques." *Encyclopaedia of Physics*, 54 (1962), 240-288.

    Comprehensive historical introduction to detectors, amplifiers, and direct telescopic applications by the pioneer of the vacuum tube amplifier in astronomy. 218 citations.

OTHER INSTRUMENTATION AND TECHNIQUES
(UV, IR, X-RAY, COSMIC RAY, SPACE BORNE)

495. Allen, David A. "Infrared Astronomy: an Assessment." *Quarterly Journal of the Royal Astronomical Society*, 18 (1977), 188-198.

    Provides a brief historical sketch of 19th-century advances in infra-red studies of astronomical objects, as an introduction to a contemporary review. 10 citations.

496. Coblentz, W.W. "Thermocouple Measurements of Stellar and Planetary Radiation." *Popular Astronomy*, 31 (1923), 105-121.

    Reviews the previous decade of studies involving the use of thermopiles and thermocouples for the determination of the characteristics of stellar radiation, primarily for the calculation of stellar temperatures for the various Harvard spectral classes. Provides details of the designs of early thermopiles and thermocouples from 1911 used by the author, a staff member of the National Bureau of Standards. 26 citations including references to pioneer studies since the 1860s by Huggins, Stone, Pfund, Nichols, Stebbins, Coblentz and others.

*   Cornell, E.S. "Early Studies in Radiant Heat."

    Cited herein as item 480.

497. Cornell, E.S. "The Radiant Heat Spectrum from Herschel
      to Melloni." *Annals of Science*, 3 (1938), 119–137.
      402–416.

      Traces studies of the maximum heating effect of the
      solar spectrum from the late 18th-century studies of
      Landriani, William Herschel and others through the mid-
      19th century by Melloni. Examines the major papers pub-
      lished and the growing realization of the similarity
      between radiant heat and light. 25+31 citations. See
      items 480, 507, and 962.

498. Curtis, Heber D. "Voyages to the Moon." *Publications of
      the Astronomical Society of the Pacific*, 32 (1920),
      145–150.

      Briefly reviews early speculation and fiction concern-
      ing space travel (1600–1900) as a prelude to a discussion
      of the possibilities of space travel, based upon recent
      claims of Robert Goddard in 1919 in his "A Method of
      Reaching Extreme Altitudes" (*Smithsonian Misc. Collections*,
      71 (1919), No. 2.).

499. Dorman, L.I. *Cosmic Rays, Variations and Space Explora-
      tions*. New York: American Elsevier, 1974. Pp. xv + 675.

      Comprehensive review of the study of variations in
      cosmic ray flux. Includes a brief 12-page historical
      discussion that identifies periods in the history of the
      study. 55 pages of references.

500. Eddy, John A. "Thomas A. Edison and Infra-Red Astronomy."
      Journal for the History of Astronomy, 3 (1972), 165–187.

      Exposition and critique of Edison's design and construc-
      tion of an infra-red sensor he used to measure the infra-
      red radiance of the solar corona during an eclipse of
      the Sun in 1878. Edison's contacts with astronomers in-
      cluding C.A. Young, S.P. Langley, and N. Lockyer are
      described. 82 citations.

501. Emme, Eugene M. *Aeronautics and Astronautics: An American
      Chronology of Science and Technology in the Exploration
      of Space, 1915–1960*. Washington, D.C.: Government
      Printing Office, 1961. Pp. xi + 240.

      Reference work for the history of recent work on ter-
      restrial and extra-terrestrial flight centering upon
      the role of NACA and the first three years of NASA. Ex-
      tensive appendices and bibliography. Reviewed in: *Tech-
      nology & Culture*, 3 (1962), 210–212.

*   Emme, Eugene M. "Aeronautics, Rocketry and Astronautics."
       Cited herein as item 14.

502.   Friedman, Herbert. "Ultraviolet and X-rays from the Sun."
       Annual Review of Astronomy and Astrophysics, 1 (1963),
       59–96.

       Description of the history of rocket-borne instrumenta-
       tion in the study of the solar spectrum from the 1930s
       by a pioneer in the field. 62 citations.

503.   Friedman, Herbert. "Rocket Astronomy." *Education in and
       History of Modern Astronomy* (item 125), 267–273.

       Review and personal account of development of scientific
       uses of rocketry centering on Friedman's work in X-ray
       astronomy.

*   Hall, R. Cargill. *Lunar Impact, A History of Project
       Ranger.*

       Cited herein as item 251.

504.   Krause, Ernst H. "High Altitude Research with V-2 Rockets."
       Proceedings of the American Philosophical Society, 91
       (1947), 430–446.

       Reviews the first few years of research in ionospheric
       studies, solar ultraviolet spectra and cosmic ray detection
       and the capabilities and limitations of V-2 rockets for
       scientific research. 42 citations. Reprinted in: *Annual
       Reports of the Smithsonian Institution*, (1948), 189–208.

*   Langley, S.P. "The New Spectrum."
       Cited herein as item 1011.

505.   Ley, Willy. *Rockets, Missiles, and Space Travel.* New
       York: Viking, 1952. Pp. xii + 436.

       General popular review, updating and expanding the
       original 1944 edition, including historical chapters on
       space flight and the plurality of worlds prior to the
       modern era; the development of rocketry through World
       War II; and post-war advances into space. Includes tech-
       nical appendices on characteristics of German rocketry
       and post-war American developments, and a 17-page bibli-
       ography listing works on rocketry and the literary history
       of "imaginative literature on Space Travel."

506. Lovell, Bernard. *The Origins and International Economics of Space Exploration*. Edinburgh: Edinburgh University Press, 1973. Pp. viii + 104.

Reprint of a lecture given at Edinburgh in 1973. Traces the history of astronautics since 1957 but examines the political and economic influences that governed the advancement of the field in the UK, Europe, the US, and USSR, since World War II. Briefly discusses the philosophical question of "good and bad science" based upon personal experiences. Reviews the applications of space borne technology to the sciences as well as society in general. 78 notes and references. Reviewed in: *Technology and Culture*, 19 (1978), 249-251.

507. Lovell, D.J. "Herschel's Dilemma in the Interpretation of Thermal Radiation." *Isis*, 59 (1968), 46-60.

Examines William Herschel's discovery of infra-red heat radiation in 1800 and his conceptual difficulties in its interpretation as invisible radiation based upon subsequent investigations of the properties of radiant heat. 62 citations. See also item 492.

508. Millikan, R.A. "High Frequency Rays of Cosmic Origin." *Popular Astronomy*, 34 (1926), 232-238.

Reviews work since 1903 on the detection of cosmic rays. 12 citations.

*   Ordway, F.I. *Annotated Bibliography of Space Science and Technology, with an Astronomical Supplement*.

Cited herein as item 46.

*   Roland, Alex. *A Guide to Research in NASA History*.

Cited herein as item 48a.

*   Rosse, Lord. "On the Radiation of Heat from the Moon, the Law of Its Absorption by Our Atmosphere, and Its Variation in Amount with Her Phases."

Cited herein as item 962.

*   Rosse, Lord. "The Radiant Heat from the Moon during the Progress of an Eclipse."

Cited herein as item 963.

509. Rossi, Bruno. *Cosmic Rays*. New York: McGraw-Hill, 1964.
     Pp. x + 268.

     Includes introductory chapters outlining the history
     of cosmic ray research. No direct citations.

510. Rynin, N.A. *Interplanetary Flight and Communication*.
     (1928-1932) 3 volumes in 9 parts. Translated by Israel
     Program for Scientific Translations, 1970-71. NASA
     Publications TT F-640-648. Springfield, Va.: National
     Technical Information Service. Pp. 1500+.

     Translations of originals written between 1928 and
     1932. TT-640 is titled "Dreams, Legends and Early Fan-
     tasies" and includes a history of speculation on space
     flight. TT-641, "Space Craft in Science Fiction," reviews
     19th- and early 20th-century writings. Reviewed in:
     Technology and Culture, 14 (1973), 317-323.

511. Spitzer, Lyman. "The Beginnings and Future of Space
     Astronomy." *American Scientist*, 50 (1962), 473-484.

     Lecture given at the Third International Space Science
     Symposium (COSPAR) held in Washington, D.C., in 1962.
     Reviews limitations of ground based astronomy, the be-
     ginnings of observational astronomy from space, and
     possible future programs, including the existence of
     extra-solar planets. 10 citations.

*   Swenson, Loyd S., James M. Grimwood, and Charles C.
     Alexander. *This New Ocean: A History of Project Mercury*.

     Cited herein as item 255.

*   Tousey, R. "The Spectrum of the Sun in the Extreme Ultra-
     violet."

     Cited herein as item 1031.

512. Venable, W.M. Henry, Sr., and William H. Venable, Jr.
     "Samuel P. Langley's Bolometer, the Quantitative
     Measurement of Energy Distribution in Radiation and
     the Discovery of the Far Infra-Red." *XIIe Congrès
     International d'Histoire des Sciences Tome V (Paris,
     1968)*. Paris: Blanchard, 1971. Pp. 107-110.

     Reviews Langley's infra-red studies of the radiation
     of the Sun in the late 19th century including a descrip-
     tion of Langley's bolometer. 18 citations.

513. Von Braun, Wernher, and Frederich I. Ordway. *History of
      Rocketry and Space Travel*. Revised Edition. New York:
      Thomas Y. Crowell, 1969. Pp. xi + 254.

      Update of 1966 descriptive text reviewing the pre-war,
      wartime, and post-war progress of astronautics. Includes
      historical chapters on the plurality of the worlds and
      the lure for speculation; rocketry from the Chinese
      through modern times; studies of pioneer rocketeers in-
      cluding Robert Goddard, Konstantin Tsiolkovsky, and Her-
      mann Oberth; the growth of rocketry during World War II.
      Extensive bibliography. Reviewed in: *Technology and
      Culture*, 9 (1968), 250-251; *Isis*, 61 (1970), 404-405.

                    COMPUTATIONAL DEVICES

*   Beer, Arthur. "Astronomical Dating of Works of Art."

      Cited herein as item 162.

514. Cotter, Charles H. "George Biddell Airy and His Mechanical
      Correction of the Magnetic Compass." *Annals of Science*,
      33 (1976), 263-274.

      Reviews Airy's invention in 1838 during period when
      iron ships were becoming commonplace. Reviews history
      of ship magnetism and Airy's system but concentrates upon
      his role, as Astronomer Royal at Greenwich, in establishing
      system of magnetic corrections. 40 citations and notes.

*   Gingerich, Owen. "Applications of High-Speed Computers
      to the History of Astronomy."

      Cited herein as item 166.

515. Goldstine, Herrman H. *The Computer from Pascal to von
      Neumann*. Princeton: Princeton University Press, 1972.
      Pp. vii + 378.

      In three parts: (1) To World War II; (2) During World
      War II; (3) After World War II. The second and third
      parts center on Goldstine's personal involvement but
      part one deals with both the development of computing
      devices and the scientific background and technological
      expertise available at the time. Specifically, Goldstine
      provides commentary on the astronomical needs for comput-
      ing power including discussions of Forest Ray Moulton's

ballistics studies and the collaborative work of Wallace J. Eckert, Gerald Clemence, and Jan Schilt of Columbia and the Naval Observatory in developing improved lunar ephemerides. Reviewed in: *Isis*, 67 (1976), 295-297.

* Huffer, C.M., and G.W. Collins. "Computation of Elements of Eclipsing Binary Stars by High-Speed Computing Machines."

Cited herein as item 1073.

* Mayall, Margaret W., ed. *Centennial Symposia*.

Cited herein as item 141.

516. Merzbach, Uta C. *George Scheutz and the First Printing Calculator*. Washington, D.C.: Smithsonian Institution Press, 1977. Pp. iv + 74.

History of the first operational "difference engine" or calculator. Illustrates the first use of machines for the computation of atmospheric refraction tables at the Dudley Observatory in Albany, N.Y. Illustrated, with bibliography.

517. Wrubel, Marshal H. "The Electronic Computer as an Astronomical Instrument." *Vistas in Astronomy*, 3 (1959), 107-116.

Reviews the possible applications and potential for use of electronic computers in astronomy, to "automatize all the standard computations that are being done at an observatory." Provides an example through the computation of line profiles by first presenting the method in its algebraic form, and then in its coded form. Written at a time when computers were still not at the research front, Wrubel notes: "Naturally, astronomers cannot be expected to change their methods overnight. It will be the young astronomer, accustomed to learning new things, who will accept electronic computation in his stride. It is to these men, especially, that this article is addressed."

CLOCKS AND TIMEKEEPING DEVICES

518. Antiquarian Horological Society. *Pioneers of Precision Timekeeping*. London: Thanet Printing Works, 1965. Pp. 117.

Illustrated collection of 6 papers derived from a
symposium and exhibition by the Antiquarian Horological
Society of British timekeeping held in London in 1955.
Describes in detail the workings of important devices
including John Harrison's chronometer of 1770 and another
by Larum Kendall used on the H.M.S. Bounty. Reviewed in:
Isis, 58 (1967), 265-266.

*   Brown, Basil. *Astronomical Atlases, Maps and Charts: An
        Historical and General Guide.*

        Cited herein as item 828.

519.   Davies, Alun C. "The Life and Death of a Scientific
        Instrument: The Marine Chronometer." *Annals of Science*,
        35 (1978), 509-525.

        Traces the success of Harrison's chronometers which
        stimulated a small craft industry lasting well into the
        20th century and which was only replaced when alternative
        modes of determining Greenwich Time became available
        through radio broadcast. Includes detailed listings of
        the rate of production of chronometers. 69 citations.

520.   Forbes, Eric Gray. "The Origin and Development of the
        Marine Chronometer." *Annals of Science*, 22 (1966),
        1-25.

        Reviews the method of determining local mean time and
        the need for accurate marine chronometers with which
        longitude could be determined at sea. Centers upon the
        quest for the £20,000 award, offered by the Parliament
        of Great Britain for the production of a suitable device
        in 1714, by John Harrison and other clock makers, and
        Harrison's final successes in the 1760s. Includes accounts
        of the testing of the chronometers, the conditions of the
        award, and claims for the award by others. 50 citations.

*   Fox, Philip. *Adler Planetarium and Astronomical Museum
        of Chicago.*

        Cited herein as item 395.

521.   Gould, Rupert Thomas. *The Marine Chronometer. Its History
        and Development.* London: J.D. Potter, 1923. Reprinted,
        1971. Pp. xvi + 287.

        Traces the development of the marine chronometer as
        a central theme in the solution to the problem of the
        determination of longitude at sea. Provides background

on alternatives to the use of the chronometer, notably
the employment of the observed position of the Moon
compared to its calculated position, a method long suggested and eventually developed by N. Maskelyne circa
1763, ironically at the same time as the successful trials
of John Harrison's chronometers in 1761 and 1764. Centers
on Harrison's work and his successors. Includes bibliographical material. A detailed synopsis by George Sarton
is available in the review: *Isis*, 6 (1924), 122-129.

522. Horsky, Zdenek. "Astronomy and the Art of Clockmaking
      in the Fourteenth, Fifteenth and Sixteenth Centuries."
      New Aspects in the History and Philosophy of Astronomy
      (item 123), 25-34.

Reviews the development of geared clocks useful for
astronomical measurements. Shows that at first, their
utility was only as demonstration devices rather than
for accurate measurement. 19 citations.

523. Irwin, John B. "The Case of the Carpenter's Chronometers."
      The Griffith Observer, 37 (1973), 2-6.

Reviews the testing of Harrison's chronometers aboard
the ship Centurion, and the gradual acceptance of
Harrison's work. No direct citations.

524. King, H.C., and J.R. Millburn. *Geared to the Stars: The
      Evolution of Planetariums, Orreries and Astronomical
      Clocks*. Toronto: University of Toronto Press, 1978.
      Pp. xvii + 442.

Definitive large scale study of astronomical timepieces,
both useful, pedagogical, and ornamental including mechanical orreries (planetaria named after an early patron)
capable of showing the heliocentric relationship of all
known planetary bodies, and their evolution into modern
projection planetariums. Extensive illustrations, diagrams,
notes and bibliography. Reviewed in: *Annals of Science*,
36 (1979), 419-420; *Isis*, 71 (1980), 160-161.

*   Marguet, F. *Histoire Générale de la Navigation du XVe
      au XXe siècle*.

Cited herein as item 792.

525. Mayall, R. Newton. "The Inventor of Standard Time."
      Popular Astronomy, 50 (1942), 204-209.

Brief narrative concluding that C.F. Dowd was the author of the principle of Standard Time. Describes the original principle and plan, and other claims of priority. No direct citations.

526. Milham, W.I. *Time and Timekeepers*. New York: Macmillan, 1923. Pp. xix + 609.

Detailed historical account of clocks and clockmaking from the early 17th century to date. Annotated bibliography with 518 citations.

527. Millburn, John R. "Benjamin Martin and the Development of the Orrery." *British Journal for the History of Science*, 6 (1972-73), 378-399.

Concludes that Martin was primarily responsible for the popularization of the orrery by demonstrating its great facility as an educational demonstration device. 91 citations.

528. Millburn, John R. "William Stukeley and the Early History of the Orrery." *Annals of Science*, 31 (1974), 511-528.

Reviews the early history of mechanical planetaria centering upon the question of their origin. Examines the possibility that John Rowley's famous model, made about 1713 for Charles Boyle, fourth Earl of Orrery (hence the name), was "developed from a model said to have been made about 1705 by Stephen Hales, and recorded in a drawing by William Stukeley." Concludes that Hale's device was a true orrery. 56 citations.

*   Quill, Humphrey. *John Harrison: The Man Who Found Longitude*.

Cited herein as item 1357.

529. Sadler, D.H. "Astronomical Measures of Time." *Quarterly Journal of the Royal Astronomical Society*, 9 (1968), 281-293.

Examines the use of Ephemeris Time and Atomic Clock Time after a brief historical introduction to the "relationship between astronomical determinations of time and 'mechanical' time-keepers or clocks." Reviews the improvement in the determination of the length of the day from Harrison's clock of 1760 to atomic clocks.

*   Taylor, E.G.R. *The Haven-Finding Art.*
              Cited herein as item 865.

*   Thomson, Malcolm M. *The Beginning of the Long Dash: A
              History of Timekeeping in Canada.*
              Cited herein as item 239.

DESCRIPTIVE ASTRONOMY

GENERAL

530. Armitage, Angus. "The Cosmology of Giordano Bruno."
    Annals of Science, 6 (1948), 24-31.

    Reviews Bruno's combination of the heliocentric hypothesis and the concept of an infinite universe in the late 16th century.

531. Baum, Richard. *The Planets: Some Myths & Realities*. New York: John Wiley, 1973. Pp. 200.

    Explores historical episodes in telescopic astronomy that remain controversial. Included are the search for lunar satellites; mountains on Venus; rings around Uranus and Neptune; lost planets, and other unexplained observations. Numerous citations and extensive bibliography.

532. Bienkowska, Barbara. *Kopernik i heliocentryzm w polskiej Kulturze umyslowej do Końca xviii wieku*. (Copernicus and the Heliocentric Theory in Polish Culture until the End of the 18th Century). Warsaw, 1971. Pp. 295.

    Describes (in Polish) the gradual acceptance of heliocentric theory in Poland. Includes analysis of scientific works, general textbooks, popular periodicals, theatrical and artistic commentary. Extensive bibliography. Reviewed in: *Isis*, 63 (1972), 254-255; *JHA*, 3 (1972), 220.

533. Brewster, David. *More Worlds Than One. The Creed of the Philosopher and the Hope of the Christian*. London: John Murray, 1854. Pp. vii + 262.

    An extended criticism and rebuttal of the contemporary work *Of the Plurality of Worlds: An Essay* (see item 594). Argues for the plurality of inhabited worlds. Provides a brief history of the concept, beginning with Fontenelle's famous work in 1686, its reception and lasting influence. Some citations. See item 534.

534.   Brooke, John Hedley. "Natural Theology and the Plurality
       of Worlds: Observations on the Brewster-Whewell Debate."
       Annals of Science, 34 (1977), 221-286.

       Analyzes debate between David Brewster and William
       Whewell on the existence of life beyond the earth as a
       case study of natural theology prior to Darwin but during
       a period when Chambers' *Vestiges of the Natural History
       of Creation* was being attacked. Argues that natural theology
       was far from static during this period, as evidenced by
       the differences between Brewster and Whewell. See items
       533, 594. 391 citations and annotations.

*   Calinger, Ronald. "Kant and Newtonian Science: The Pre-
       Critical Period."

       Cited herein as item 754.

535.   Carré, Marie-Rose. "A Man Between Two Worlds: Pierre Borel
       and His *Discours nouveau prouvant la pluralité des mondes*
       of 1657." *Isis*, 65 (1974), 322-335.

       Analyzes the central theme of Borel's work, which was
       that since the universe was heliocentric, and the Earth
       a planet, then all other planets must be inhabited. Examines
       contemporary reactions to Borel's statement. 26 citations.

536.   Collier, Katherine Brownell. *Cosmogonies of Our Fathers*.
       New York: Columbia University Press, 1934. Pp. 500.

       Attempts an examination of efforts between the late
       16th century and early 19th century to reconcile the
       heliocentric universe with the Bible, as part of a general
       study of reconciliations of science and biblical doctrine.
       Reviewed in: *Isis*, 24 (1936), 167.

*   [Copernicus, N.] *Avant, avec, après Copernic*.

       Cited herein as item 127.

537.   Deisch, Noel. "The Navigation of Space in Early Specula-
       tion and in Modern Research." *Popular Astronomy*, 38
       (1930), 73-88.

       Includes commentary on the history of the doctrine of
       the Plurality of Worlds. 19 citations.

*   Dingle, Herbert. "Thomas Wright's Astronomical Heritage."

       Cited herein as item 1268.

538. Drake, Stillman, and C.D. O'Malley, eds. *The Controversy
      over the Comets of 1618*. Philadelphia: University of
      Pennsylvania Press, 1960. Pp. xxv + 380.

      Translations, with commentary, of writings by Galileo,
      H. Grassi, M. Guiducci and J. Kepler on the appearance of
      three bright comets which "inspired a large number of
      books and pamphlets, because comets were at that time
      generally regarded with superstitious dread, and because
      even among the learned it was still debated whether they
      were atmospheric or celestial phenomena." Numerous cita-
      tions.

*   Fernie, J.D. "The Historical Quest for the Nature of the
      Spiral Nebulae."

      Cited herein as item 1153.

539. Flammarion, Camille. *La Planète Mars et ses conditions
      d'habitabilité*. 2 volumes. Paris: Gauthier-Villars et
      Fils, 1892 (vol. I); 1909 (vol. II). Pp. 608.

      Detailed, well-illustrated discussion of history of
      observations of the planet from 1636 to 1892 with analysis
      and extensive speculation.

*   Flammarion, Camille. *Popular Astronomy*.

      Cited herein as item 1413.

540. Flamsteed, John. *The Gresham Lectures of John Flamsteed*.
      Eric G. Forbes, ed. London: Mansell, 1975. Pp. xviii +
      479.

      Includes 39 lectures between 1681 and 1684 on theory
      of telescope optics, solar parallax, astronomical refrac-
      tion and physical constitution of celestial objects.
      Stereographic celestial scenes are included with annota-
      tions by the editor, as well as an extensive introduction
      to the background of the lectures. Reviewed in: *Annals
      of Science*, 33 (1976), 417-420; *Isis*, 68 (1977), 644-645.

541. Fontenelle, Bernard LeBovier de. *Entretiens sur la
      pluralité des mondes. Digressions sur les anciens et
      les modernes.* [1686; 1748]. Robert Shackleton, ed.
      Oxford: Clarendon Press, 1955. Pp. 218.

      Critical edition with extensive introduction by Shackle-
      ton of Fontenelle's exposition of Cartesian cosmology.
      Identifies origins of Fontenelle's interest in Plurality

of the Worlds and lists the many editions of these works, and other related works on Fontenelle. Utilizes the 1748 edition. See also translation by John Glanvill of original 1686 edition: Nonesuch Press, 1929, with prologue by David Garnett. Reviewed in: *Isis*, 47 (1956), 452-453.

*   Forbes, Eric Gray, ed. *Human Implications of Scientific Advance*.

Cited herein as item 362.

542.   Godwin, Francis. *The Man in the Moone* [1638]. Edited, translated, and annotated by Annie Amartin. Nancy: Université de Nancy II, 1979. Pp. x + 169.

English transcription with French translation of Godwin's 1638 astronomical fantasy of journeys to the moon. Includes background remarks on 17th- and 18th-century speculations on cosmic voyages and the Plurality of the Worlds. Bibliography and numerous citations.

543.   Grant, Edward. "Medieval and Seventeenth-Century Conceptions of an Infinite Void Space beyond the Cosmos." *Isis*, 60 (1969), 39-60.

Analyzes the responses to the Aristotelian belief that "absolutely nothing, or mere privation" would be found beyond the visible cosmos. Examines medieval and anti-Aristotelian cosmological views that influenced Newton, and the difficulties of conceptualizing void space. Concludes that medieval Scholastics' acceptance of an "imaginary infinite void space" beyond the Aristotelian cosmos was based purely upon theological grounds but that with advances in physics and astronomy in the 17th century, which showed that Nature did not "abhor a vacuum," a significant change in the conception of the reality of the void took place. 91 citations.

*   Grant, Robert. *History of Physical Astronomy, from the Earliest Ages to the Middle of the Nineteenth Century*.

Cited herein as item 91.

544.   Green, A.H. *The Birth and Growth of Worlds*. London: Society for Promoting Christian Knowledge, 1890. Pp. 61.

Brief popular review of theories of the origin and structure of the Earth and cosmos since the late 17th century by a respected Oxford geologist. Short bibliography.

*   Gushee, Vera. "Thomas Wright of Durham, Astronomer."

       Cited herein as item 1294.

*   Hardin, Clyde L. "The Scientific Work of the Reverend John Michell."

       Cited herein as item 1298.

545.   Harrison, Thomas P. "Birds in the Moon." *Isis*, 45 (1954), 323-330.

       Traces the 17th-century efforts of Charles Morton to combine astronomy and ornithology into the earliest treatise on bird migration written in England. Stimulated by the 17th-century intellectual climate favoring Plurality of the Worlds, Morton suggested that birds migrated to the Moon. Examines Morton's familiarity with works on plurality, including Wilkins' *The Discovery of a New World in the Moon*, and Godwin's famous romance: *The Man in the Moone*. 17 citations.

546.   Hartner, Willy. "Mediaeval Views on Cosmic Dimensions and Ptolemy's Kitab Al-Manghurat." *Mélanges Alexandre Koyré, Volume 1*. Histoire de La Pensée, XII. Paris: Hermann, 1964. Pp. 254-282.

       Detailed exposition of methods and techniques for determining distances and dimensions of Sun, Moon and planets. 66 citations.

547.   Hastie, William, tr. *Kant's Cosmogony*. Glasgow: J. Maclehose, 1900. Pp. cix + 205. Revised with an introduction by Willy Ley. New York, 1968.

       A translation with detailed commentary of Kant's writings on the retardation of the Earth's rotation, and Kant's *Natural History and Theory of the Heavens*. Describes Kant's speculations on the origin and development of the Solar System, the Milky Way, and island universes and compares them to the works of Wright, Lambert and Laplace. Numerous citations.

548.   Hellman, C. Doris. *The Comet of 1577. Its Place in the History of Astronomy*. New York: Columbia University Press, 1944. Pp. 488.

       Exhaustive study of the 1577 passage of a comet observed by Brahe which was responsible for his condemnation of Copernican theory. Examines Brahe's measurement of the parallax of the comet, and the opinions of those who

believed with him that the comet was more distant than
the Moon. Examines also the varied reactions of scientists
"preachers and poets," and others of the time. Heavily
annotated with extensive bibliography. Reviewed in: *Isis*,
36 (1946), 266–270.

549.   Hellman, C. Doris. "The Gradual Abandonment of the
       Aristotelean Universe: A Preliminary Note on Some
       Sidelights." *Mélanges Alexandre Koyré, Volume 1.*
       Histoire de La Pensée, XII. Paris: Hermann, 1964.
       Pp. 283–293.

       Examines elements in the transition, notably the in-
       crease of accurate observations of planetary positions
       and the exploitation of the appearances of comets and
       novae in the 16th century. 36 citations.

*   Herrmann, D.B. "Karl Friedrich Zöllner und sein Beitrag
       zur Reception der naturwissenschaftlichen Schriften
       Immanuel Kants."

       Cited herein as item 1302.

550.   Hine, William L. "Mersenne and Copernicanism." *Isis*, 64
       (1973), 18–32.

       Traces Mersenne's attempts to promote Copernicanism,
       and constraints upon him requiring that he not abandon
       Aristotelianism. The result was that he presented the
       Copernican doctrine as a hypothetical theory. 86 cita-
       tions.

551.   Hoskin, M.A. "The Cosmology of Thomas Wright of Durham."
       Journal for the History of Astronomy, 1 (1970), 44–52.

       Argues that Wright's model of the form of the Universe
       was based more upon moral and religious ideas than upon
       the scientific analysis of observational data. Develops
       Wright's thinking prior to his publication of *An Original
       Theory or new Hypothesis of the Universe* (London, 1750)
       and through the writing of his *Second Thoughts* ... in
       1770. See also: *Journal for the History of Astronomy*, 2
       (1971), 208.

*   Hoskin, M.A. "Newton, Providence and the Universe of
       Stars."

       Cited herein as item 725.

552. Hoskin, M.A. "Lambert and Herschel." *Journal for the History of Astronomy*, 9 (1978), 140-142.

Demonstrates, through an examination of William Herschel's unpublished papers, that Herschel indeed critically studied Lambert's speculations upon solar motion and the structure of the universe, and though he rejected much of Lambert's speculation, it did play an important role in the development of Herschel's own thinking after 1799.

553. Hoskin, M.A. "Cosmology in the Eighteenth and Nineteenth Centuries." *Human Implications of Scientific Advance* (item 362), 544-552.

Examines some of the major concepts and methodological principles involved in early scientific cosmology. Concludes that inadequacy of observational data was primary limitation of study, and that cosmological speculation was temporarily dormant in the late 19th century. 8 citations.

*   Hoyt, William Graves. *Lowell and Mars*.

Cited herein as item 954.

554. Huygens, Christiaan. *The Celestial Worlds Discover'd* [1698]. London: F. Coss, 1968. Pp. vi + vi + 160.

Facsimile reprint of anonymous English translation of this 1698 treatise that examines the physical appearances of the planets and conjectures on the Plurality of the Worlds. Known originally as *Cosmotheros*, it includes Huygens' pioneer studies of stellar photometry. Reviewed in: *British Journal for the History of Science*, 4 (1969), 406-407.

555. Jaki, Stanley L. "Lambert: Self-taught Physicist." *Physics Today*, 30 (1977), 25-32.

Brief examination of J.H. Lambert's professional training and areas of research. Contrasts Lambert's and Kant's cosmologies and Lambert's preoccupation with the role of comets. Describes Lambert's model of the Milky Way: a ring made up of innumerable subsystems with void within the ring. Attempts to show existence of modern elements of cosmology in Lambert's thinking. 19 citations.

556. Jaki, Stanley L. "Lambert and the Watershed of Cosmology." *Scientia*, 113 (1978), 75-95.

Reviews the character of post-Newtonian and Cartesian cosmologies in the early 18th century and their change in character to speculations upon structure and limits by Thomas Wright, Immanuel Kant and Johann Heinrich Lambert at mid-century. Provides detailed analysis of William Herschel's contributions and regard for writings on cosmological speculation. 88 citations.

557.   Jaquel, Roger. *Le savant et philosophe multhousien Jean-Henri Lambert (1728-1777): Études critiques et documentaires*. Paris: Éditions Ophrys, 1977. Pp. 170.

Collected papers by Jaquel on Lambert's contributions to symbolic logic, cartography and cosmogony, with an introduction to his background and life. Examines in some detail Lambert's theory of comets as ejecta from stellar volcanoes, and Lambert's statements on the Plurality of Worlds. Reviewed in: *Annals of Science*, 35 (1978); 543-544; *Isis*, 70 (1979), 178.

558.   Jones, H.W. "Leibniz' Cosmology and Thomas White's *Euclides Physicus*." *Archives Internationale d'Histoire des Sciences*, 25 (1975), 277-303.

Examines Leibniz' use of Thomas White's work in the development of his cosmology showing its influence, through Leibniz' notes on White, on Leibniz' comcepts of extent, motion and force. 12 citations.

559.   Jones, Kenneth Glyn. "The Observational Basis for Kant's 'Cosmogony': A Critical Analysis." *Journal for the History of Astronomy*, 2 (1971), 29-34.

Argues that Kant's *Cosmogony* largely was based upon philosophical considerations and that its supposed observational basis is illusory. Reviews early observations of nebulae and the suggestion of the existence of nebulae external to the Milky Way. 26 citations.

560.   Kant, E. *Allgemeine Naturgeschichte und Theorie des Himmels* [1755]. Translated by W. Hastie as *Kant's Cosmogony*, with new introduction by G.J. Whitrow. New York: Johnson Reprint, 1970. Pp. xl + 205.

A softbound edition is available, with introduction by M.K. Munitz. Arbor Paperbacks, 1969. Pp. xxiv + 180. Reviewed in: *JHA*, 3 (1972), 68.

561.   Koyré, Alexandre. *From the Closed World to the Infinite Universe*. Baltimore: Johns Hopkins Press, 1957. Re-

printed by Harper Torchbooks, New York: Harper and
Row, 1958. Pp. x + 312.

Examines role of the concepts of space in cosmological
speculation during the 16th to 18th centuries. Latter
half of work centers upon Newtonian cosmology and its
adherents. Centers upon the rejection of christianized
Aristotelianism in cosmological thought and concepts of
the structure and hierarchy of space, which ended in the
reestablishment of the concept of void space by the end
of the 18th century. Reviewed in: *Isis*, 49 (1958), 363-
366; *Annals of Science*, 25 (1969), 357-359.

*   Labrousse, Elisabeth. *L'entrée de Saturne au Lion*.

        Cited herein as item 625.

562.   Lambert, J.H. *Cosmological Letters on the Arrangement of
        the World-Edifice* [1761]. Stanley L. Jaki, translator
        and editor. New York: Science History Publications,
        1976. Pp. 245.

        Translation, with a 54-page introduction by Jaki, of
        Lambert's 1761 work which speculates upon the hierarchical
        organization of a collision-free static universe safe
        for life. Jaki provides extensive commentary and bibliog-
        raphy. Lambert's work takes the form of correspondence
        with a peer, but was written entirely by Lambert. Examines
        comets, the Milky Way and the motions of fixed stars,
        under the general assumption that perfection is achieved
        in the universe because all possible variations in form
        and substance which can occur do occur. Essay review in:
        JHA, 9 (1978), 134-139. Reviewed also in: *Centaurus*, 22
        (1978), 74-76; *Isis*, 70 (1979), 316-317.

*   Ley, Willy. *Rockets, Missiles, and Space Travel*.

        Cited herein as item 505.

*   Lohne, J. "The Fair Fame of Thomas Harriott."

        Cited herein as item 1333.

*   Lowell, Percival. *Mars*.

        Cited herein as item 955.

*   Lowell, Percival. *Mars and Its Canals*.

        Cited herein as item 956.

*   Lowell, Percival. *Mars as the Abode of Life*.

      Cited herein as item 957.

*   Lowell, Percival. *The Evolution of Worlds*.

      Cited herein as item 958.

563. MacKlem, Michael. *The Anatomy of the World. Relations
      between Natural and Moral Law from Donne to Pope*.
      Minneapolis: The University of Minnesota Press, 1958.
      Pp. x + 139.

      General synthetic review of historical studies center-
      ing upon 18th-century essays on cosmological thinking.
      Reviewed in: *Isis*, 50 (1959), 506-507.

564. Marsak, Leonard M. "Cartesianism in Fontenelle and French
      Science, 1686-1752." *Isis*, 50 (1959), 51-60.

      Argues that Fontenelle "rejected the methodology and
      metaphysics of Descartes," though he accepted Cartesian
      cosmology. 42 citations.

565. McColley, Grant. "The Seventeenth-Century Doctrine of
      a Plurality of Worlds." *Annals of Science*, 1 (1936),
      385-430.

      A general study beginning with an exposition of Greek
      thought, through Saint Augustine, Albertus Magnus, and
      through the post-Copernican period and the plurality-
      infinity doctrine. Examines the reaction to telescopic
      discoveries and notes the surviving influence of the
      "principle of Plentitude." 202 citations.

566. McColley, Grant. "The Second Edition of *The Discovery
      of a World in the Moone*." *Annals of Science*, 1 (1936),
      330-334.

      Examines two separate editions of this famous work by
      John Wilkins circa 1638-1640 to show that indeed two
      distinct editions were produced, thus increasing the
      historical influence of this work, an important document
      arguing for the acceptance of the new astronomy and a
      more rational and less literal approach to Scripture.
      19 citations.

567. McColley, Grant. "Nicholas Reymers and the Fourth System
      of the World." *Popular Astronomy*, 46 (1938), 25-31.

Reviews this late 16th-century cosmology which placed
the Sun at the center of the solar system, and a rotating
Earth at the center of the fixed stars. Examines Reymers'
astronomical work. 25 citations.

568. McColley, Grant. "The Ross-Wilkins Controversy." *Annals
of Science*, 3 (1938), 153-189.

Recounts the nature and influence of this famous controversy in the 1630s between science and scriptural
literalism, of interest for its role in the 17th-century
acceptance of scientific cosmology, especially in John
Wilkins' writings on the new cosmology. 114 citations.

569. McColley, Grant. "Christopher Scheiner and the Decline
of Neo-Aristotelianism." *Isis*, 32 (1940), 63-69.

Reviews the contents of the *Rosa Ursina* (1626-1630)
by Christopher Scheiner, a book that contained his many
early solar observations but which also concentrated an
attack upon the Aristotelian belief in solid crystalline
orbital spheres and the incorruptibility of celestial
bodies. Examines the background and influence of the
14 chapters in the book devoted to the attack upon
Aristotle. Scheiner's strong argument was based upon
documentation showing "that a large portion, if not the
weight, of ancient and modern authority either was opposed
to the doctrine of solid spheres and immutable heavens,
or supported the conception that the heavens are liquid
and the globes within them corruptible." 14 citations.

570. McColley, Grant. "J.H. and the *Astronomia Crystallina*."
Annals of Science, 4 (1940), 319-321.

Describes a small pamphlet by an unknown writer in
the late 17th century which follows Nicolas Reymers'
1588 hypothesis of a "Fourth System of the World," a
compromise between Ptolemaic and Copernican cosmology.
See also item 567. 10 citations.

571. McColley, Grant. "Nathanael Carpenter and the 'Philos-
ophia Libra.'" *Popular Astronomy*, 48 (1940), 143-146.

Reviews the contributions of this 17th-century English
cleric who was an early advocate of the "New Astronomy"
and who advocated the Fourth System of the World (Sun
in center of planetary orbits, Earth at center of stellar
sphere and Earth in rotation). 16 citations.

572. McColley, Grant. "George Valla: An Unnoted Advocate of
          the Geo-Heliocentric Theory." *Isis*, 33 (1941), 312-314.

      Examines Valla's role in the late 16th- and early 17th-
      century struggle for a suitable astronomical system when
      five systems vied for acceptance: revised Ptolemaic;
      semi-Ptolemaic; geo-heliocentric; Fourth System of the
      World; and heliocentric. See also item 567.

*   Meadows, A.J. "The Discovery of an Atmosphere on Venus."

      Cited herein as item 960.

*   Meadows, A.J. *The High Firmament*.

      Cited herein as item 157.

573. Michel, Paul Henri. *La Cosmologie de Giordano Bruno*.
          Paris: Hermann, 1962. Pp. viii + 344.

      Well-documented exposition (in French) of Bruno's
      cosmology. Argues that Bruno did create a systematic
      philosophy of Nature. Reviewed in: *Annals of Science*,
      17 (1961), 63.

574. Mitchell, S.A. "The Moon Hoax." *Popular Astronomy*, 8
          (1900), 256-267.

      Argues that a French astronomer, M.J.N. Nicollet, was
      the true author of the celebrated "Moon Hoax" of the
      mid-19th century, and not Richard Adams Locke. Examines
      the hoax, a popular article recording John Herschel's
      "observations" of life on the Moon, and the origins of
      Nicollet's efforts. No direct citations.

575. Mullen, Richard D. "The Undisciplined Imagination:
          Edgar Rice Burroughs and Lowellian Mars." *Science
          Fiction: The Other Side of Realism*. Thomas D. Clareson,
          ed. Bowling Green: Bowling Green University Popular
          Press, 1971. Pp. 229-247.

      Contrary to conceptions of the popular influence of
      Percival Lowell's writings claiming evidence for the
      existence of highly civilized and technological life on
      Mars, Mullen argues that one writer at least, Burroughs,
      was not so influenced in his fiction and remained free
      to speculate. 15 citations.

576. Nicolson, Marjorie Hope. *Voyages to the Moon*. New York:
          Macmillan, 1948. Pp. xiii + 297. Reprinted: New York:
          Macmillan Paperback, 1960.

Describes supernatural voyages, flight aided by birds, use of artificial wings and flying chariots in 17th- and 18th-century literature and thought, as modes of travelling beyond the Earth. Extensive 30-page bibliography. Reviewed in: *Isis*, 40 (1949), 286-287.

* Nicolson, Marjorie H. "English Almanacs and the 'New Astronomy.'"

Cited herein as item 862.

577. Nicolson, Marjorie Hope, and Nora Mohler. "The Scientific Background of Swift's Voyage to Laputa." *Annals of Science*, 2 (1937), 299-334; 405-430.

Demonstrates that contemporary scientists, primarily members of the Royal Society, were the subjects of Swift's satire. The second paper examines the sources of Swift's "Flying Island" showing that it was "drawn" carefully and thoughtfully from contemporary science. 230 citations.

578. Numbers, Ronald L. "The Nebular Hypothesis of Isaac Orr." *Journal for the History of Astronomy*, 3 (1972), 49-51.

Exposition of cosmological thought of an early 19th-century American, his scientific isolation, and the reception of his ideas. Indicates that Laplace's Nebular Hypothesis was not widely known in America in the early 19th century. 11 citations.

579. Numbers, Ronald L. *Creation by Natural Law: Laplace's Nebular Hypothesis in American Thought*. Seattle: Univ. of Washington Press, 1977. Pp. xi + 184.

Documents the scientific, lay, and theological reactions to Laplace's Nebular Hypothesis in mid-19th-century America and shows that the general acceptance of Laplace aided the American reception of Darwinian evolution. Provides detailed analysis of the development of Laplace's cosmogony, its modification by Daniel Kirkwood in America, and the extent to which Americans (such as Isaac Orr and Benjamin Franklin) indulged in cosmological speculation. Heavily annotated with detailed bibliographical listings. Reviewed in: *Isis*, 69 (1978), 312; *JHA*, 9 (1978), 70-71; *Sky and Telescope*, 55 (1978), 73.

580.   Pizor, Faith K., and T. Allen Comp, eds. *The Man in
        the Moone and Other Lunar Fantasies*. New York: Praeger,
        1971. Pp. xx + 230.

        Collection of nine literary essays from the 17th through
        19th centuries dealing with travelling to the Moon and
        life seen there. Brief introductions provided for each
        selection. Reviewed in: *Isis*, 63 (1972), 108.

581.   Reaves, Gibson. "The Great Moon Hoax of 1835." *The
        Griffith Observer*, 18 (1954), 126-134.

        Examines the state of popular astronomy at the time of
        the hoax, and the reception of this famous article that
        appeared in the New York "Sun" reporting upon John
        Herschel's "observations" of life on the moon. Few in-
        direct citations. See also item 574.

582.   Robison, Wade L. "Galileo on the Moons of Jupiter."
        Annals of Science, 31 (1974), 165-169.

        Argues that Galileo did not personally compare the
        Jovian system of satellites to the solar system as a
        small-scale version of the Copernican system. 9 citations.

583.   Roger, Jacques. *Mémoires du Muséum national d'Histoire
        naturelle. Serie C: Sciences de la terre. Tome X:
        Buffon, Les Époques de la nature*. Paris: Éditions du
        Museum, 1962. Pp. clii + 343.

        Annotated critical edition of Buffon's 1778 work which
        includes much of his astronomical work and interests.
        Roger's introduction examines the evolution of Buffon's
        cosmology and religious beliefs, as well as his contri-
        butions to geological thought. Reviewed in: *Isis*, 55
        (1964), 39-94.

584.   Rosen, Edward. "Kepler's Defense of Tycho Against Ursus."
        Popular Astronomy, 54 (1946), 405-412.

        Reviews Kepler's involvement (circa 1595-1601) in the
        dispute between Tycho Brahe and Nicholas Reymers ("Ursus").
        Provides an annotated facsimile of Kepler's *Defense of
        Tycho*, and examines the possible date of its writing.
        40 citations.

*   Rosen, Edward, tr. *Kepler's Somnium: The Dream, or
        Posthumous Work on Lunar Astronomy*.

        Cited herein as item 710.

585. Rousseau, G.S. "Poiesis and Urania: The Relation of Poetry and Astronomy in the English Enlightenment." *XIIᵉ Congrès International d'Histoire des Sciences, Tome IIIB (Paris, 1968)*. Paris: Blanchard, 1971. Pp. 113-116.

Examines the influence of science upon Milton and Pope concluding that while Milton was directly affected by Galileo, Pope learned of Newtonian astronomy only through the popular lectures of William Whiston. 6 citations.

586. Russell, John L. "Cosmological Teaching in the Seventeenth-Century Scottish Universities." *Journal for the History of Astronomy*, 5 (1974), 122-132; 145-154.

Traces resistance to Newtonian, Galilean and Cartesian thinking as illustrated in theses completed during the period. 71 citations.

587. Rybka, Eugeniusz. *Four Hundred Years of the Copernican Heritage*. Krakow: Jagellonian University Press, 1964. Pp. 235.

Constitutes a biographical study (in English) of Copernicus set within the intellectual climate of his time. Includes chapters on Brahe, the fight for the heliocentric system, Newtonian cosmology, the revival of astronomical research in Poland, and the influence of Copernicus upon modern cosmology. 52-item bibliography.

* Sarton, George. "Laplace's Religion."

Cited herein as item 805.

588. Schaffer, Simon. "The Phoenix of Nature: Fire and Evolutionary Cosmology in Wright and Kant." *Journal for the History of Astronomy*, 9 (1978), 180-200.

Examines the difficulties in establishing a truly rational theory of cosmical evolution in the mid-18th century. "The idea of development contained within itself a fundamental contradiction: the system of the world had been ordained by God, and must therefore be perfect, while any change in this state would imply a departure from divine perfection." Analyzes Wright's use of a "philosophy of fire" as the basis for the action of God in maintaining a "moral and vital evolution" and his dependence upon the earlier writings of William Whiston who developed a cosmology that "distinguished between

a 'power' of God and a 'law' of Nature." The "Phoenix of
Nature," or the regeneration through fire, is used by
Kant as a universal explanatory mechanism, and Schaffer
shows how its use, blended with Kant's concept of gravi-
tationally active matter, subject always to God's will,
was finally able to break "the tyranny of the stable
state" and produce a truly evolutionary cosmology. 91
citations.

589.   Schaffer, Simon. "'The Great Laboratories of the Uni-
       verse': William Herschel on Matter Theory and Planetary
       Life." *Journal for the History of Astronomy*, 11 (1980),
       81-111.

       Examines the connection between Herschel's studies of
       the structure and evolution of stellar systems and his
       interest in the nature of light. Develops in detail
       Herschel's conclusion that stellar systems are evolving
       through the action of gravity and relates this conclusion
       to his speculations on Plurality of the Worlds. 107
       citations.

590.   Skabelund, Donald. "Cosmology on the American Frontier:
       Orson Pratt's Key to the Universe." *Centaurus*, 11
       (1965), 190-204.

       Describes the background and contents of a small
       monograph written by a Mormon apostle in 1879 which in-
       cluded a number of unique empirical laws.

*   Sticker, Bernhard. *Bau und Bildung des Weltalls*.

       Cited herein as item 1222.

591.   Stimson, Dorothy. *The Gradual Acceptance of the Copernican
       Theory of the Universe*. New York: Baker & Taylor,
       1917. Pp. 147.

       In three major sections: exposition of the state of
       astronomical thought to 1400; Copernicus and his times;
       the reception of Copernican theory in the 16th through
       early 18th centuries. Appendices provide translations
       of relevant documents. Includes a 15-page bibliography
       and numerous citations.

592.   Tipler, Frank J. "Extraterrestrial Intelligent Beings
       Do Not Exist." *Quarterly Journal of the Royal Astro-
       nomical Society*, 21 (1980), 267-281.

Provides, within a contemporary review of the question, a brief but cogent bibliography to historical literature on the Pluality of Worlds. 81 citations.

* Tumanian, B.E. *The Geocentric and Heliocentric Systems in Armenia.*

Cited herein as item 241.

* Van Helden, Albert. "The Telescope in the Seventeenth Century."

Cited herein as item 458.

* Von Braun, Wernher, and Frederich I. Ordway. *History of Rocketry and Space Travel.*

Cited herein as item 513.

593. Wallace, Alfred Russel. *Man's Place in the Universe.* New York: McClure, Phillips, 1904. Pp. viii + 338.

Applies biological knowledge to synthesis of contemporary astronomical thought on the existence of life in the Universe. Concludes that Earth is unique for life in our Solar System, and that our central position in the Universe (the Milky Way) is crucial for stability and hence for the existence and maintenance of life. Offers valuable insight into speculations on Plurality of the Worlds circa 1900. Lucid review of astronomy circa 1900. Numerous citations. Reviewed by Simon Newcomb in: *Nation,* 78 (Jan. 14, 1904), 34-35.

594. Whewell, William. *Of the Plurality of Worlds: An Essay.* London: Parker, 1853; 3rd Edition: 1854. Pp. v + 279.

Concludes, using religious arguments, that "if there be, in any other planet, intellectual and moral beings, they must not only be like men, but must be men, ..." (p. 260), and since the Earth has been shown to be in a unique condition for man, there is little to support the existence of life elsewhere. Argues that further speculation on the matter, from the astronomical,geological, or biological sciences, "must, we think, be held to be eminently rash and unphilosophical." The author of this important essay, which does provide a good review of astronomical knowledge relating to plurality at the time, was a pivotal figure in the philosophy of science and an important commentator on astronomical matters, especially the Nebular Hypothesis, and how these matters must be viewed within the tenets of theology. This work was critically reviewed by David Brewster (see item 533). Some citations, no index.

*   Whiston, William. *Astronomical Lectures*.

Cited herein as item 1385.

*   Whitney, C.A. *The Discovery of our Galaxy*.

Cited herein as item 116.

595.   Whitrow, G.J. "Kant and the Extragalactic Nebulae." *Quarterly Journal of the Royal Astronomical Society*, 8 (1967), 48-56.

Provides a brief biographical sketch of Kant as an introduction to a discussion of his 1755 work *Natural History and Theory of the Heavens*. Examines origin and object of Kant's production of this work and the uniqueness of his speculation on the extragalactic nature of nebulae. Provides brief review of theories of the Milky Way by Galileo, Kepler, and Wright, and early allied studies of nebulae. 14 citations.

596.   Whitrow, G.J. "The Nebular Hypothesis of Kant and Laplace." *XIIe Congrès International d'Histoire des Sciences, Tome IIIB (Paris, 1968)*. Paris: Blanchard, 1971. Pp. 175-180.

Descriptive review of the cosmogonies of Kant, Laplace, and Swedenborg emphasizing the differences between the first two. Notes features of Kant's theory reflected in modern work. 9 citations.

597.   Whittaker, Sir Edmund. *From Euclid to Eddington, A Study of Conceptions of the External World*. Cambridge: Cambridge University Press, 1949. Pp. ix + 212.

Contains the Tarner Lectures for 1947. Provides a general philosophical narrative tracing the concept of space and extension from Plato and Aristotle through Descartes, Gassendi, Newton and Leibniz to modern conceptions of the dimensions of space, "extra-galactic" geometry, and aspects of relativity. Includes a summary of his "The Beginning and End of the World" lecture from 1942 on cosmical processes, what is permanent in Nature, and the degradation of the energy of the Universe. No direct citations.

598. Wilkins, John. *The Discovery of a World in the Moone* [1638]. Barbara Shapiro, ed. New York: Scholars Facsimiles and Reprints, 1973. Pp. x + 209.

   Facsimile reprint of 1638 edition which was the first English study of plurality based upon the telescopic observations of Galileo, Scheiner and others of the time. Includes annotation and an introduction. Reviewed in: *Isis*, 67 (1976), 645-646.

599. Wright, Thomas. *Clavis coelestis: Being the Explication of a Diagram entitled A Synopsis of the Universe or, the Visible World Epitomised by Thomas Wright* (1742). Michael A. Hoskin, ed. London: Dawson, 1967. Pp. xi + 78.

   A facsimile edition with a preface by Hoskin, of a 6x4-foot chart illustrating Wright's theory of the Universe, with detailed discussions of its various aspects. Provides insight into the development of Wright's cosmology between his *A Theory of the Universe* of 1734 and his *An Original Theory or New Hypothesis of the Universe* of 1750. Essay review by H. Woolf in: *Isis*, 63 (1972), 235-241.

600. Wright, Thomas. *Second or Singular Thoughts upon the Theory of the Universe*. Michael A. Hoskin, ed. London: Dawson, 1968. Pp. 93.

   Facsimile of post-1750 unpublished revision of Thomas Wright's cosmology. Wright's cosmology is also discussed by S. Schaffer in: *JHA*, 9 (1978), 181; an essay review by Harry Woolf is in: *Isis*, 63 (1972), 235-241.

601. Wright, Thomas. *An Original Theory or New Hypothesis of the Universe*, 1750. A Facsimile reprint together with the first publication of *A Theory of the Universe* (1734). M.A. Hoskin, ed. London: MacDonald, 1971. Pp. xxviii + 178.

   Facsimile reprinting of Wright's first cosmological essay, and his major work, which was an exploration of the structure of the Milky Way. Wright was deeply influenced by theology but removed the Sun from the center of the Universe and made it a star like other stars within an ordered universe subject to moral and physical law. Essay review by Harry Woolf in: *Isis*, 63 (1972), 235-241; see also: *JHA*, 2 (1971), 208.

602. Yourgrau, Wolfgang, and Allan D. Breck, eds. *Cosmology,
      History, and Theology*. New York: Plenum Press, 1977.
      Pp. xvi + 416.

      24 articles from a 1974 symposium of scientists, the-
      ologians, philosophers and historians. Historical papers
      on the role of time in cosmology; Thomas Wright; William
      Herschel; Laplace and the Nebular Hypothesis and geology.

603. Zirkle, Conway. "The Theory of Concentric Spheres: Edmund
      Halley, Cotton Mather, John Cleves Symmes." *Isis*, 37
      (1947), 155-159.

      Describes speculations on the "hollow earth" hypothesis
      wherein the Earth is composed of a series of concentric
      spheres, more or less habitable. Concludes that much
      early speculation was derived from the serious hypothesis
      of Halley in 1692 to explain the drift of the Earth's
      magnetic poles by the existence of a magnetic nucleus
      within the Earth that moved with respect to the Earth's
      outer shell. Includes excerpts from the writings of
      Halley, Mather and Symmes on this subject. Numerous
      citations to primary and secondary literature.

                    TERRESTRIAL AND ATMOSPHERIC

*   Biermann, Kurt-R., ed. *Briefwechsel zwischen Alexander
      von Humboldt und Carl Friedrich Gauss*.

      Cited herein as item 942.

604. Biermann, Kurt-R. "Alexander von Humboldt als Initiator
      und Organisator Internationaler Zusammenarbeit auf
      Geophysikalischen Gebeit." *Human Implications of
      Scientific Advance* (item 362), 126-138.

      Examines Humboldt's role in international studies of
      geomagnetism by looking at the development of inter-
      national cooperation, the origins of Humboldt's interest,
      and the background of the history of geomagnetic studies.
      51 citations with commentary by J.A. Cawood.

605. Boyer, Carl B. "The Tertiary Rainbow: An Historical
      Account." *Isis*, 49 (1958), 141-154.

      General account, from Aristotelean theory through the
      18th century, of speculation over the existence of a

third rainbow. Shows that the Aristotelean belief in
the invisibility of the third rainbow was contested
during the Middle Ages, and was a popular topic of
scholastic philosophers. Reviews Cartesian and Newtonian
theories of the rainbow noting that Newton's provided
predictions for the radius of the third bow. Follows
later theories and studies by Halley, Bernoulli, Thomas
Young, and observes that even at present, "the interaction of the solar rays with the particles of the media
through which the light passes" is imperfectly understood.
58 citations.

606. Briggs, J. Morton, Jr. "Aurora and Enlightenment: Eighteenth-Century Explanations of the Aurora Borealis."
Isis, 58 (1967), 491-503.

Examines increased study of Aurorae after a spectacular
display in 1716. Reviews Halley's and Jacques Philippe
Maraldi's observations and deductions and how their work
"set the style" for later reports. Examines the search
for a cause, including the exhalation of fumes from the
Earth's interior as a result of earthquakes; Halley's
modification of the "fume" theory wherein the Earth's
magnetic field was the cause; Mairau's belief in the
role of the Zodiacal Light in 1732; and Euler's electrical
theory based upon his work on the origin of cometary
tails. 29 citations.

607. Brush, Stephen G. "Nineteenth-Century Debates about the
Inside of the Earth: Solid, Liquid or Gas." *Annals of
Science*, 36 (1979), 225-254.

Includes review of astronomical arguments against the
molten core/thin crust theory of the Earth by William
Hopkins, on the basis of calculations of the Earth's
precessional constant. 136 citations.

*   Burstyn, Harold L. "The Deflecting Force of the Earth's
Rotation from Galileo to Newton."

Cited herein as item 701.

*   Burstyn, Harold L. "Early Explanations of the Role of
the Earth's Rotation in the Circulation of the Atmosphere and the Ocean."

Cited herein as item 753.

608. Cawood, John. "Terrestrial Magnetism and the Development
     of International Collaboration in the Early Nineteenth
     Century." *Annals of Science*, 34 (1977), 551-587.

     Traces the origins of international cooperation in the
     study of geomagnetism from the late 18th century through
     the 1830s in France, Britain and Germany. Centers on the
     efforts of Humboldt, his collaborations with Biot and
     Gay-Lussac; the relationship of astronomical interests
     with problems in geomagnetism, and the resultant partici-
     pation of F. Arago. Examines also the development of a
     theoretical understanding of the cause of terrestrial
     magnetism. 112 citations.

609. Cawood, John. "The Magnetic Crusade: Science and Politics
     in Early Victorian Britain." *Isis*, 70 (1979), 493-518.

     Examines the origins and structure of "the geomagnetic
     lobby" led by Edward Sabine, Humphrey Lloyd, John Herschel
     and others in the 1830s that convinced the British Associ-
     ation and the Royal Society to support a world-wide sea
     and land based study of the Earth's magnetism. Explores
     the role the "Magnetic Crusade" played in the politics
     of Victorian science and shows that its prosecution
     "revealed the importance of the military in the develop-
     ment of the geophysical sciences." Reviews early 19th-
     century background to the Crusade and reveals the close
     connection it maintained with astronomical institutions.
     95 citations.

610. Chapman, Sydney. "Aurora and Geomagnetic Storms." D.P.
     LeGalley and A. Rosen, eds. *Space Physics*. New York:
     Wiley, 1964. Pp. 226-269.

     Historical review, from the 17th century to the present,
     of observations and speculations about aurorae and mag-
     netic disturbances on Earth. 90 citations.

611. Chapman, Sydney. "Alexander von Humboldt and Geomagnetic
     Science." *Archive for History of Exact Sciences*, 2
     (1962), 41-51.

     Reviews the history of geomagnetic science identifying
     Humboldt's contributions. Identifies the main problems
     of early geomagnetism including the origin of the mag-
     netic field, causes for secular variations, daily varia-
     tions, and the origins of Humboldt's magnetic storms.
     Reviews 20th-century studies, the "World Magnetic Survey"
     and contemporary modes of magnetic surveying. 2 citations.

612. Chapman, Sydney. "The Earth." *Quarterly Journal of the Royal Astronomical Society*, 11 (1970), 382-395.

General review of theories of the Earth from earliest times, including Roger Bacon's theory of tides; geomagnetism beginning with Gilbert, Halley, and the variations in the needle observed by George Graham; the study of aurorae; the figure of the Earth after Newton; the motions of the Earth; the nature of the atmosphere. 23 citations.

* Eather, Robert H. *Majestic Lights, The Aurora in Science, History and the Arts*.

Cited herein as item 945a.

* Ferraro, V.C.A. "The Solar-Terrestrial Environment: An Historical Survey."

Cited herein as item 946.

612a. Goldstein, Bernard R. "Refraction, Twilight, and the Height of the Atmosphere." *Vistas in Astronomy*, 20 (1976), 105-107.

Examines early speculation and later observational methodology regarding the structure of, and extent of, the Earth's atmosphere from the Middle Ages to the mid-17th century and the contributions of Kepler and Hooke. 19 citations.

613. Hughes, Arthur. "Science in English Encyclopaedias, 1704-1875--IV." *Annals of Science*, 11 (1955), 74-92.

Examines the development of the "deluge controversy" as reflected in the treatment of the history of the Earth provided in encyclopedias. Notes Buffon's development of a theory of the Earth and its origin. 44 citations.

614. Joy, A.H. "Refraction in Astronomy." *Astronomical Society of the Pacific Leaflet No. 220* (June 1947). Pp. 8.

Brief introduction to atmospheric color and refraction. Centers on recognition of the existence of atmospheric refraction in astronomy by Ptolemy and its treatment through Newton to present knowledge.

615. Link, Frantisek. "On the History of the Aurora Borealis." *New Aspects in the History and Philosophy of Astronomy* (item 123), 297-306.

Abstracts observations since antiquity and attempts
a statistical time analysis. 27 citations.

616.   Mahan, A.I. "Astronomical Refraction: Some History and
       Theories." *Applied Optics*, 1 (1962), 497-511.

Detailed technical review and analysis of theories of
the Earth's atmospheric structure as revealed by studies,
both observational and theoretical, of the phenomenon
of refraction, or the progressive displacement of the
positions of celestial objects with increasing zenith
distance. Covers all periods, emphasizing modern developments. 63 citations.

617.   Malin, S.R.C. "British World Magnetic Charts." *Quarterly
       Journal of the Royal Astronomical Society*, 10 (1969),
       309-316.

Reviews the history of the series beginning with the
first by Edmond Halley in 1702 used for navigation,
Edward Sabine's organization of magnetic observatories,
and advances in magnetic cartography to the present era.
13 citations.

*   McKinley, D.W. *Meteor Science and Engineering*.

Cited herein as item 983.

617a. Mendillo, Michael, and John Keady. "Watching the Aurora
       from Colonial America." *EOS: Transactions, American
       Geophysical Union*, 57, No. 7 (July 1976), 485-491.

Traces early American interest in astronomical and
geophysical phenomena including excerpts from the writings
of John Winthrop, Benjamin Franklin, and Ezra Stiles.
16 citations.

618.   Middleton, W.E. Knowles. "Bouguer, Lambert, and the
       Theory of Horizontal Visibility." *Isis*, 51 (1960),
       145-149.

Technical examination of 18th-century theories of
atmospheric optics and the theory of seeing through the
atmosphere. Argues for the priority of Pierre Bouguer's
pioneering studies in practical photometry which established an exponential law for the attenuation of light
in transparent media in 1729, normally attributed to
Lambert and Beer. 23 citations.

618a. Minnaert, M. *Light and Colour in the Open Air*. London:
      G. Bell, 1940. Pp. xl + 362.

      Standard classic descriptive review of atmospheric
      optics and atmospheric phenomena including rainbows,
      halos and coronae. Bibliographical footnotes.

619. Ockenden, R.E. "Marco Antonio de Dominis and His Explana-
      tion of the Rainbow." *Isis*, 26 (1936), 40-49.

      Reviews the controversy over the first correct inter-
      pretation of the rainbow. Examines Dominis' independent
      rediscovery of a theory of the rainbow in 1611, after
      its original interpretation by Qutbal-din and Theodoric
      of Freiburg in the early 14th century as due to two re-
      fractions and one reflection of the sun's rays (for the
      primary bow) and an extra reflection for the secondary
      bow. Concludes that these early theories were not correct,
      whereas Descartes in 1637 did provide a correct theory,
      even though his priority was contested by Newton. 30
      citations.

*   Richardson, Robert S. "Sunspot Problems Old and New."

      Cited herein as item 1027.

*   Ruse, Michael. "The Scientific Methodology of William
      Whewell."

      Cited herein as item 803.

620. Schröder, W. *Entwicklungsphasen der Erforschung der
      Leuchtenden Nachtwolken*. Berlin: Akademie-Verlag,
      1975. Pp. vi + 64.

      Traces history of observations and speculations upon
      noctilucent clouds, a rare high-altitude form of cloud.
      Centers upon Otto Jesse's recognition of clouds as a
      true form after the eruption of Krakatoa in 1883 stimu-
      lated interest in studying atmospheric phenomena. Con-
      siders the questions of the cloud's origin and nature
      still open. Reviewed in: *JHA*, 10 (1979), 69-70.

*   Zirkle, Conway. "The Theory of Concentric Spheres:
      Edmund Halley, Cotton Mather, John Cleves Symmes."

      Cited herein as item 603.

THE SUN

*   Airy, G.B. "On the Results of Recent Calculations on
       the Eclipse of Thales and Eclipses Connected with It."

       Cited herein as item 745.

621.   Alter, Dinsmore. "Sunspot Observations Before the Inven-
       tion of the Telescope." *The Griffith Observer*, 11
       (1947), 137-139.

       Based upon secondary sources compiled from the work
       of Grant and Wolf, this brief review provides a concise
       listing of all known pre-telescopic sightings of sunspots.
       Two direct citations. See item 623.

622.   Bartholomew, C.F. "The Discovery of the Solar Granula-
       tion." *Quarterly Journal of the Royal Astronomical
       Society*, 17 (1976), 263-289.

       Reviews 18th- and 19th-century theories of the nature
       of the sun's visible surface. Examines A. Wilson and
       W. Herschel's sunspot theory, J. Nasmyth's "willow-
       leaves," the contrary observations of W.R. Dawes, and
       the ensuing controversy circa 1864 over the nature of
       the solar surface. 121 citations.

623.   Clark, David H., and F. Richard Stephenson. "An Inter-
       pretation of the Pre-Telescopic Sunspot Records from
       the Orient." *Quarterly Journal of the Royal Astronomical
       Society*, 19 (1978), 387-410.

       Detects and confirms two major historical periods in
       addition to the Maunder Minimum when sunspot activity
       was at a minimum: 1280-1350 ("Medieval Minor Minimum")
       and 1400-1600 (the "Sporer Minimum"). 15 citations.

624.   Curtis, H.D. "Rosa Ursina, sive Sol, A Retrospect."
       Popular Astronomy, 20 (1912), 561-568.

       Short examination of C. Scheiner's study of the Sun
       emphasizing scientific style. No direct citations.

*   Eddy, J.A. "The Maunder Minimum."

       Cited herein as item 1002.

*   Fernie, Donald. *The Whisper and the Vision*.

       Cited herein as item 282.

* Kilgour, Frederick G. "Professor John Winthrop's Notes on Sun Spot Observations (1737)."

  Cited herein as item 635.

625. Labrousse, Elisabeth. *L'entrée de Saturne au Lion. (l'éclipse de soleil du 12 âout 1654)*. The Hague: Nijhoff, 1974. Pp. iv + 115.

  Comparative study of how various European cultures reacted to the solar eclipse of 12 August 1654; how some were still in the grip of astrological fear and how others, in accepting heliocentrism, met the event with scientific interest and emotional indifference. Extensive documentation. Reviewed in: *Annals of Science*, 32 (1975), 604-605.

626. Lerner, Michel-Pierre. "'Sicut Nodus in Tabula': De la rotation propre du soleil au seizième siècle." *Journal for the History of Astronomy*, 11 (1980), 114-129.

  Investigates 16th-century speculation on the rotation of the Sun, before Galileo's observations. 70 citations.

* Lockyer, J. Norman. "On the Recent Solar Eclipse."

  Cited herein as item 1013.

* McColley, Grant. "Christopher Scheiner and the Decline of Neo-Aristotelianism."

  Cited herein as item 569.

* Mitchell, S.A. "Eclipses of the Sun."

  Cited herein as item 1020.

* Mitchell, S.A. *Eclipses of the Sun*.

  Cited herein as item 1021.

627. Mitchell, Walter M. "The History of the Discovery of the Solar Spots." *Popular Astronomy*, 24 (1916), 22-31; 82-96; 149-162; 206-218; 290-303; 341-354; 428-440; 488-499; 562-570.

  Traces history from Galileo's first observations to the publication of Scheiner's "Rosa Ursina" some twenty years later and provides detailed historical commentary by later authors on the relative priorities of these two pioneers. Numerous citations.

*   Richardson, Robert S. "Sunspot Problems Old and New."
         Cited herein as item 1027.

627a.   Sakurai, Kunitomo. "The Solar Activity in the Time of
         Galileo." *Journal for the History of Astronomy*, 11
         (1980), 164-173.

         Analyzes Galileo's sketches of sunspots to conclude
         that sunspot activity decreased during a period, leading
         to the era of the Maunder Minimum (1645-1715), when the
         rotation rate of the Sun increased. Work is based upon
         that of Eddy, Gilman and Trotter, who analyzed sketches
         of sunspots by Hevelius and Scheiner. 22 citations.

628.   Schove, D. Justin. "Sunspot Epochs 188 A.D. to 1610 A.D."
         Popular Astronomy, 56 (1948), 247-252.

         Includes brief bibliography on sunspot data.

629.   Shea, William R. "Galileo, Scheiner, and the Interpreta-
         tion of Sunspots." *Isis*, 61 (1970), 498-519.

         Examines the methodology that directed the work of
         Galileo and the Jesuit Christopher Scheiner to the nature
         of sunspots. Reviews Scheiner's observations in late
         1611 and their differences with Galileo's that led to
         a controversy culminating in Galileo's publication of
         Letter on the Sunspots in 1613. Argues that his interest
         in the debate between Scheiner and Galileo "lies not so
         much in its subject matter as in the method of falsifica-
         tion and confirmation developed by Galileo on this oc-
         casion." 107 citations.

630.   Tsu, Wen Shion. "A Statistical Survey of Solar Eclipses
         in Chinese History." *Popular Astronomy*, 42 (1934),
         136-141.

         Surveys records of 920 solar eclipses from 2000 B.C.
         through the modern era. Notes that records for total
         eclipses are far more complete than for partial solar
         eclipses.

631.   Weiss, J.E., and N.O. Weiss. "Andrew Marvell and the
         Maunder Minimum." *Quarterly Journal of the Royal
         Astronomical Society*, 20 (1979), 115-118.

         Demonstrates that the lack of sunspots was well known
         by 1670 through the exposition of a satirical poem, *The
         Last Instructions to a Painter* by Marvell, which "con-
         tains what is apparently the first reference in English

to the disappearance of sunspots in the late seventeenth
century." 16 citations.

*   Young, Charles A. *The Sun*.

Cited herein as item 1032.

THE MOON

632. Beer, W., and J.H. Mädler. *Der Mond nach seinen kosmischen
und individuellen Verhältnissen, oder Allegemeine
vergleichende Selenographie*. Berlin: Simon Schropp,
1837. Pp. xviii + 412.

General study of the physical aspects of the Moon as
a companion volume to their landmark map of the lunar
surface. Classic case of volcanic interpretation of lunar
craters.

632a. Enright, J.T. "The Moon Illusion Examined from a New
Point of View." *Proceedings of the American Philosophical
Society*, 119 (1975), 87-107.

Provides a review of the long history, literally since
Aristotle, of attempts to explain the Moon illusion, or
the apparent increase in the angular size of the Moon
when it is observed close to the horizon. Identifies
inconsistencies and errors in all suggested explanations
of the illusion, provides new data on the phenomenon,
and argues for physiological rather than psychological
causes. Extensive citations to historical literature.

*   Forbes, Eric Gray. "The Life and Work of Tobias Mayer
(1723-62)."

Cited herein as item 1280.

633. Home, Roderick W. "The Origin of the Lunar Craters: An
Eighteenth-Century View." *Journal for the History of
Astronomy*, 3 (1972), 1-10.

Describes the beginning epoch of speculation on the
origins of lunar features. Examines the mechanisms of
meteoritic impact and volcanic activity, the debate that
ensued over which one dominated, and pioneer studies
based upon comparison with terrestrial geological forma-
tions. 33 citations.

634. Hopmann, J. "Ermittelung von Höhen auf dem Monde."
      Mitteilungen der Universitäts--Sternwarte, Wien, 12,
      No. 8 (1963), 103-134.

      Reviews attempts since Galileo's time to determine
      lunar mountain heights. In German with English summary.
      30 citations.

635. Kilgour, Frederick G. "Professor John Winthrop's Notes
      on Sun Spot Observations [1737]." *Isis*, 29 (1938),
      355-361.

      Brief discussion of Winthrop's astronomical work with
      a reproduction of his unpublished manuscript notes on
      sunspot observations. 23 citations.

636. Kopal, Zdeněk, and R.W. Carder. *Mapping of the Moon:
      Past and Present*. Boston: D. Reidel, 1974. Pp. viii +
      234.

      Outlines the history of lunar cartography from pre-
      telescopic maps of William Gilbert to mapping by the
      U.S. Army Map Service, the U.S. Geological Survey, and
      Lowell Observatory in preparation for Apollo. Includes
      chapters on the technical aspects of lunar cartography.
      Reviewed in: *Isis*, 67 (1976), 297-298.

637. Wilkins, H.P., and P. Moore. *The Moon*. New York: Mac-
      millan, no date: circa 1960. Pp. 388.

      Reviews the origins of the names of lunar features.
      Includes listing of historical lunar maps from 1610
      to 1952, and a general description of Wilkins' 300-inch
      lunar map. Appendix provides brief biographical sketches
      of major lunar observers since Galileo. Brief bibliog-
      raphy.

                        SOLAR SYSTEM OBJECTS

638. Alexander, A.F.O'D. *The Planet Saturn. A History of
      Observation, Theory and Discovery*. London: Faber and
      Faber, 1962. Pp. 474.

      Covers the history of studies of Saturn from antiquity
      to the present touching upon all aspects of the study
      of the motions of the planet; early telescopic observa-
      tions and the recognition of the existence of its ring

system by Huygens; theories of the ring system by Kant and Laplace. The bulk of the study details post-18th-century advances in knowledge about Saturn, its rings, and its moons. Reviewed in: *Isis*, 56 (1965), 370-371.

639. Alexander, A.F.O'D. *The Planet Uranus: A History of Observation, Theory and Discovery*. London: Faber and Faber, 1965. Pp. 316.

Descriptive history of the discovery of the planet by William Herschel in 1781 and subsequent studies of its orbit, figure, moons, and spectroscopic features. Numerous citations and bibliography.

640. Antoniadi, E.M. *La Planète Mars (1659-1929)*. Paris: Hermann, 1930. Pp. 239.

General review of the nature of the Martian surface. Primarily a heavily annotated collection of observations of surface features examined since 1659. Emphasizes later work, especially that of the author.

641. Armitage, Angus. "Master Georg Dörffel and the Rise of Cometary Astronomy." *Annals of Science*, 7 (1951), 303-315.

Examines the role of Dörffel, a Saxon clergyman, who was the first to suggest in 1681 that cometary orbits were parabolic. Includes a general narrative on the history of cometary studies from the passage of the comet of 1577 and first attempts to measure cometary parallax. 36 citations.

* Armitage, Angus. "The Pilgrimage of Pingré."

Cited herein as item 1232.

642. Austin, R.H. "Uranus Observed." *British Journal for the History of Science*, 3 (1967), 275-284.

Demonstrates that William Herschel's first observations of the planet, which showed it to be increasing in apparent size, were a result of his expectations that he had discovered a new comet and were, in fact, spurious observations. 12 citations.

643. Blunck, Jurgen. *Mars and Its Satellites. A Detailed Commentary on the Nomenclature*. New York: Exposition Press, 1977. Pp. 200.

Traces the origins of the names of the features of the
surface of Mars dating from the time of Huygens to the
1976 establishment of nomenclature by the IAU. Includes
a list of maps and globes of Mars using older nomenclature
noting over 100 items.

644. Chambers, George F. *The Story of the Comets*. Oxford:
      Clarendon Press, 1909. Pp. xiii + 256.

Comprehensive popular account covering descriptive and
dynamical aspects of comets; cometary spectra; "lost"
comets; methods of discovery; ancient records; major
comets of the second half of the 19th century; comets
in history and poetry; and cometary statistics. Includes
list of recent comets, 1888–1908; a short bibliography
of major historical works on comets; and an ephemeris
for the passage of Halley's Comet, January–July, 1910.
Numerous citations.

645. Chandler, S.C. "Historical Note on the Rotation of Venus
      and Mercury." *Popular Astronomy*, 4 (1897), 393–397.

Reviews attempts to determine rotation periods since
the time of J.D. Cassini in 1666, concluding with G.B.
Schiaparelli's recent studies. No direct citations.

646. Curtis, H.D. "The Comet-Seeker Hoax." *Popular Astronomy*,
      46 (1938), 71–75.

Retells famous hoax perpetrated on E.E. Barnard in
1891 wherein a newspaper article alleged that this
astronomer possessed a machine that could detect comets
automatically. Includes a transcript of original article,
and notes reactions to hoax. No direct citations.

*   Drake, Stillman, and C.D. O'Malley, eds. *The Controversy
       over the Comets of 1618*.

Cited herein as item 538.

*   Flammarion, C. *La Planète Mars et ses Conditions d'Habi-
       tabilité*.

Cited herein as item 539.

647. Fontenrose, Robert. "In Search of Vulcan." *Journal for
      the History of Astronomy*, 4 (1973), 145–158.

Examines search for intramercurial planet from Leverrier's theory of the motion of Mercury in 1859 through
early 20th-century conclusions of its non-existence.

Emphasizes reception of observations claiming to have found Vulcan. 64 citations.

* Forbes, Eric G. "Gauss and the Discovery of Ceres."

Cited herein as item 765.

* Forbes, Eric G. "Early Astronomical Researches of John Flamsteed."

Cited herein as item 1284.

648. Gingerich, Owen. "The Satellites of Mars: Prediction and Discovery." *Journal for the History of Astronomy*, 1 (1970), 109-115.

Reviews Jonathan Swift's prediction of the existence of two Martian moons in *Gulliver's Travels* (London, 1726). Suggests that Swift was influenced by Kepler's work in making the prediction. Examines the actual discovery of the moons in 1877 by Asaph Hall.

649. Goldstein, Bernard R. "Some Medieval Reports of Venus and Mercury Transits." *Centaurus*, 14 (1969), 49-59.

Reprints newspaper report of Venus transit observed visually in 1874 that demonstrates that pretelescopic observations were possible. Notes several instances where early reports of spots are probably due to Venus transits. Includes reports of al-Kindi, Avicenna, Averres and Ilan Bāya, none of which are considered to provide definite evidence. 29 citations.

* Grosser, Morton. *The Discovery of Neptune*.

Cited herein as item 773.

* Hellman, C. Doris. *The Comet of 1577*.

Cited herein as item 548.

650. Hellman, C. Doris. "The Role of Measurement in the Downfall of a System: Some Examples from Sixteenth Century Comet and Nova Observations." *New Aspects in the History and Philosophy of Astronomy* (item 123), 43-52.

Provides a new interpretation of Tycho Brahe's well-known observations of the distance to the Comet of 1577 and the Nova of 1572. Examines the techniques of observing celestial positions and the accuracies obtained. 11 bibliographical citations.

651. Jervis, Jane L. "Vögelin on the Comet of 1532: Error
      Analysis in the 16th Century." *Centaurus*, 23 (1980),
      216-229.

      Reviews J. Vögelin's calculations and arguments in
      the 1530s that the comet of 1532 passed less than one
      earth radius above the surface of the Earth, and examines
      the cosmological context within which this work was
      accepted concluding with Tycho Brahe's dismissal of it.
      Shows that Vögelin's methods, based upon Regiomontanus'
      geometry, were correct and that his calculations for
      the distance of the comet fell within expected ranges
      at the time. 8 citations.

652. Lankford, John. "A Note on T.J.J. See's Observations of
      Craters on Mercury." *Journal for the History of Astronomy*
      11 (1980), 129-132.

      Argues that recent claims that See observed craters
      on Mercury cannot be supported and that See apparently
      doctored his observational data to fit his speculations
      on the origin of the solar system. Provides important
      commentary on See's controversial life and how it must
      be studied by historians. 17 citations.

653. Laves, Kurt. "Three Hundred Years of Research on the
      Motions of the Satellites, 1610-1910." *Popular
      Astronomy*, 21 (1913), 279-291; 332-345.

      Reviews observations of motions of planetary satellites
      and the rise of theoretical techniques used to describe
      them. No direct citations.

654. Maunder, E. Walter. "The 'Canals' of Mars." *Scientia*,
      7 (1910), 253-269.

      Reviews 250 years of visual studies of the surface of
      Mars but concentrates on Lowell's recent study of the
      canals of Mars. Reviews the controversy over the nature
      of the canals, the opposing views of Lowell and Antoniadi,
      the problems of interpreting both visual and photographic
      observations, and concludes that "there is still con-
      troversy as to the true nature of these straight lines
      and round spots."

655. Moore, Patrick. *The Planet Venus*. London: Faber and
      Faber, 1956. Pp. 151.

      Popular descriptive history of knowledge about Venus
      from the 17th century to date. Includes discussion of

spurious observations of polar caps, canals, and a satellite, together with a review of contemporary knowledge. 155 citations.

656. Partridge, E.A., and H.C. Whitaker. "Galileo's Work on Saturn's Rings--A Historical Correction." *Popular Astronomy*, 3 (1896), 408-414.

Argues that Arago's treatment of Galileo was unfair and, in recounting his work on Saturn, incorrect. 21 citations.

657. Pascu, D. "A History of the Discovery and Positional Observation of the Martian Satellites; 1877-1977." *Vistas in Astronomy*, 22 (1978), 141-148.

Reviews the motivations behind the discovery of the Martian satellites, the visual techniques utilized, and later techniques for their continued observation. 23 citations.

658. Ronchi, L.R., and G. Abetti. "Psycho-physiological Effects in Visual Astronomical Observations--The Planet Mars." *Atti della Fondazione 'Giorgio Ronchi' e Contributi dell'Istituto nazionale di Ottica*, 19 No. 1 (Jan.-Feb. 1964), 1-34.

Study of the physiological and psychological factors involved in the interpretation of visual observations made under extreme conditions and where a certain degree of expectation is involved. Centers upon a comparison of visual and photographic observations of Mars. 20 citations.

659. Shapley, Dora. "Pre-Huygenian Observations of Saturn's Ring." *Isis*, 40 (1949), 12-17.

Reviews observations by Galileo, Scheiner, Gassendi, Fontana, Riccioli, Hevelius and others and their interpretation of Saturn's anomalous figure, finally explained successfully by Huygens as a ring. 19 citations.

660. Streeter, John W. "John Winthrop, Junior, and the Fifth Satellite of Jupiter." *Isis*, 39 (1948), 159-163.

Analysis of Winthrop's alleged observation of a fifth Jovian satellite in 1664 with a 3 1/2-foot telescope. Argues that it was a spurious observation and had no influence upon the actual discovery of the moon by E.E. Barnard in 1892. Concludes that Winthrop had observed a faint star, identified as HR7128. 21 citations.

661.   Tsu, Wen Shion. "The Observations of Halley's Comet in
       Chinese History." *Popular Astronomy*, 42 (1934), 191-
       201.

       Records 30 passages dating from 240 B.C. together with
       notation from Chinese records identifying the character
       of each event.

662.   Turner, Herbert Hall. "Halley's Comet." *Royal Institution
       Library of Science: Astronomy* (item 140), 137-148.

       Lecture delivered February 18, 1910. Reviews Halley's
       prediction of the return of a comet and provides a brief
       history of comet lore as an introduction to a detailed
       examination of its pending return in 1910.

663.   Van Helden, Albert. "Saturn and His Anses." *Journal for
       the History of Astronomy*, 5 (1974), 105-121; 155-174.

       The first paper provides a detailed descriptive review
       and analysis of the visible nature of Saturn's rings
       from Galileo to Gassendi in the 1640s before Huygens'
       realization that they were, in fact, a ring system. The
       second paper, entitled "Annulo Cingitur: The Solution
       to the Problem of Saturn," examines Huygens' prediction
       of the nature of the ring system and observations by
       Campani and others. Emphasizes conceptual as well as
       technical difficulties in coming to the ring model.
       60 + 117 citations.

664.   Van Helden, Albert. "The Importance of the Transit of
       Mercury of 1631." *Journal for the History of Astronomy*,
       7 (1976), 1-10.

       Argues that the importance of the event was in provid-
       ing "an indisputable quantitative measure of the apparent
       magnitude of a planetary disc." Provides useful back-
       ground for understanding the role of planetary dimensions
       in 17th-century speculations on the construction of the
       solar system. 50 citations.

665.   Watson, James C. *A Popular Treatise on Comets*. Phila-
       delphia: James Challen, 1861. Pp. x + 363.

       General exposition reviewing the history of comet
       lore; major passages of comets since the 16th century;
       general descriptions of comets; the computation of the
       orbits of periodic comets; speculation on the physical
       constitution of comets; the origins of comets and the
       Nebular Hypothesis; and the question of the existence

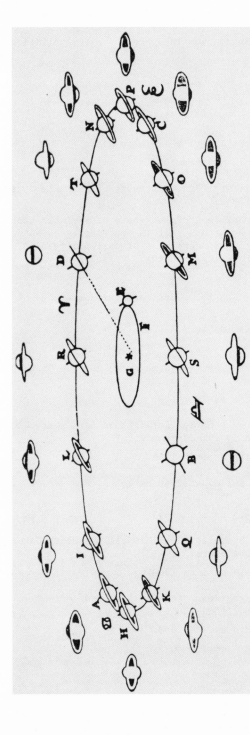

Figure used by Huygens to show how a ring can generate the various appearances of Saturn, from *Oeuvres complètes*, xv, 309. (From: Albert van Helden, "'Annulo Cingitur': The Solution of the Problem of Saturn" (Item 663).)

of a resisting medium in space. No direct citations or
index.

STARS

*   Aitken, Robert Grant. *The Binary Stars.*

       Cited herein as item 1067.

666.   Bobrovnikoff, N.T. "The Discovery of Variable Stars."
       Isis, 33 (1942), 687-689.

       Observes that lack of awareness of variable stars by
       ancients is a curious fact in the history of astronomy.
       Reviews history of detection of variability of Algol
       and Mira concluding that their variability was not known
       to the ancients.

*   Forbes, Eric G. "Early Astronomical Researches of John
       Flamsteed."

       Cited herein as item 1284.

*   Hellman, C. Doris. "The Role of Measurement in the
       Downfall of a System: Some Examples from Sixteenth
       Century Comet and Nova Observations."

       Cited herein as item 650.

*   Herrmann, D.B. "Aus der Entwicklung der Grössenklassen--
       Definition im 19. Jahrhundert."

       Cited herein as item 1058.

*   Hoskin, Michael A. *William Herschel, Pioneer of Sidereal
       Astronomy.*

       Cited herein as item 1315.

*   McCormmach, Russell. "John Michell and Henry Cavendish:
       Weighing the Stars."

       Cited herein as item 1340.

*   Warner, Deborah Jean. *The Sky Explored, Celestial
       Cartography, 1500-1800.*

       Cited herein as item 853.

STELLAR SYSTEMS AND NEBULAE

*   Clerke, Agnes. *The Herschels and Modern Astronomy*.

Cited herein as item 1261.

*   Fernie, J.D. "The Historical Quest for the Nature of the
Spiral Nebulae."

Cited herein as item 1153.

667.   Hagen, J.G. *William Herschel's Fifty-Two Nebulosities*.
Specola Vaticana, 1927. Pp. 30.

Bound collection of Hagen's papers from the *Monthly
Notices of the Royal Astronomical Society* (Volumes 86-
87)on Herschel's early observations of extended nebulous
fields.

*   Holden, E.S., and C.S. Hastings. *A Synopsis of the
Scientific Writings of Sir William Herschel*.

Cited herein as item 1314.

*   Hoskin, Michael A. *William Herschel, Pioneer of Sidereal
Astronomy*.

Cited herein as item 1315.

668.   Hoskin, Michael A. *William Herschel and the Construction
of the Heavens*. New York: W.W. Norton & Co., 1964.
Pp. 199.

Examines Herschel's techniques for gauging the structure of the visible universe, the assumptions he was
restricted to using, and how these assumptions were
received by his contemporaries. Includes excerpts from
selected papers by Herschel dealing with his cosmology,
arranged chronologically by topic. Extensive annotation
and introductory material provided, as well as footnote
discussions of scientific points by David Dewhirst. Essay
reviews: *British Journal for the History of Science*,
2 (1965), 356-358; *History of Science*, 3 (1964), 91-101.
Reviewed also in: *Isis*, 55 (1964), 452-454; *Annals of
Science*, 18 (1962), 130.

*   Hoskin, Michael A. "Apparatus and Ideas in Mid-nineteenth-
century Cosmology."

Cited herein as item 399.

669. Hoskin, Michael A. "William Herschel's Early Investiga-
      tions of Nebulae: A Reassessment." *Journal for the
      History of Astronomy*, 10 (1979), 165-176.

      Clarifies Herschel's first years of activity in study-
      ing nebulae and star clusters and his speculations on
      their character after his receipt of Messier's *Catalogue*
      in 1781. Shows, through an examination of Herschel's
      observing logs, that he at first believed in the existence
      of two types of nebulae: true nebulae and nebulae that
      could eventually be resolved into stars. When, by 1784,
      he found nebulae that exhibited the characteristics of
      both types he therefore decided, until the early 1790s,
      that all nebulae were composed of stars. 56 citations.

670. Jaki, Stanley L. *The Paradox of Olbers' Paradox: A Case
      History of Scientific Thought*. New York: Herder and
      Herder, 1969. Pp. 269.

      Examines Wilhelm Olbers' attempt in 1823 to explain
      the darkness of the night sky by assuming the existence
      of absorption in space. Jaki re-examines Olbers' use
      of the 1744 work of Jean-Philippe Loys de Cheseaux in:
      "New Light on Olbers' Dependence on Cheseaux." *Journal
      for the History of Astronomy*, 1 (1970), 53-55. Text
      includes general history of resolution of the paradox.
      Reviewed in: *JHA*, 1 (1970), 83.

671. Jaki, Stanley L. *The Milky Way: An Elusive Road for
      Science*. New York: Science History Publications,
      1972. Pp. xi + 352.

      General review of cosmological speculation on the nature
      of the Milky Way from Greek thought to modern times and
      the discoveries of Shapley and Hubble. Extensive cita-
      tions. Summaries have appeared in *JHA*, 2 (1971), 161-
      167; 3 (1972), 199-204. Essay review in: *History of
      Science*, 12 (1974), 299-306. Reviewed also in: *Annals
      of Science*, 31 (1974), 580-581; *JHA*, 4 (1973), 200;
      Isis, 66 (1975), 115-116.

*   Jaki, Stanley L. "Lambert and the Watershed of Cosmology."

      Cited herein as item 556.

*   Jaki, Stanley L. "Johann Georg Soldner and the Gravita-
      tional Bending of Light, with an English Translation
      of His Essay on It Published in 1801."

      Cited herein as item 782.

672. Jones, Kenneth Glyn. *Messier's Nebulae and Star Clusters*.
       London: Faber and Faber, 1968. Pp. 480.

       Popular descriptive introduction to the 104 star
       clusters, nebulae and galaxies listed by Messier in his
       original catalogues of 1771 through 1784. Short biographies
       of Messier and other contemporary observers. Includes
       four-page bibliography.

*   Jones, Kenneth Glyn. "The Observational Basis for Kant's
       'Cosmogony': A Critical Analysis."

       Cited herein as item 559.

673. Jones, Kenneth Glyn. *The Search for the Nebulae*. New
       York: Science History Publications, 1975. Pp. 84.

       Revised series of papers dealing with the early history
       of the search for nebulae prior to the work of the late
       18th-century comet hunter Charles Messier. Covers period
       from Greeks to Mechain. Includes list of all nebulae
       and clusters of stars detected as of 1781. Reviewed
       in: *Annals of Science*, 34 (1977), 98-99.

674. Mallas, John H., and Evered Kreimer. *The Messier Album*.
       Cambridge: Sky Publishing Corp., 1978. Pp. viii + 248.

       Photographs and drawings of all objects catalogued by
       Charles Messier in the late 18th century, with a brief
       historical introduction by O. Gingerich.

*   Nichol, J.P. *The Architecture of the Heavens*.

       Cited herein as item 1156.

*   Schaffer, Simon. "'The Great Laboratories of the Uni-
       verse': William Herschel on Matter Theory and Plane-
       tary Life."

       Cited herein as item 589.

THEORETICAL ASTRONOMY

GENERAL

675. Brown, E.W. *An Introductory Treatise on the Lunar Theory*. Cambridge: Cambridge University Press, 1896. Reprinted, New York: Dover, 1960. Pp. xvi + 292.

Technical review of major investigations on the lunar theory, or the problem of three bodies, covering both practical studies and theoretical studies of periodic solutions, solutions by infinite series, and the general principles underlying the practical methods. Provides detailed analysis of the methods of earlier investigators from Clairaut, d'Alembert, Laplace, and others to Hill, Poincaré, and Gyldén. Numerous citations.

* Chauvenet, W. *Manual of Spherical and Practical Astronomy*.

Cited herein as item 820.

676. Clemence, G.M. "The Concept of Ephemeris Time: A Case of Inadvertent Plagiarism." *Journal for the History of Astronomy*, 2 (1971), 73-79.

Brief review of recognition that the rotation of the Earth is not constant and that consequently there is a need to establish an independent standard of uniform time. Examines steps taken to establish "Ephemeris Time" and the roles of various astronomers involved in its definition from the 1920s through the 1960s. 8 citations.

677. Dziobek, Otto. *Mathematical Theories of Planetary Motions*. New York: Register, 1892. Reprinted, New York: Dover, 1962. Pp. vi + 294.

Mathematical review of two-body motion and the theory of perturbations. Includes historical review sections on two- and three-body motion; the development of analytical integration techniques in the 18th and 19th centuries;

and the development of the theory of perturbations.
Original translation by M.W. Harrington and W.J. Hussey.
Numerous citations.

678. Frischauf, Johannes. *Grundriss der theoretischen Astronomie
      und der Geschichte der Planetentheorien*. Leipzig:
      Engelmann, 1903. Pp. xv + 199.

      Mathematical development of historical aspects of
      celestial mechanics from Copernicus through Gauss, Olbers
      and Encke. Some citations to primary and secondary mate-
      rial.

679. Gregory, David. *The Elements of Physical and Geometrical
      Astronomy* [1726]. 2 volumes. I.B. Cohen, ed. New York:
      Johnson Reprint Corporation, 1972. Pp. xxvi + xiv +
      893.

      Reprint of 1726 work representative of immediate Post-
      Newtonian British astronomy after the publication of the
      Principia. Reviewed in: *JHA*, 4 (1973), 204-205.

680. Klinkerfues, W. *Theoretische Astronomie*. Braunschweig:
      F. Vieweg, 1912. Pp. xxxviii + 1070.

      Comprehensive review of all aspects of celestial
      mechanics, originally published in 1872 and here greatly
      revised and expanded into a major review textbook. In-
      cludes historical commentary on recent research of W.
      Gibbs, P. Harzer and A. Leuschner. Numerous citations.

681. Moulton, Forest Ray. *An Introduction to Celestial
      Mechanics*. New York: Macmillan, 1902; 1914. Pp. xvi +
      437.

      Comprehensive textbook introduction that includes
      historical commentary and bibliographical essays arranged
      topically in each chapter.

682. Moulton, Forest Ray. *Periodic Orbits*. Washington, D.C.:
      Carnegie Institution Publication No. 161, 1920. Re-
      printed, New York: Johnson Reprint Corporation, 1964.
      Pp. xiii + 524.

      Comprehensive review with detailed historical develop-
      ment. 101 citations.

683. Newcomb, Simon. *A Compendium of Spherical Astronomy with
      its applications to the determination and reduction
      of positions of the fixed stars*. New York: Macmillan,
      1906. Reprinted, New York: Dover, 1960. Pp. xviii + 444.

Often reprinted standard work including detailed historical review (pp. 339-351) of methods of deriving stellar positions and motions. Reviewed in: *The Observatory*, 29 (1906), 366-368; *Nature*, 74 (Aug. 16, 1906), 379-380.

684. Plummer, H.C. *An Introductory Treatise on Dynamical Astronomy*. Cambridge: Cambridge University Press, 1918. Reprinted, New York: Dover, 1960. Pp. xix + 343.

Mathematical review of two-body motion; techniques for the determination of orbits; planetary theory; restricted problem of three bodies; the Lunar Theory; motions of the Earth and Moon. Limited to exposition of mathematical techniques without examination of formalism.

685. Seeliger, Hugo. "General Problems of Celestial Mechanics." *Popular Astronomy*, 4 (1897), 407-414; 475-481; 545-549.

Translation of an address in 1892 to the Royal Bavarian Academy of Sciences. Reviews the development of Newtonian mechanics and contemporary studies in stellar astronomy. No direct citations.

686. Smart, William M. *Text-Book on Spherical Astronomy*. Cambridge: Cambridge University Press, 1931. 5th Edn., 1962. Pp. xii + 430.

Standard mathematical introduction to all aspects of practical astronomy including: spherical trigonometry; the celestial sphere; theory of refraction; the description, use and reduction of meridian circle observations; gravitational equations of planetary motion; definitions and the measurement of time; planetary phenomena; the law of aberration and its reduction; geocentric, lunar, planetary, solar and stellar parallax; precession and nutation; stellar proper motions; techniques and the reduction of astronomical photography; celestial navigation including observations by sextant and circles of position; orbits of visual, eclipsing and spectroscopic binary stars; occultations and eclipses. Includes appendices on the method of dependences for the reduction of photographic plates; the determination of stellar magnitudes; the use of a coelostat for the production of a stationary image; and tables of astronomical data. No direct citations.

*   Stephenson, F.R., and D.H. Clark. *Applications of Early Astronomical Records*.

Cited herein as item 175.

687. Stumpff, K. *Himmelsmechanik*. 2 volumes. Berlin: Deutscher
      Verlag der Wissenschaften, 1959. Pp. 508 + 682.

      Comprehensive mathematical text with introductory
      chapters on historical topics including the theory of
      planetary motion from the Greeks through Newton. Post-
      Newtonian methods are treated topically and are presented
      at an introductory mathematical level. Extensive bibliog-
      raphy.

688. Tisserand, Francois Felix. *Traité de Méchanique Céleste*.
      4 volumes. Paris: Gauthier-Villars et Fils, 1896. Re-
      printed: Gauthier-Villars, 1960. Pp. 2000+.

      Comprehensive four-volume work published between 1889
      and 1896 that represents an update of Laplace's *Méchanique
      Céleste*. Constitutes an important review and analysis
      of 19th-century celestial mechanics.

689. Watson, J.C. *Theoretical Astronomy*. Philadelphia: J.B.
      Lippincott, 1868. Reprinted, New York: Dover, 1964.
      Pp. vii + 662.

      Comprehensive textbook on celestial mechanics repre-
      sentative of knowledge and techniques of the mid-19th
      century. Includes brief historical preface and new in-
      troduction by G.M. Clemence.

690. Whitrow, G.J. "The Laws of Motion." *British Journal for
      the History of Science*, 5 (1971), 217-234.

      Presidential address at July 1970 meeting of the
      Society in Leicester. Reviews concepts of motion from
      Greek thought through Galileo, Descartes, Newton, Euler,
      Lagrange, d'Alembert, Mach, Poincaré, Jacobi, Milne,
      and Einstein. 22 citations.

691. Wintner, Aurel. *The Analytical Foundations of Celestial
      Mechanics*. Princeton: Princeton University Press,
      1941. Pp. xii + 448.

      Within this contemporary mathematical text is a detailed
      appendix of "Historical Notes and References" on impor-
      tant points in 18th through 20th-century celestial
      mechanics. Reviewed in: *Isis*, 34 (1943), 230.

PRE-NEWTONIAN

692. Aiton, E.J. "Galileo's Theory of the Tides." *Annals of Science*, 10 (1954), 44-57.

   Describes Galileo's unsuccessful theory based upon the motion of the Earth. Compares Galileo's theory to that of Descartes. 31 citations.

693. Aiton, E.J. "Descartes' Theory of the Tides." *Annals of Science*, 11 (1955), 337-348.

   Examines Descartes' contributions, which were based on his theory of vortices, and which were superseded by the Newtonian theories of Maclaurin, Euler and Bernoulli. 51 citations.

694. Aiton, E.J. "The Cartesian Theory of Gravity." *Annals of Science*, 15 (1959), 27-49.

   Argues that Descartes' introduction of an elastic aether was "the first hypothesis of the aether worthy of the title mathematical physics." 66 citations.

695. Aiton, E.J. *The Vortex Theory of Planetary Motions*. London: MacDonald, 1972. Pp. x + 282.

   Examines Newton's criticism of the Cartesian vortex theory and continues with an exposition of Leibniz's harmonic vortex theory. Notes early Cartesians' ignorance of Kepler's laws and provides substantial discussion of the physical basis for Kepler's laws. Extension of a series of papers in the *Annals of Science*, 13 (1957), 249-264; 14 (1958), 132-147; 157-172. Includes bibliography. Reviewed in: *British Journal for the History of Science*, 7 (1974), 91-92; *Centaurus*, 18 (1973-1974), 86-87.

696. Aiton, E.J. "Johannes Kepler in the Light of Recent Research." *History of Science*, 14 (1976), 77-100.

   Examines recent studies and provides an extensive bibliography within 134 reference notes.

697. Ariotti, Piero E. "Toward Absolute Time: The Undermining and Refutation of the Aristotelian Conception of Time in the Sixteenth and Seventeenth Centuries." *Annals of Science*, 30 (1973), 31-50.

Examines factors that contributed to the introduction
of an absolute system of time independent of external
motion. Reviews contributions of Copernicus, Kepler,
Galileo, Huygens and Hooke within the framework of
Newtonian mechanics. 35 citations.

698.   Armitage, Angus. "Borelli's Hypothesis and the Rise of
       Celestial Mechanics." *Annals of Science*, 6 (1950),
       268-282.

       Reviews G.A. Borelli's mechanical theory of planetary
       motions published in 1666 and its influence upon Hooke,
       Newton and others. Concludes with an exposition of Hooke's
       relations with Newton, and a comparison of Hooke's and
       Borelli's views. Numerous citations.

*   Beer, Arthur, and Peter Beer, eds. "Kepler, Four Hundred
       Years."

       Cited herein as item 1243.

699.   Bochner, Salomon. "Mathematical Background Space in
       Astronomy and Cosmology." *Vistas in Astronomy*, 19
       (1977), 133-161.

       Argues that the developments that led from Nicholas
       of Cusa's and Copernicus' work to that of Newton and
       the resultant astronomical revolution were due to the
       "infusion, and even irruption of a radically new mathe-
       matics, namely of mathematical analysis, into the texture
       and structure of Astronomy." 107 citations.

700.   Boyer, Carl B. "Note on Epicycles & the Ellipse from
       Copernicus to LaHire." *Isis*, 38 (1947), 54-56.

       Traces knowledge of a theorem that shows how ellipitical
       motion can be generated by a combination of a circle
       rolling with no slippage on the interior surface of a
       circle twice its radius. Shows that use of circular
       motion persisted after Kepler. 12 citations.

701.   Burstyn, Harold L. "The Deflecting Force of the Earth's
       Rotation from Galileo to Newton." *Annals of Science*,
       21 (1965), 47-80.

       Traces the development of studies of relative motion
       in the 17th century within the context of the gradual
       shift to the Copernican theory. Shows that theoretical
       arguments put forth to prove the Earth's motion, while
       not immediately fruitful, did lay essential groundwork

for the quantitative developments in later centuries.
Concludes that no significant advance over Galileo's
initial work was made during the century. 93 citations.

702. Dobrzycki, Jerzy, ed. *The Reception of Copernicus' Helio-*
     centric Theory. Dordrecht: D. Reidel, 1972. Pp. 368.

     Proceedings of a 1973 IUHPS symposium organized by
     the Copernicus Committee. Eleven papers on the diffusion
     of Copernican thought in Europe, Japan, Great Britain
     and America through the 18th century. Extensive cita-
     tions.

*   Dreyer, J.L.E. *A History of Astronomy from Thales to*
     Kepler.

     Cited herein as item 89.

703. Gingerich, Owen. "Johannes Kepler and the New Astronomy."
     Quarterly Journal of the Royal Astronomical Society,
     13 (1972), 346-373.

     The George Darwin lecture delivered on 10 December
     1971. Reviews Kepler's scientific life and development.
     Includes a translation by Gingerich and William Walder-
     man of Kepler's "Preface" to his *Rudolphine Tables.* 9
     citations and short bibliography.

704. Kepler, J. *Neue Astronomie* [1609]. Max Casper, editor.
     München-Berlin: R. Oldenbourg, 1929. Pp. 416.

     German edition, with extensive annotations and intro-
     duction, of Kepler's exposition of 1609 outlining his
     first two laws of planetary motion based upon his study
     of Mars.

705. Kepler, J. *Kepler's Conversation with Galileo's Sidereal*
     Messenger. Edward Rosen, translator. New York: Johnson
     Reprint Corporation, 1965. Pp. xix + 164.

     Translation by Rosen with an introduction and 403
     notes which comprise the bulk of the work.

706. Koyré, Alexandre. "A Documentary History of the Problem
     of Fall from Kepler to Newton." *Transactions of the*
     American Philosophical Society, N.S., 45 (1955), 329-
     395.

     Examines the development of attempts to solve the
     problem of determining the trajectories of falling ob-
     jects under the influence of a gravitational force on

a rotating earth. Bulk of text contains detailed excerpts
and statements from major writers during the period.
370 citations.

707. Koyré, Alexandre. *La Révolution Astronomique: Copernic,
      Kepler, Borelli*. Paris: Herrmann, 1961. Pp. 525.

      Three essays on the contributions of these astronomers
      to the astronomical revolution of the 16th and early
      17th centuries. Reviewed in: *History of Science*, 3 (1964),
      134-139; *Annals of Science*, 16 (1960), 271.

*   Krafft, Fritz, Karl Meyer, and Bernhard Sticker, eds.
      Internationales Kepler-Symposium.

      Cited herein as item 1325.

*   Laves, Kurt. "Three Hundred Years of Research on the
      Motions of the Satellites, 1610-1910."

      Cited herein as item 653.

708. Maeyama, Y. "The Historical Development of Solar Theories
      in the Late Sixteenth and Seventeenth Centuries."
      Vistas in Astronomy, 16 (1974), 35-60.

      Provides mathematical analysis of planetary theories
      during the period compared against the background of
      the then available observational error range. Much of
      the discussion centers upon Tycho's observations and
      planetary theory, and the importance of the solar paral-
      lax and knowledge of the eccentricity of the Sun's
      motion. 12 citations.

709. Pannekoek, A. "The Planetary Theory of Kepler." *Popular
      Astronomy*, 56 (1948), 63-75.

      Brief exposition derived from Kepler's "Astronomia
      Nova" outlining Kepler's derivation of elliptical motion.
      No direct citations.

710. Rosen, Edward, tr. *Kepler's Somnium: The Dream, or
      Posthumous Work on Lunar Astronomy*. Madison: University
      of Wisconsin Press, 1967. Pp. xxiv + 255.

      Translation, with extended commentary, of Kepler's
      1634 fantasy written to support Copernicanism. Reviewed
      in: *Annals of Science*, 23 (1967), 245-246.

711. Russell, J.L. "Kepler's Laws of Planetary Motion: 1609-
     1666." *British Journal for the History of Science*, 2
     (1964), 1-24.

     Argues that while Kepler's laws were not universally
     recognized during the period, they certainly were well
     known and studied, more than some historical treatments
     suggest. 30 citations and bibliography.

712. Small, Robert. *An Account of the Astronomical Discoveries
     of Kepler* [1804]. Madison: U. of Wisconsin Press,
     1963. Pp. xvi + 386.

     Facsimile reprint of 1804 text with a foreword by
     W.D. Stahlman. Reviews the derivations of Kepler's first
     two laws of planetary motion as expressed in his "Astro-
     nomia Nova" of 1609. The first four chapters review
     planetary theory from Ptolemy through Tycho and the
     remaining text concentrates on Kepler's derivations.
     Reviewed in: *Annals of Science*, 17 (1961), 271-272.

713. Thoren, Victor E. "Kepler's Second Law in England."
     British Journal for the History of Science, 7 (1974),
     243-256.

     Concludes that even though equant theories continued
     in use in England through the 17th century, Kepler's
     elliptical models "entered England with Jeremiah Horrox"
     but lay dormant through the 1640s, coming out only after
     the 1660s. 63 citations.

714. Wilson, Curtis A. "Kepler's Derivation of the Elliptical
     Path." *Isis*, 59 (1968), 5-25.

     Detailed examination of how Kepler dealt with the
     problem of observational error in the development of
     his first law of planetary motion. 107 citations.

715. Wilson, Curtis A. "From Kepler's Laws, So-called, to
     Universal Gravitation: Empirical Factors." *Archive
     for History of Exact Sciences*, 6 (1969), 89-170.

     Reviews the empirical relations of Kepler's laws of
     planetary motion and argues "that the Keplerian revolu-
     tion rests finally upon conjecture, and that its two
     main principles--the first two Keplerian laws, socalled--
     while yielding a revolutionary improvement in predictive
     accuracy when taken together, cannot be verified sepa-
     rately with satisfactory precision." Examines Newton's
     arguments for universal gravitation and their development.
     273 citations.

NEWTONIAN

716. Aiton, E.J. "The Contributions of Newton, Bernoulli and
      Euler to the Theory of the Tides." *Annals of Science*,
      11 (1955), 206-223.

      Reviews the history of studies of the causes of the
      tides beginning with Kepler's association of tidal
      phenomena with the Moon's attraction, and Newton's veri-
      fication in his development of the principle of universal
      gravitation. Follows extensions of Newton's work by
      Daniel Bernoulli and Euler. Concludes that Bernoulli's
      treatment was more analytical than that of Newton and,
      in being the first to provide tide tables based upon
      theory, were of the greatest influence on later develop-
      ments. Provides an analysis of Euler's chief contribu-
      tion--the recognition of the importance of the horizontal
      component of the disturbing force as the primary cause
      of the tides. 57 citations.

717. Aiton, E.J. "The Celestial Mechanics of Leibniz."
      Annals of Science, 16 (1960), 65-82.

      Concludes that Leibniz's chief contribution was to
      apply the differential calculus to the theory of plane-
      tary motion. Includes an examination of Leibniz's deriva-
      tions and his contemporary influence. 56 citations.

718. Aiton, E.J. "The Celestial Mechanics of Leibniz in the
      Light of Newtonian Criticism." *Annals of Science*, 18
      (1962), 31-40.

      Argues that Newton's criticism of Leibniz's *Tentamen*
      are unfounded but do "illuminate the distinctive contri-
      bution that Leibniz made to the theory of planetary
      motion." 29 citations.

719. Aiton, E.J. "An Imaginary Error in the Celestial Mechan-
      ics of Leibniz." *Annals of Science*, 21 (1965), 169-173.

      Argues that Newtonian criticisms of Leibniz's theory
      of planetary motions lack foundation and that recent
      studies by A. Koyré (*Newtonian Studies*, see item 726)
      constitute "a new misinterpretation." Specific matter
      considered in this paper was Leibniz's derivation of
      Kepler's harmonic law utilizing a harmonic vortex. 22
      citations.

720. Aiton, E.J. "Newton's Aether-Stream Hypothesis and the Inverse Square Law of Gravitation." *Annals of Science*, 25 (1969), 255-260.

    Examines Newton's apparent attempt to suggest a general explanation of gravity descriptively based upon his elastic aether stream hypothesis for the transmission of light as discussed in his *Opticks*. Concludes that this attempt was not intended to account for the inverse square law. 35 citations.

*   Aiton, E.J. *The Vortex Theory of Planetary Motions*.

    Cited herein as item 695.

*   Armitage, Angus. "Master Georg Dörffel and the Rise of Cometary Astronomy."

    Cited herein as item 641.

721. Beer, Peter, ed. "Newton and the Enlightenment." *Vistas in Astronomy*, 22 (1978), 367-557.

    Texts of 21 papers presented at a symposium held at Cagliari, Italy, on 3-5 October 1977. Covers many aspects of Newton's science and philosophy; comparative analyses of his work with that of Buffon, R.G. Boscovitch, Locke and Lambert, and the influence of Newton's work in France.

*   Bochner, Salomon. "Mathematical Background Space in Astronomy and Cosmology."

    Cited herein as item 699.

722. Cohen, I. Bernard. *Introduction to Newton's 'Principia.'* London: Cambridge University Press, 1971. Pp. xxi + 380.

    Introduction to the variorum edition of the *Principia* edited by Cohen and Alexandre Koyré. Constitutes a documentary biography of the origins and production of the *Principia*. Reviewed in: *British Journal for the History of Science*, 6 (1972-1973), 94-95.

722a. Gaythorpe, S.B. "On Horrocks' Treatment of the Evection and the Equation of the Centre, with a Note on the Elliptic Hypothesis of Albert Curtz and Its Correction by Boulliau and Newton." *Monthly Notices of the Royal Astronomical Society*, 85 (1925), 858-865.

A technical review comparing historical studies on the
lunar theory to modern knowledge. Emphasizes Horrocks'
original methodology in the use of successive approximations and corrections. Numerous citations.

723. Herivel, John. *The Background to Newton's 'Principia.'*
London: Clarendon Press, 1965. Pp. xvi + 337.

Presents development of Newton's thought based heavily
on manuscript sources. Reviewed in: *British Journal for
the History of Science*, 3 (1967), 298-299.

724. Hiscock, W.G., ed. *David Gregory, Isaac Newton and Their
Circle. Extracts from David Gregory's "Memoranda," 1677-
1708*. Oxford: Private Printing, 1937. Pp. x + 48.

Collection of letters and manuscript excerpts found in
Christ Church, Oxford, pertaining to Newton's activities
and interests as viewed through Gregory's correspondence
and diaries. Manuscripts include the second edition of
the *Principia* (1713), one on the dispute between Newton
and Leibniz, and Gregory's diary of activities during
this period. Reviewed in: *Isis*, 28 (1938), 105-106.

725. Hoskin, M.A. "Newton, Providence and the Universe of Stars."
Journal for the History of Astronomy, 8 (1977), 77-101.

Examines evolution of Newton's recognition of the problem of the stability of the Universe from the year of the
printing of his *Principia* through the early 18th century.
At first, 1685-87, Newton was apparently unaware of the
problem and by 1694 he was ready to invoke divine intervention to prevent gravitational collapse. To Newton,
"... the system of the stars, like the system of planets
and comets, has been constructed with providential ingenuity
to minimize the tendency to gravitational collapse." An
Appendix reprints Newton's analysis, in Latin. 61 citations.

726. Koyré, Alexandre. *Newtonian Studies*. London: Chapman and
Hall, 1965. Pp. viii + 288.

Collection of essays on various aspects of Newton's
work, with a major chapter on Newton and Descartes. Aiton's
review (*Annals of Science*, 21 (1965), 204-205) criticizes
Koyré's treatment of Leibniz. See also item 719.

727. Pannekoek. A. "The Planetary Theory of Newton." *Popular
Astronomy*, 56 (1946), 177-192.

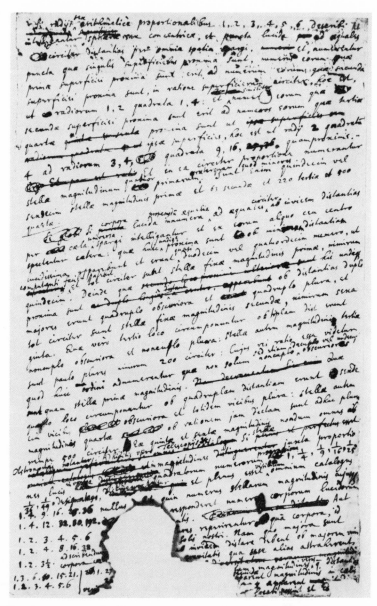

"Further attempts by Newton to express the relationship between predictions from his geometrical model and data in the star catalogues." *(From M.A. Hoskin, "Newton, Providence and the Universe of Stars" (Item 725); original source: University Library Cambridge ULC Add. MS 3965, f74ʳ.)*

Brief exposition based upon F. Cajiori's translation of the *Principia*. No direct citations.

728. Patterson, Louise Diehl. "Hooke's Gravitation Theory and Its Influence on Newton. I: Hooke's Gravitation Theory." *Isis*, 40 (1949), 327-340.

Reevaluates Hooke's contribution to gravitation theory based upon an analysis of Hooke's surviving papers. Argues that 19th-century studies of Hooke's role and influence in the development of Newtonian gravitational theory were biased and incomplete. 86 citations.

729. Rigaud, Stephen Peter. *Historical Essay on the First Publiction of Sir Isaac Newton's Principia*. Oxford, 1838. Reprinted with an introduction by I.B. Cohen. New York: Johnson Reprint, 1972. Pp. xi + 108 + 80.

Facsimile reprint of Rigaud's classic text based upon primary sources. Cohen's introduction provides useful citations to Newton bibliographies.

730. Rosenfeld, Leonard. "Newton and the Law of Gravitation." *Archive for History of Exact Sciences*, 2 (1965), 365-386.

Traces development of Newton's discovery of the law of Universal Gravitation centering upon the apparent 20-year hiatus between the time he was aware of the identity of terrestrial and celestial gravitational forces and his announcement of the law. Provides general review of Newtonian studies, including a 30-item bibliography. 72 citations.

731. Rufus, W. Carl. "David Rittenhouse as a Newtonian Philosopher and Defender." *Popular Astronomy*, 56 (1948), 122-130.

Reviews Rittenhouse's writings on Newton, his observations of transits, and general astronomical activities. Few citations.

732. Voltaire, F.M.A. *Letters Concerning the English Nation*. London: Davis and Lyon, 1733. Pp. xvi + 253.

Voltaire's account of Newtonian physics, including the first printed statement of Newton and the apple. Translation predates French edition by one year. Includes chapter on Locke, Bacon and Descartes. Frequently reprinted.

733. Waff, Craig B. "Newton and the Motion of the Moon."
     Centaurus, 21 (1977), 64–75.

     Extensive review of I.B. Cohen's edition of Isaac Newton's
     Theory of the Moon's Motion (1702), published in 1975. 29
     citations.

734. Westfall, Richard S. "Hooke and the Law of Universal Gravi-
     tation." *British Journal for the History of Science*, 3
     (1967), 245–261.

     Argues that Hooke contributed both to the mechanics and
     conception of universal gravitation and examines his in-
     fluence upon Newton within the context of the priority
     dispute between the two. 40 citations.

735. Whiteside, D.T. "The Expanding World of Newtonian Research."
     History of Science, 1 (1962), 16–29.

     Examines general aspects of the historical problem of
     understanding Newton, influences upon him, and his influ-
     ence upon the modern scientific age. Identifies significance
     of recent historical work that extends beyond secondary
     works to original documentation. 39 citations.

736. Whiteside, D.T. "Newton's Early Thoughts on Planetary
     Motion: A Fresh Look." *British Journal for the History
     of Science*, 2 (1964), 117–138.

     Detailed exposition and analysis of the influences upon
     Newton that led him through to his world-system identified
     in his *De Motu* and elaborated in the *Principia*. Emphasizes
     the roles and influence of Hooke and Halley. 60 citations.

*   Whiteside, D.T., ed. *The Mathematical Papers of Isaac
     Newton*.

     Cited herein as item 1386.

737. Whiteside, D.T. "Before the Principia: The Maturing of
     Newton's Thoughts on Dynamical Astronomy." *Journal for
     the History of Astronomy*, 1 (1970), 5–19.

     Traces the intricate and difficult path from Newton's
     first conception of universal gravitation to its elucida-
     tion in the *Principia*. Examines the chief influences upon
     Newton, including the works of René Descartes and contact
     with Halley and Hooke. 45 citations.

738. Whiteside, D.T. "The Mathematical Principles Underlying Newton's *Principia Mathematica*." *Journal for the History of Astronomy*, 1 (1970), 116-138.

     Examines why only a handful of Newton's contemporaries "achieved a working knowledge of the *Principia*'s technical content." 63 citations.

739. Whiteside, D.T. "Newton's Lunar Theory: From High Hope to Disenchantment." *Vistas in Astronomy*, 19 (1977), 317-328.

     Short narration on the development of the theory which was, to Newton, never fully satisfactory. 54 citations.

740. Wilson, David B. "Herschel and Whewell's Version of Newtonianism." *Journal of the History of Ideas*, 35 (1974), 79-97.

     Examines the scientific friendship of John Herschel and William Whewell in the first half of the 19th century, and the similarities in their thought, notably their common position regarding Newtonian mechanics, astronomy, natural theology, and rules of reasoning. 109 citations.

                              POST-NEWTONIAN

741. Abalakin, Victor K. "The Development of Theoretical Astronomy in the U.S.S.R." *Vistas in Astronomy*, 19 (1977), 163-177.

     Traces the history of celestial mechanics in Russia since the time of L. Euler at the Academy of Sciences at St. Petersburg but centers upon late 19th- and 20th-century studies. Extensive citations to Russian literature.

*   Adams, John Couch. *The Scientific Papers of John Couch Adams*.

     Cited herein as item 1224.

742. Agostinelli, C. "Sul problema de tre Corpi." *Rendiconti del Seminario matematico i fisico di Milano*, 21 (1951), 165-195.

     Surveys the three-body problem from Newton to date concentrating upon the research of Lagrange, Sundman, Hill and Poincaré. 40 citations.

743. Airy, G.B. *Gravitation: An Elementary Explanation of the Principal Perturbations in the Solar System*. London: Charles Knight, 1834. Pp. xxiii + 215.

Descriptive exposition of perturbation theory for general audiences. Includes general discussion of perturbations by many bodies; theory of the Moon's motion; theory of Jupiter's satellites; changes in orbital characteristics; and perturbations by non-spherical bodies. No direct citations.

744. Airy, G.B. "Account of Some Circumstances Historically Connected with the Planet Exterior to Uranus." *Monthly Notices of the Royal Astronomical Society*, 7 (1846), 121-144.

After brief introductory remarks, reprints 23 extracts of correspondence with associates including T.J. Hussey, M.E. Bouvard, Challis, J.C. Adams and U. LeVerrier. Argues that the discovery "is the effect of a movement of the age," and was being pursued equally by many, both observers and theoreticians. Airy's discussion is followed (pp. 145-152) by remarks extracted from Challis' presentation, and an account of the actual discovery of the planet at Berlin.

745. Airy, G.B. "On the Results of Recent Calculations on the Eclipse of Thales and Eclipses Connected with It." *The Royal Institution Library of Science: Astronomy* (item 140), 8-15.

Lecture date: February 4, 1853. Discusses the historical attempts to date an eclipse, reportedly observed by Thales, as 585 B.C. Reviews Laplace's discovery of the secular change in the Moon's mean motion which greatly increased the accuracy of eclipse predictions, and then the application by Francis Baily of improved French lunar tables to predict date of Thales' eclipse. Continues with later, more refined determinations and the importance of the Greenwich Lunar Observations, 1750-1830.

746. Andoyer, Henri. *L'oeuvre scientifique de Laplace*. Paris: Payot, 1922. Pp. 162.

Concise summary of Laplace's works, beginning with a discussion of his *Traité de Mécanique Céleste* and *L'Exposition du système du Monde*, and notes on Laplace's philosophical background. Includes biographical and bibliographical commentary providing background for his contributions to celestial mechanics; the theory of probabilities; the solutions to differential equations; and to his interests in moral science, finance and politics. Reviewed in: *Isis*, 5 (1923), 159-160.

*   Arago, F. "Laplace."

       Cited herein as item 1229.

747.   Backlund, Oskar. "The Development of Celestial Mechanics
       During the Nineteenth Century." *International University
       Lectures at the Congress of Arts and Science Universal
       Exposition, Saint Louis (1904)*. Volume VI. New York:
       University Alliance, 1909. Pp. 83-95.

       Brief general review of major accomplishments in theory
       and observation. Concludes that all branches, save for the
       lunar theory, had kept pace with observation, especially
       in the cases of the computations of orbits of asteroids;
       the discovery of Neptune and the solution of the inequalitie
       of the motions of Jupiter, Saturn and Uranus; and the study
       of the motions of newly discovered satellites of Mars and
       Jupiter. Notes that in the past 30 years, America had
       moved into prominence in this field. No direct citations.
       Brief bibliography of general and related works appears
       on pp. 96-101.

*   Beer, Arthur, and Peter Beer, eds. "The Origins, Achieve-
       ment and Influence of the Royal Observatory, Greenwich:
       1675-1975."

       Cited herein as item 124.

748.   Boss, Valentin. *Newton and Russia: The Early Influence,
       1698-1796*. Cambridge: Harvard University Press, 1972.
       Pp. xviii + 310.

       Detailed examination of the spread of Newtonianism in
       Russia centering upon the role of Daniel Bruce in introduc-
       ing it in the early 18th century. Identifies the central
       place of the St. Petersburg Academy of Sciences, composed
       largely of non-Russians, in its acceptance at mid-century
       after chemist Michael V. Lomonosov, a critic of Newtonianism
       failed to maintain a successful opposition. Claims that
       Euler continued to oppose Newton in the late 18th century,
       a conclusion disputed by R. Calinger in his review of this
       work (*Science*, 180 (1973), 623-624). Also reviewed in:
       Isis, 65 (1974), 537-538.

749.   Brill, Dieter R., and Robert C. Perisho. "Resource Letter
       GR-1 on General Relativity." *American Journal of Physics*,
       36 (1968). 8p.

       Annotated bibliography of 77 papers, texts, and other
       general bibliographies including elementary expositions,

general texts, works reviewing current research, the experimental verification of general relativity, equations of motion in general relativity, cosmology, gravitational collapse, Mach's Principle, and gravitational waves. In addition to the 77 primary citations, extensive citations to general reviews in each subject matter area are included.

750. Brown, Ernest W. "The Problem of the Moon's Motion." *Publications of the Astronomical Society of the Pacific*, 32 (1920), 93-104.

General review of the history of studies of the Moon's motion. Introduces the problem and how it was examined in three stages: the gathering of observations; the formulation of Newton's laws in their generality; and the application of these laws to deduce their consequences "in a form in which they may be most easily compared with observation." Brown notes that Hansen was the only worker "to have accomplished the whole of it with any degree of completeness." Describes the construction of his own theory of the Moon, and the tables derived from it. Examines recent history of study of secular acceleration of Moon's mean motion.

751. Brown, Lloyd A. *The Story of Maps*. Boston: Little, Brown & Co., 1949. Pp. 397.

Traces progress of map making from earliest times. Includes chapters on navigation; the determination of longitude by the observation of the Jovian moons and by lunar tables; the development of the chronometer and the geodetic surveys of Picard and the Cassinis; and the designation of Greenwich time. Reviewed in: *Isis*, 41 (1950), 243-244.

752. Brunet, Pierre. *L'introduction des théories de Newton en France au XVIIIᵉ siècle (1738)*. Paris: Blanchard, 1931. Pp. vii + 355.

Examines the gradual reception of Newtonian thought in France between 1700 and 1738 when Voltaire published his *Les Eléments de la philosophie de Newton*, a popular essay on Newton's gravitation theory with an excerpt on his *Opticks*. Notes the earlier role of Maupertius circa 1730-1732 as first astronomer in France to examine Newtonian and Cartesian theories equally, and to favor Newton. The general text deals also with receptions in areas of chemistry, physics, and the realm of speculation on the origin of the Earth. Reviewed in: *Isis*, 17 (1932), 433-435.

* Brush, Stephen G. "Poincaré and Cosmic Evolution."

   Cited herein as item 1036.

753. Burstyn, Harold L. "Early Explanations of the Role of the
     Earth's Rotation in the Circulation of the Atmosphere
     and the Ocean." *Isis*, 57 (1966), 167-187.

     Discusses the work of George Hadley and Colin Maclaurin
     in 1735 and 1740 respectively in revealing the "deflecting
     effect of the earth's rotation on winds and currents, the
     basis for scientific understanding of the motions in the
     earth's fluid envelope." Traces the background to Hadley's
     and Maclaurin's synthesis in 16th- and 17th-century examina
     tions of the phenomenon; Newton's exposition in the *Prin-
     cipia*; and Mariotte's study of the rotations of fluids.
     88 citations.

754. Calinger, Ronald. "Kant and Newtonian Science: The Pre-
     Critical Period." *Isis*, 70 (1979), 349-362.

     Examines Kant's first contacts with, early response to,
     and gradual mastery of Newtonian science before his studies
     leading to his *Critique of Pure Reason* in the 1770s. Traces
     Kant's developing zeal for Newtonianism through his first
     essays, and his *Theory of the Heavens* in 1755 which include
     his concept of external island universes based upon the
     work of "two prominent Newtonians ... James Bradley and
     Pierre Maupertius." 54 citations.

755. Campbell, W.W. "The Closing of a Famous Astronomical
     Problem." *Publications of the Astronomical Society of
     the Pacific*, 21 (1909), 103-115.

     Provides historical commentary on the work of Leverrier
     on the motion of Mercury and its stimulus in the search
     for Vulcan. Reviews eclipse results, studies of the motions
     of the inner planets, studies of the Zodiacal Light, and
     claims of direct detection of an intra-Mercurial planet
     dubbed "Vulcan." Concludes that the supposed object does
     not exist.

756. Chandler, Philip. "Clairaut's Critique of Newtonian
     Attraction: Some Insights into His Philosophy of Science."
     Annals of Science, 32 (1975), 369-378.

     Examines the controversy between Clairaut and Buffon
     that arose when Clairaut tentatively proposed, in 1749,
     a revision in Newton's inverse square law to account for
     discrepancies in the theory of motion of the Moon. Even

though Clairaut rejected Buffon's defense of the inverse
square law, a subsequent analysis brought him back to it,
and also to the first successful theory of the motion of
the apsides of the lunar orbit. 59 citations.

757. Clemence, Gerald M. "The Motion of Mercury, 1765–1937."
     *Astronomical Papers of the American Ephemeris and Nautical
     Almanac*, 11, Pt. 1 (1943), 9–221.

     Evaluation of observations since the time of James
     Bradley originally collected by Simon Newcomb. No historical
     discussion but of general historical interest for discussion
     of instrumental techniques and reduction procedures.

758. Cotter, Charles H. *A History of Nautical Astronomy*. New
     York: American Elsevier, 1968. Pp. xii + 387.

     A general history covering all periods from the Babylonians
     through modern times including the problems of time measure-
     ment at sea; changing instrumentation; observational methods;
     techniques of finding latitude and longitude; and naviga-
     tion tables. Includes a 375-item bibliography. Reviewed
     in: *Isis*, 64 (1973), 560–563.

759. Cotter, Charles H. *Studies in Maritime History*. 3 volumes.
     London: Mansell, 1977. 25 fiches.

     Comprehensive study of navigation. Volume 1 reviews
     nautical astronomical tables; Volume 2 examines the history
     of ship magnetism; Volume 3 covers the history of naviga-
     tion by dead reckoning. Microfiche of typescript. Includes
     extensive bibliography.

760. Cousin, J. *Introduction à l'étude de l'astronomie physique*.
     Paris: Veuve Dessaint, 1787. Pp. xv + 323.

     Mathematical review and analysis of 18th-century physical
     astronomy.

*   Delauney, C.E. *Rapport sur les progrès de l'astronomie*.

     Cited herein as item 128.

761. Dubyago, A.D. *The Determination of Orbits*. New York:
     Macmillan, 1961. Pp. xiv + 431.

     Includes a chapter on the history of orbit determination,
     concentrating upon Gauss and Olbers. 72 citations.

*   Dunnington, G. Waldo. *Carl Friedrich Gauss: Titan of
     Science*.

     Cited herein as item 1274.

761a. Earman, John, and Clark Glymour. "Relativity and Eclipses: The British Eclipse Expeditions of 1919 and Their Pre- decessors." *Historical Studies in the Physical Sciences*, 11 (1980), 49-85.

Examines in detail the motives and influences behind the British Eclipse expeditions conducted at a time when the British were largely disinterested in relativity and when, during the planning stages circa 1917, the British were still bitterly at war with Germany. Identifies the great influence of A.S. Eddington and F.W. Dyson. Reviews attempts prior to 1919 to measure the deflection of star- light at the limb of the Sun. 71 citations.

761b. Earman, John, and Clark Glymour. "The Gravitational Red Shift as a Test of General Relativity: History and Analysis." *Studies in the History and Philosophy of Science*, 11 (1980), 175-214.

Detailed technical and historical review of the interplay of theoretical predictions of, and continued failures by observations to confirm, the relativistic red shift of the spectrum of the Sun, from Einstein's first prediction in 1907 through the 1920s when some consensus was believed to have been achieved. Argues that "the red shift is a litmus, and its coloring reveals most of the major themes that dominated the development and reception of general relativity." 118 citations.

*   Euler, Leonhard. *Manuscripta Euleriana*.

Cited herein as item 1277.

762.   Euler, Leonhard. *Opera Omnia. Briefwechsel. Series Quarta A: Commercium epistolicum*. Volume 1: A. Juskevic, V. Smirnov, and W. Habicht, eds. Basel: Birkhauser, 1975. Pp. xviii + 666.

Provides a synthesis of Euler's correspondence selected from the projected six volumes in the series. Includes correspondence with Tobias Mayer on lunar motion and with Clairaut on the three-body problem. Reviewed in: *Isis*, 67 (1976), 617-619.

*   Fontenrose, Robert. "In Search of Vulcan."

Cited herein as item 647.

*   Forbes, Eric G. "The Life and Work of Tobias Mayer (1723- 62)."

Cited herein as item 1280.

763. Forbes, Eric G. "The Correspondence between Carl Friedrich Gauss and the Rev. Nevil Maskelyne (1802-05)." *Annals of Science*, 27 (1971), 213-237.

    Reviews their letters which began in response to Gauss' development of an efficient method to determine asteroid orbits. Reviews the background to their correspondence and reprints 18 letters with detailed annotations.

764. Forbes, Eric G. "Schultz's Proposal for Finding Longitude at Sea." *Journal for the History of Astronomy*, 2 (1971), 35-41.

    Presentation of Heinrich Schultz's derivation in 1762 of a method of longitude determination by observations of the Moon's meridian transits. His relation with a co-author, G.F. Stender, is discussed, and a critical evaluation of his method is provided. 16 citations.

765. Forbes, Eric G. "Gauss and the Discovery of Ceres." *Journal for the History of Astronomy*, 2 (1971), 195-199.

    Describes Giuseppe Piazzi's discovery of the first minor planet which he named Ceres. Relates how Carl Friedrich Gauss, upon hearing of Piazzi's discovery, created a new method of orbit computation and how the elements of Ceres' orbit derived from this new method became generally accepted. 23 citations. See item 770.

766. Forbes, Eric G. "Tobias Mayer's Method for Calculating the Circumstances of a Solar Eclipse." *Annals of Science*, 28 (1972), 177-189.

    Detailed analysis of one of Mayer's six lectures in his *Opera Inedita* translated by Forbes. 23 citations.

767. Forbes, Eric G. *The Euler-Mayer Correspondence (1751-1755): A New Perspective on Eighteenth Century Advances in the Lunar Theory*. New York: American Elsevier, 1972. Pp. x + 118.

    Translations provide insight to Euler's influence in the development of Mayer's lunar theory and his praise for Mayer's lunar tables based upon improved observing technique. Detailed annotation is included for the 31 letters, which also include the development of Euler's interests in studies of atmospheric refraction, the motions of bodies in a resisting aethereal medium, and the possibility of the existence of a lunar atmosphere. Reviewed in: *Annals of Science*, 30 (1973), 222-224; *Isis*, 64 (1973), 552-554; *JHA*, 4 (1973), 205-206.

*   Forbes, Eric G., ed. *The Unpublished Writings of Tobias Mayer*.

Cited herein as item 1282.

768.   Forbes, Eric G. *The Birth of Scientific Navigation. The Solving in the 18th Century of the Problem of Finding Longitude at Sea*. Greenwich: National Maritime Museum, 1974. Pp. vii + 25.

Brief account beginning with the foundation of the Board of Longitude, the construction of Hadley's reflecting quadrant, Tobias Mayer's repeating circle, and finally Harrison's marine chronometer--the first acceptable time keeping device accurate enough for navigational needs. Reviewed in: *Centaurus*, 19 (1975), 149; *Isis*, 66 (1975), 579-580.

769.   Forbes, Eric G. "Die Entwicklung der Navigationswissen-schaft im 18. Jahrhundert." *Rete*, 2 (1975), 307-321.

Reviews 18th-century advances leading to an effective and reliable mode of navigation, notably John Hadley's 1731 invention of an accurate octant; Tobias Mayer's 1754 publication of solar and lunar ephemerides that allowed for the precise comparison of observed and predicted posi-tions of the Moon--the deviations between prediction and observation being a measure of terrestrial longitude; and John Harrison's successful construction of a precise marine chronometer, tested in 1764. 44 citations.

*   Forbes, Eric G. "Tobias Mayer's Contributions to Observa-tional Astronomy."

Cited herein as item 1285.

770.   Gauss, Karl Friedrich. *Theoria Motus: Theory of the Motion of the Heavenly Bodies Moving About the Sun in Conic Sections* (1809). Translated with an Appendix by Charles Henry Davis. New York: Little, Brown and Co., 1857. Reprinted, New York: Dover, 1963. Pp. xvii + 326.

Translation of original 1809 work by Gauss with an appendix on later work by Encke and Peirce. This famous text provided for the first time a quick, efficient method for the determination of an orbit from only a few observa-tions taken over a short period of time, a problem stimu-lated by the discovery of asteroids. See item 765.

771.   Gautier, Alfred. *Essai historique sur le problème des trois
       corps*. Paris: 1817. Pp. xii + 283.

       Provides historical development of the application of
       the classic three-body problem to the theories of motion
       of the Moon, planets, and their satellites.

*   Gould, Rupert Thomas. *The Marine Chronometer*.

       Cited herein as item 521.

*   Grant, Robert. *History of Physical Astronomy, from the
       Earliest Ages to the Middle of the Nineteenth Century*.

       Cited herein as item 91.

772.   Grosser, Morton. "The Search for a Planet beyond Neptune."
       Isis, 55 (1964), 163-183.

       Reviews the predictions by P. Lowell and W.H. Pickering
       and earlier workers for the position of a planet beyond
       Neptune and the search for that planet at the Lowell
       Observatory. Compares the predictions of Lowell, Pickering
       and five earlier workers and concludes, with V. Kourganoff
       and in opposition to E.W. Brown, that the predictions of
       Lowell and Pickering were genuine. 58 citations.

773.   Grosser, Morton. *The Discovery of Neptune*. Cambridge:
       Harvard University Press, 1962; New York: Dover, 1979.
       Pp. 172.

       Traces progress in observational planetary astronomy
       through William Herschel's discovery of Uranus in 1781
       and the rapid realization that the motion of Uranus was
       being perturbed by an unknown mass. Examines in detail
       the work of Gauss, E. Bouvard, G.B. Airy and others in
       discovering asteroids, calculating planetary tables,
       and in general providing the observational and theoretical
       background for the independent predictions of the place
       of Neptune by Urbain J.J. Leverrier and John Couch Adams.
       The acrimonious priority dispute that followed the observa-
       tional verification of the existence of Neptune at Berlin
       based upon Leverrier's predictions is reviewed in detail.
       Includes short glossary of terms, extensive citations,
       and a bibliography of archival, primary and secondary
       sources. Reviewed in: *Isis*, 54 (1963), 413-414; *Centaurus*,
       10 (1964-65), 54.

*   Grünbaum, Adolf. "*Ad Hoc* Auxiliary Hypotheses and Falsi-
       ficationism."

       Cited herein as item 167a.

774.   Hahn, Roger. *Laplace as a Newtonian Scientist*. University
       of California: William Andrews Clark Memorial Library,
       1967. Pp. iv + 26.

       Argues that Laplace was not, in fact, a pure Newtonian.
       Examines Laplace's early training in theology and philosophy
       and his initial interest in Descartes, Leibniz and Male-
       branche. Reviewed in: *Annals of Science*, 25 (1969), 175.

*   Hall, Tord. *Carl Friedrich Gauss*.

       Cited herein as item 1295.

*   Hankins, Thomas L. *Jean d'Alembert*.

       Cited herein as item 1297.

775.   Hanson, Norwood Russell. "Leverrier: The Zenith and Nadir
       of Newtonian Mechanics." *Isis*, 53 (1962), 359-378.

       Analyzes Leverrier's establishment of the existence of
       an unseen planetary body beyond Uranus, and notes that
       while Adams assumed the existence of such a body, Leverrier
       demonstrated the necessity for its existence. Argues that
       Leverrier's development and application of Newtonian
       celestial mechanics provided it with its greatest success
       in the discovery of Neptune, and also initiated its eventual
       demise with his analysis of the orbit of Mercury in 1859.
       Reviews late 19th-century studies of Mercury's orbit and
       their failure to explain its motion. 23 citations and an
       appendix "On the Impossibility of a Straight Line Solution
       to the Three-Body Problem."

776.   Harvey, A.L. "Brief Review of Lorentz-Covariant Scaler
       Theories of Gravitation." *American Journal of Physics*,
       33 (1965), 449-460.

       Reviews work of Poincaré, Kottler, Whitrow and others.

777.   Hill, G.W. "Remarks on the Progress of Celestial Mechanics
       Since the Middle of the Century." *Bulletin of the
       American Mathematical Society*, Series 2 (1896), 125-136.

       Presidential address delivered on December 27, 1895.
       Reviews causes for the high degree of success of the
       application of mathematics to the solution of the motions
       of celestial objects. Examines the lunar and planetary
       theories of Delaunay, Gyldén and Poincaré, noting the
       general influence of P.A. Hansen. Notes the lack of
       primary scientific texts in American libraries, and the
       lack of comprehensive histories of the subject at the
       time. No direct citations.

* Hodghead, Beverly L. "Address of the Retiring President of the Society, in Awarding the Bruce Medal to Ernest W. Brown, Professor of Mathematics at Yale University."

  Cited herein as item 1311.

778. Howse, Derek. *Greenwich Time and the Discovery of the Longitude*. Oxford: Oxford University Press, 1980. Pp. 254.

  General popular review of all aspects of the problem of navigation at sea including the designation of Greenwich Time as Universal Time after the 1884 International Meridian Conference held in Washington. Appendices on technical subjects. Numerous citations. Reviewed in: *Sky and Telescope*, 60 (Dec. 1980), 513-514.

779. Hoyt, William Graves. "W.H. Pickering's Planetary Predictions and the Discovery of Pluto." *Isis*, 67 (1976), 551-564.

  Examines W.H. Pickering's speculation on the existence of at least seven additional planets during the years 1909-1932 and his contribution to the discovery of Pluto. 77 citations.

780. Hoyt, William Graves. *Planets X and Pluto*. Tucson: U. of Arizona, 1979. Pp. 352.

  Surveys the discovery of Uranus, the asteroids and Neptune, and centers upon the long search for "Planet X" beyond Neptune prosecuted at the Lowell Observatory. Examines the controversy over the role of predictions calculated by Lowell in search of the planet prior to 1916. Extensive bibliography and archival documentation. Reviewed in: *Isis*, 72 (1981), 148.

781. Jaki, Stanley L. "The Original Formulation of the Titius-Bode Law." *Journal for the History of Astronomy*, 3 (1972), 136-138.

  Attempts to clarify origins of numerical relation that describes ratios of planetary distances from the Sun. Emphasizes role of mid-18th-century workers including Christian Wolff and Charles Bonnet, as well as Johann Daniel Titius and Johann Elert Bode. 14 citations.

* Jaki, Stanley L. "The Five Forms of Laplace's Cosmogony."

  Cited herein as item 1044.

782. Jaki, Stanley L. "Johann Georg Soldner and the Gravita-
      tional Bending of Light, with an English Translation
      of His Essay on It Published in 1801." *Foundations of
      Physics*, 8 (1978), 927-950.

      Examines the influences upon Soldner that led him to
      derive relations for the gravitational bending of star-
      light due to the gravitational field of the Sun, according
      to Newtonian principles. Evaluates secondary references
      to Soldner's work by relativists in the early 20th cen-
      tury. 44 citations.

783. Jeffreys, Harold. "Tidal Friction." *Quarterly Journal of
      the Royal Astronomical Society*, 16 (1975), 145-151.

      Brief technical review of all sources of knowledge from
      astronomy, geology and dynamical theory relating to the
      acceleration of the Moon's mean motion and deceleration
      of the Earth's rotation. Includes historical introduction
      noting work of Halley, Laplace, J.C. Adams, J.R. Mayer,
      and G.H. Mayer. 16 citations.

784. Jones, Harold Spencer. *John Couch Adams and the Discovery
      of Neptune*. Cambridge: Cambridge University Press, 1947.
      Pp. 42.

      Short essay attempting to clarify the roles played by
      British scientists during the period when Neptune was
      discovered. Centers upon J.C. Adams' solution of the orbit
      of Uranus that predicted the existence of a planet farther
      out in the solar system. No direct citations.

785. Klein, F., M. Brendel and L. Schlesinger. "Über die
      astronomischen Arbeiten von Gauss." *Materialien für
      eine wissenschaftliche Biographie von Gauss*, Heft VII.
      Leipzig: B.G. Teubner, 1919. Pp. 106.

      Reviews Gauss' studies in celestial mechanics including
      his lunar theory and techniques for determining the orbits
      of asteroids.

786. Kuiper, G.P. "Limits of Completeness." *The Solar System*.
      Volume 3. Chicago: University of Chicago Press, 1961.
      Pp. 575-591.

      Describes searches for intra-Mercurial planets, planetary
      satellites and planetary rings. Hints at possible rings
      around other Jovian planets and discusses the McDonald
      Observatory Satellite Survey. Notes earlier photographic
      and visual discoveries of faint satellites. 22 citations.

787. Laplace, Pierre Simon, Marquis de. *Traité de Méchanique Céleste*. 4 volumes. Paris, 1799–1805. Translated as *Méchanique Céleste* by N. Bowditch. Reprinted, New York: Chelsea, 1966.

 Includes a 148-page memoir on the life of Nathaniel Bowditch in Volume 1. In the last volume, Laplace reviews the development of physical astronomy concentrating on the period since Newton.

788. Lecat, Maurice. *Bibliographie de la Relativité*. Bruxelles: M. Lamertin, 1924. Pp. xii + 337.

 Alphabetical listing of over 3000 primary papers on all aspects of relativity. Provides journal cross-index and chronological summary of publication histories.

789. Levy, J. "Trois siècles de mécanique céleste à l'Observatoire de Paris." *L'Astronomie*, 82 (1968), 381–393.

 Transcription of a lecture presented to the French Astronomical Society on 21 June 1967. Brief popular review of the work of major figures associated with the Paris Observatory from Clairaut, D'Alembert, Lagrange and Laplace through Leverrier, Delaunay, Tisserand and Poincaré. No direct citations.

* Loria, Gino. "Nel secondo centenario della nascita di G.L. Lagrange, 1736–1936."

 Cited herein as item 1334.

790. Lyttleton, R.A. "A Short Method for the Discovery of Neptune." *Monthly Notices of the Royal Astronomical Society*, 118 (1958), 551–559.

 Provides mathematical basis for a new method that provides a reasonable measure of accuracy using Bode's Law.

791. Lyttleton, R.A. "The Rediscovery of Neptune." *Vistas in Astronomy*, 3 (1960), 25–46.

 Presents a simplified method for predicting the position of Neptune using fewer calculations than the methods of Adams and Le Verrier. Reviews the original methods, and the then available observational data. 6 citations.

791a. Lyttleton, R.A. "History of the Mass of Mercury." *Quarterly Journal of the Royal Astronomical Society*, 21 (1980), 400–413.

Reviews determinations since the late 18th century
beginning with Lagrange and Laplace. Examines early studies
based upon the conjecture that the mean density of a planet
was inversely proportional to its solar distance. Analyzes
19th-century dynamical studies through Newcomb and the
generally favorable reception of Newcomb's value in 1896,
in spite of the fact that Newcomb's determination was
arbitrary and well outside the limits of error set by his
own analysis. 30 citations.

792. Marguet, F. *Histoire générale de la navigation du XVe au
      XXe siècle*. Paris: Société d'éditions géographique,
      maritimes et coloniales, 1931. Pp. 301.

      A general technical review of the history of the art
      of navigation. Includes discussions of the growth of
      instrumentation, notably chronometers; major voyages for
      mapping, and the detection of the drift of magnetic declina-
      tion. Describes the various astronomical techniques and
      tables developed after Tobias Mayer's work culminating
      in refinements by P.A. Hansen. Includes the use of Jovian
      satellites, lunar tables, Harrison's chronometer. Reviewed
      in: *Isis*, 19 (1933), 235-237.

*   McCrea, W.H. "The Royal Greenwich Observatory, 1675-1975."

      Cited herein as item 316.

*   McCrea, W.H. "Einstein: Relations with the Royal Astronomic
      Society."

      Cited herein as item 369.

793. Morgan, H.R. "Motions in the Solar System." *Popular Astron-
      omy*, 54 (1946), 2-16.

      Concise review of theory and observations of planetary,
      lunar and solar positions and motions since the 18th
      century but concentrating upon the late 19th and 20th
      centuries. Examines computation of tables, fundamental
      reference systems, the variation of latitude, and the
      lunar inequality. 78 citations.

794. Moulton, F.R. "On Certain Rigorous Methods of Treating
      Problems in Celestial Mechanics." *Publications of the
      Yerkes Observatory*, 2 (1903), 117-142.

      Reviews methods of convergence that allow for rigorous
      proofs of the validity of computational methods in celestial
      mechanics. Includes historical discussions within context

of problem. Contends that originators of many of the early computational methods (Lagrange and Laplace) did not have the means available to test the validity of their methods. Hence the conclusions derived from them, which included the problem of the stability of the solar system, were invalid.

795. Moulton, F.R. "The Problem of Three Bodies." *Popular Astronomy*, 22 (1914), 197-207.

Provides a brief historical introduction to the problem beginning with Newton, Clairaut, d'Alembert and Euler. No direct citations.

* Munk, W.H. "Polar Wandering: A Marathon of Errors."

Cited herein as item 898.

796. Newcomb, Simon. "On the Present State of the Theories of the Celestial Motions." *Sidereal Messenger*, 2 (1883), 11-17; 33-39.

Reprinting of introduction to Newcomb's classic "On the Recurrence of Solar Eclipses, with tables of eclipses from B.C. 700 to A.D. 2300." Washington, D.C.: Bureau of Navigation, Navy Department, 1879. Pp. 55.

797. Nieto, Michael Martin. *The Titius-Bode Law of Planetary Distances: Its History and Theory*. Oxford: Pergamon, 1972. Pp. xii + 161.

Begins with historical development from Kepler to Titius and Bode and then reviews modifications to the law in the 19th and 20th centuries. Reviews theories forwarded to explain the law including those of modern times. Attempts an evaluation of the validity of the law and a contemporary assessment based upon theories of the origin of the solar system. 200+ citations and extensive bibliography. Reviewed in: *Centaurus*, 18 (1973-74), 87; *JHA*, 4 (1973), 141-142.

* Nordenmark, N.V.E. *Pehr Wilhelm Wargentin*.

Cited herein as item 1349.

798. Numbers, Ronald L. "The American Kepler: Daniel Kirkwood and His Analogy." *Journal for the History of Astronomy*, 4 (1973), 13-21.

Examines the role of Kirkwood's analogy, or his empirical relationship between the ratios of planetary rotation periods and the diameter of their "spheres of attraction"

(the width of the original nebulous ring that formed the
planet), in the American acceptance of Laplace's Nebular
Hypothesis in the mid-19th century. 30 citations.

799. Pannekoek, A. "The Planetary Theory of Laplace." *Popular
      Astronomy*, 56 (1948), 300-312.

      Brief technical review of Laplace's *Traité de Mécanique
      Céleste* as a culmination of 18th-century celestial mechanic
      No direct citations.

800. Pannekoek, A. "The Discovery of Neptune." *Centaurus*, 3
      (1953-54), 126-137.

      Analyzes events leading to the prediction and discovery
      of the planet, within the social context of 19th-century
      British and French science. Argues that the controversy
      that arose over the question of the reliability of the
      predictions was more an event in the European "social-
      cultural struggle," reflecting the author's Marxist
      philosophy. 2 citations.

801. Picard, Emile. *Un double centenaire, Newton et Laplace.*
      Discours prononcé à la Sorbonne, le mercredi 4 mai
      1927. Paris: Palais de l'Institut, 1927. Pp. 26. Re-
      printed: *Revue génèrale des sciences*, 38 (1927), 357-
      366.

      An address written on the bicentenary of Newton's death
      recalling France's elaboration and refinement of Newtonian
      mechanics. Reviewed in: *Isis*, 11 (1928), 387-393.

802. Poynting, J.H. "Some Astronomical Consequences of the
      Pressure of Light." *Royal Institution Library of Science:
      Astronomy* (item 140), 75-77.

      Abstract of Poynting's May 1906 address reviewing the
      physics of the pressure due to radiation and its possible
      effect upon the orbits of meteoroids and the stability
      of the rings of Saturn.

*   Ronan, Colin A. *Their Majesties' Astronomers--A Survey
      of Astronomy in Britain between the two Elizabeths.*

      Cited herein as item 226.

*   Rufus, W. Carl. "David Rittenhouse as a Newtonian Philos-
      opher and Defender."

      Cited herein as item 731.

803. Ruse, Michael. "The Scientific Methodology of William Whewell." *Centaurus*, 20 (1976), 227-257.

Examines Whewell's philosophy of science and his contributions to various areas of science including astronomy. Whewell's major contributions to astronomy centered on the theory of tides and the search for empirical relations in tidal data. Analyzes how Whewell's philosophy of science can be found in the scientific methodology he employed in the analysis of tides. 104 citations.

* Sadler, D.H. "The Bicentenary of the Nautical Almanac."

Cited herein as item 386.

804. Sarton, George. "Discovery of the Main Nutation of the Earth's Axis." *Isis*, 17 (1932), 333-383.

Traces development of realization that luni-solar precession was not sole long-term motion of Earth's axis but that there was another second-order term that was finally observed by James Bradley. Examines details of Bradley's work and the development of his technique and rationale. Includes facsimile reproduction of James Bradley's account of his discovery, originally in the *Philosophical Transactions*, 45 (1748), 1-43. 10 citations.

805. Sarton, George. "Laplace's Religion." *Isis*, 33 (1941), 309-312.

Short commentary on Laplace's lack of need for divine intervention as exhibited in his *Mécanique Céleste*, Napoleon's quip regarding this attitude, and the general attitude of rejection of theological investigation.

806. Schaffer, Simon. "John Michell and Black Holes." *Journal for the History of Astronomy*, 10 (1979), 42-43.

Briefly describes late 18th-century speculation by Michell on the attraction of light by gravity and his discussion of bodies massive enough to capture their own light and any light falling upon them. Notes relationship of this early speculative thought by Michell, Laplace, Herschel and others to the acceptance of the corpuscular theory of light. 8 citations.

* Schaffer, Simon. "'The Great Laboratories of the Universe': William Herschel on Matter Theory and Planetary Life."

Cited herein as item 589.

*   See, T.J.J. "The Services of Benjamin Peirce to American
        Mathematics and Astronomy."

        Cited herein as item 1366.

807.   Shenynin, O.B. "Laplace's Theory of Errors." *Archive for
        History of Exact Sciences*, 17 (1977), 1-61.

        While not dealing directly with Laplace's celestial
        mechanics, provides useful background to his work on the
        principles of the adjustment of observations and their
        mathematical treatment. 20 footnotes and 83 citations.

808.   Smart, W.M. "John Couch Adams and the Discovery of Neptune.
        Popular Astronomy, 55 (1947), 301-311.

        Brief review centering upon the priority dispute and
        Airy's lack of familiarity with Adams' work. A rebuttal
        to Smart's review was provided by H. Spencer-Jones (item
        810).

809.   Somerville, Mary. *The Mechanism of the Heavens*. London:
        John Murray, 1831. Pp. lxx + 621.

        Provides in part a translation and detailed summary of
        Laplace's *Mécanique Céleste* within a general technical
        exposition of planetary theory and physical astronomy.

810.   Spencer-Jones, Harold. "G.B. Airy and the Discovery of
        Neptune." *Popular Astronomy*, 55 (1947), 312-316.

        Rebuttal to Smart's negative portrayal of Airy's role
        (item 808), including Smart's printed reply.

*   Stearns, Raymond Phineas. "The Course of Capt. Edmond
        Halley in the Year 1700."

        Cited herein as item 901.

811.   Stigler, Stephen M. "Laplace's Early Work: Chronology
        and Citations." *Isis*, 69 (1978), 234-254.

        Detailed examination and chronological accounting of
        Laplace's first papers presented to the French Academie
        des Sciences between 1770 and the end of 1773. Includes
        several papers on topics in celestial mechanics. Examines,
        by a citation study and general commentary, the direct
        influences upon Laplace by d'Alembert and many others.
        45 citations.

*   Stoney, G. Johnstone. "The Story of the November Meteors."

Cited herein as item 989.

*   Swenson, Loyd S. *The Ethereal Aether: A History of the Michelson-Morley-Miller Aether-Drift Experiments, 1880-1930.*

Cited herein as item 886.

812. Szanser, Adam J. "Marian Kowalski: A Little Known Pioneer in Stellar Statistics." *Quarterly Journal of the Royal Astronomical Society*, 11 (1970), 341-350.

Recalls the work of this mid-19th-century Russian/Polish astronomer, chiefly his priority as the first to produce a theory of the motion of Neptune. Examines his improved lunar and solar theory for predicting eclipses; improved theory of atmospheric refraction; and his priority in the statistical study of stellar proper motions. 17 citations.

813. Todhunter, I. *History of the Mathematical Theories of Attraction and the Figure of the Earth, from the Time of Newton to that of Laplace.* 2 volumes. London: Constable, 1873. Reprinted, New York: 1962. Pp. xviii + 476 + 508.

Comprehensive history tracing the analyses which led to the elaboration of Newtonian celestial mechanics in the 18th century. Discusses in turn the works of Newton, Huygens, Maupertius, and major figures through the 19th century. Extensive textual citations.

814. Tombaugh, C.W. "The Trans-Neptunian Planet Search." *The Solar System*, Volume 3. Chicago: University of Chicago Press, 1961. Pp. 12-30.

Recalls the methods used in his discovery of Pluto and the general Lowell Observatory project searching for the planet. Examines project searching for the existence of additional terrestrial satellites and general sky coverage of photographic patrol plates used in the Pluto search. 14 citations.

*   Turner, H.H. *Astronomical Discovery.*

Cited herein as item 887.

815. Von Kluber, H. "The Determination of Einstein's Light-Deflection in the Gravitational Field of the Sun." *Vistas in Astronomy*, 3 (1960), 47-77.

Detailed review of the determination of light-deflection
in the gravitational field of the Sun with a historical
introduction to predictions from classical theory and
Einstein's modifications to the mathematical treatment of
the general problem. Includes an outline of all known
observational efforts from 1919 through 1952. 94 citations.

*   Waters, David Watkins. "The Finding of the Longitude of
       Places, 1415-1767."

       Cited herein as item 866.

816.   Whitrow, G.J., and G.E. Morduch. "Relativistic Theories
       of Gravitation: A Comparative Analysis with Particular
       Reference to Astronomical Tests." *Vistas in Astronomy*,
       6 (1965), 1-67.

       Examines results of Lorentz-invariant theories and com-
       pares them to predictions from general relativity as tested
       by astronomical observations. Numerous citations.

817.   Wilson, Curtis A. "Perturbations and Solar Tables from
       Lacaille to Delambre: The Rapprochement of Observation
       and Theory." *Archive for History of Exact Sciences*,
       22 (1980), 53-188; 189-304.

       Comprehensive examination of how lunar and planetary
       perturbations upon the motions of the Earth were incor-
       porated into tables representing the apparent motion of
       the Sun. Provides general background information on 17th-
       and 18th-century observational astronomy necessary to an
       understanding of limitations to the development of the
       Solar Theory. Extensive partially annotated 517-item
       bibliography of primary and secondary sources.

*   Wilson, David B. "George Gabriel Stokes on Stellar
       Aberration and the Luminiferous Ether."

       Cited herein as item 891.

                           COMPUTATIONAL METHODS

*   Brown, Ernest W. "The Problem of the Moon's Motion."

       Cited herein as item 750.

*   Clemence, Gerald M. "The Motion of Mercury, 1765-1937."

       Cited herein as item 757.

*   Duncombe, Raynor L. "Personal Equation in Astronomy."

        Cited herein as item 821.

*   Forbes, Eric G. "Gauss and the Discovery of Ceres."

        Cited herein as item 765.

*   Huffer, C.M., and G.W. Collins. "Computation of Elements of Eclipsing Binary Stars by High-Speed Computing Machines."

        Cited herein as item 1073.

*   Shenynin, O.B. "Mathematical Treatment of Astronomical Observations."

        Cited herein as item 159.

*   Shenynin, O.B. "Laplace's Theory of Errors."

        Cited herein as item 807.

818.   Speiser, David. "The Distance of the Fixed Stars and the Riddle of the Sun's Radiation." *Mélanges Alexandre Koyré*. Volume 1. Histoire de la Pensée, XII. Paris: Hermann, 1964. Pp. 541-551.

       Centers upon L. Euler's research on the brightnesses of the stars compared to the solar brightness, and his derivation of formulae yielding estimates of stellar distances. Notes that Euler's estimates of about 100,000 times the Sun's distance approached present-day measurements. 10 citations.

POSITIONAL ASTRONOMY

GENERAL

819. Baillaud, Benjamin. *Histoire de l'astronomie de position.*
3 volumes. n.p., 1933. Pp. 228 + 474.

Comprehensive study of all aspects of the history of
positional astronomy including itemized review of major
photographic studies since the mid 19th century. Emphasizes the role of France in the development of photographic astrometry and reviews the major institutional
and international efforts to produce a comprehensive
photographic map of the sky. Numerous citations and chapter bibliographies.

* Campbell, William Wallace. *Stellar Motions.*

Cited herein as item 1164.

820. Chauvenet, W. *Manual of Spherical and Practical Astronomy.*
2 volumes. Philadelphia, 1863; 5th edn., 1891. Reprinted,
New York: Dover, 1960. Pp. 1340.

A standard work. Volume One contains spherical astronomy
with worked examples pertaining to problems in nautical
astronomy. Volume Two reviews the theory and use of
astronomical instruments in use during period of publication circa 1860-1890.

* Dewhirst, D.W. "Observatories and Instrument Makers in
the Eighteenth Century."

Cited herein as item 276.

821. Duncombe, Raynor L. "Personal Equation in Astronomy."
Popular Astronomy, 53 (1945), 2-13; 63-76; 110-121.

A review of the history of the personal equation in
astronomy from the late 18th century. Identifies psycho-

logical factors but emphasizes how it was treated during
the period by astronomers who wished to understand the
nature of observational errors, especially in early
meridian transit work. Includes general bibliography
citing 115 sources from 1779 to 1944.

*   Forbes, Eric G. "Dr. Bradley's Astronomical Observations."

      Cited herein as item 284.

*   Forbes, Eric G. "Tobias Mayer's Contributions to the
      Development of Lunar Theory."

      Cited herein as item 859.

822. Forbes, Eric G. *Tobias Mayer's Opera Inedita. The First
      Translation of the Lichtenberg Edition of 1755*. London:
      Macmillan, 1971. Pp. ix + 166.

      Six lectures by Mayer at Göttingen on astronomical topics
      including the problems of determining atmospheric refrac-
      tion, calculations to correct observed meridian transits
      of stars, predictions of eclipses, theory of color, a
      catalogue of zodiacal stars and a study of stellar proper
      motions. The last was utilized by William Herschel to
      derive the solar motion. Extensive annotations and intro-
      ductory material. See also: Eric G. Forbes. "Georg Chris-
      toph Lichtenberg and the *Opera Inedita* of Tobias Mayer."
      Annals of Science, 28 (1972), 31–42. Reviewed in: *Annals
      of Science*, 30 (1973), 222–224; *JHA*, 3 (1972), 219.

*   Forbes, Eric G., ed. *The Unpublished Writings of Tobias
      Mayer*.

      Cited herein as item 1282.

*   Forbes, G. *David Gill: Man and Astronomer*.

      Cited herein as item 1286.

*   Gill, David. "The Application of Photography in Astronomy."

      Cited herein as item 468.

*   Grant, Robert. *History of Physical Astronomy, from the
      Earliest Ages to the Middle of the Nineteenth Century*.

      Cited herein as item 91.

823. Herrmann, D.B. "Some Aspects of Positional Astronomy from
      Bradley to Bessel." *Vistas in Astronomy*, 20 (1976),
      183–186.

Examines Bessel's use of Bradley's long series of observations and argues "that the positional astronomy of the 18th and 19th century essentially represented the history of the effects of Bradley's outstanding work and that from among the direct successors of Bradley, it was Bessel who achieved the greatest advance." Statistically analyzes the growth of attainable accuracy through the 18th and 19th centuries compared to the number of stellar positions determined, the number of observatories involved and the number of catalogues produced. 15 citations.

*   Herrmann, D.B. "K.F. Zöllner in seinen Beziehungen zu O.W. Struve und Russland."

Cited herein as item 364.

*   Smart, William M. *Text-Book on Spherical Astronomy.*

Cited herein as item 686.

*   Tucker, R.H. "Transit Circles Today."

Cited herein as item 454.

824. Vasilevskis, S. "Meridian Astronomy." *Astronomical Society of the Pacific Leaflet No. 274* (Feb. 1952), Pp. 8.

Briefly explores positional astronomy both before and after the application of the telescope but emphasizes 18th- and 19th-century advances and present state of precision.

STELLAR POSITIONS AND MOTIONS--CATALOGUES

825. Armitage, Angus. "The Astronomical Work of Nicolas-Louis De Lacaille." *Annals of Science*, 12 (1956), 163-191.

Traces Lacaille's life and contributions to mid-18th-century astronomy, specifically his studies of the size and shape of the Earth; solar, lunar, and planetary geocentric parallaxes; star positions, atmospheric refraction; and the Sun's apparent motion. Examines the results of Lacaille's scientific expeditions to the Cape of Good Hope. Provides brief comparison of Lacaille's work to that of James Bradley. 72 citations.

826. Baldwin, Florence L. "Flamsteed's Numbers and Bayer's
      Greek Letters." *Popular Astronomy*, 20 (1912), 82-86.

      Provides tables for translating Bayer's Greek alpha-
      betical letters into appropriate Flamsteed numbers to
      facilitate identifications of bright stars found in both
      works.

827. Ball, Robert S. "The Distances of the Stars." *Royal
      Institution Library of Science: Astronomy* (item 140),
      185-190.

      Lecture dated February 11, 1881. Reviews the first
      measurements of stellar parallax and progress in parallax
      studies. Includes cases: 61 Cygni, Groombridge 1830,
      Groombridge 1618.

828. Brown, Basil. *Astronomical Atlases, Maps and Charts: An
      Historical and General Guide*. London: Search Publishing
      Co., 1932. Reprinted, 1968. Pp. 200.

      Includes both historical and contemporary star atlases,
      solar and stellar spectroscopic atlases, lunar maps and
      items of interest to navigation and atmospheric phenomena.
      Globes, planispheres and mechanical orreries are also
      listed and discussed. General coverage dates from the
      late 18th century. Numerous citations.

829. Burwell, Cora G. "The Astronomer's Most Useful Chart."
      Astronomical Society of the Pacific Leaflet No. 271
      (Nov. 1951). Pp. 8.

      Brief history and description of the *Bonner Durchmusterun*
      the premier 19th-century visual sky map created by Fried-
      rich Argelander.

830. Coronelli, Vincenzo Maria. *Libro dei Globi*. Facsimile
      edition with introduction by Helen Wallis. Chicago:
      Rand McNally, 1969. Pp. xxii + 139.

      Facsimile of 1693 volume of globe gores produced by
      Coronelli that ranged in size from a few inches to 15
      feet in diameter. Detailed review by D.J. Warner in *Isis*,
      62 (1971), 390-394.

*   Czenakal, V.L. "The Astronomical Instruments of the Seven-
      teenth and Eighteenth Centuries in the Museums of the
      U.S.S.R."

      Cited herein as item 393.

831. Delporte, E. *Délimitation scientifique des constellations*. Cambridge: Cambridge University Press, 1930. Pp. 41 + Maps.

Official discussion of the constellation boundaries as established under the auspices of the IAU between 1925 and 1928. Includes tables and maps.

832. Dyson, Frank Watson. "The Stars around the North Pole." *Royal Institution Library of Science: Astronomy* (item 140), 193-213.

Lecture delivered April 24, 1914. Reviews methods of determining stellar distances noting that at the time there "are not more than 100 or 150 stars whose distances have been measured with any degree of accuracy." Examines in particular statistical methods for determining distances to groups of stars.

833. Dyson, Frank Watson. "Advances in Astronomy." *Royal Institution Library of Science: Astronomy* (item 140), 239-242.

Lecture delivered April 29, 1921. Reviews ten years' progress in determining stellar distances noting: F. Schlesinger's photographic parallax programs at Yerkes and Allegheny; W.S. Adams' and A. Kohlschütter's spectroscopic parallax method; and methods of "dynamical parallaxes" for double star systems developed independently by H.N. Russell and E. Hertzsprung. 2 citations.

* Eddington, A.S. *Stellar Movements and the Structure of the Universe*.

Cited herein as item 1168.

834. Fricke, W. "Systems and Catalogues of Proper Motions." L.N. Mavridis, ed. *Structure and Evolution of the Galaxy*. Holland: D. Reidel, 1971. Pp. 17-33.

Reviews the principles of the measurement of stellar positions and the evolution of the system of positions and proper motions derived from the fundamental reference systems of A. Auwers in 1879 (the FK Catalogue); J. Peters in 1907 (the NFK system); A. Kopff in 1938 (the FK3 system); and the 1956 system of Fricke (the FK4). Compares instrumental series from different observatories and the characteristics of several fundamental proper motion systems in comparison to the FK4 system. Discusses the determination of the constants of Precession and the

determination of the fundamental kinematical properties
of the galaxy (Oort's Constant B). 20 citations.

835. Gefwert, Christoffer. "F.W.A. Argelander in Finland."
      Journal for the History of Astronomy, 6 (1975), 209-211.

      Brief sketch of Argelander's early work in positional
      astronomy, chiefly his technique and rationale for deter-
      mining the value and character of the solar motion, the
      production of his catalogue of fundamental positions
      (the Åbo catalogue of 1835), and work which eventually
      led to his later organization of the *Bonner Durchmusterung*
      in 1862. 16 citations.

836. Gill, D., and J.C. Kapteyn. "Cape Photographic Durch-
      musterung Volume 1." *Annals of the Cape Observatory*,
      3 (1896), ix-xxxi; 5-95.

      Includes a detailed historical discussion of the develop-
      ment of a pioneer project to photographically map the
      southern sky for accurate stellar positions and magnitudes.

837. Grant, Robert. "The Proper Motions of the Stars." *Royal
      Institution Library of Science: Astronomy* (item 140),
      203-206.

      Lecture dated May 5, 1882. Reviews early ideas about
      stars and the history of the detection and measurement
      of proper motions since Halley's first work in 1717.

*   Heinemann, K. *Verzeichnis von Sternkatalogen 1900-1962*.

      Cited herein as item 17.

838. Hendrie, M.J. "Star Atlases for the Amateur." *Journal of
      the British Astronomical Association*, 74 (1964), 20-24.

      Tabulates many atlases and charts that are generally
      available, including some of historical interest.

839. Hetherington, Norriss S. "The First Measurements of Stel-
      lar Parallax." *Annals of Science*, 28 (1972), 319-325.

      Reviews the long search for stellar parallaxes from
      Galileo through the Herschels. Examines the final successes
      of Struve, Bessel and Henderson, which were made possible
      by significant improvements in instrumentation. 30 cita-
      tions.

*   Hevelius, Johannes. *Johannes Hevelius and His Catalog of
      Stars*.

      Cited herein as item 1307.

840. Hoffleit, Dorrit. "The Quest for Stellar Parallax."
      Popular Astronomy, 57 (1949), 259-273.

      Provides a general accounting of the search for stellar
      parallax from antiquity through James Bradley; Bradley's
      discovery of aberration; the detection of solar motion;
      the role of atmospheric refraction and instrumental errors;
      the first successes of Bessel, Struve and Henderson; the
      era of the heliometer; early photographic work and Frank
      Schlesinger's contributions. Numerous citations.

841. Hoffleit, Dorrit. "H.N.R. as a Pioneer in Trigonometric
      Parallaxes." *In Memory of Henry Norris Russell* (item
      1356), 51-54.

      Reviews parallel work of Russell and Frank Schlesinger
      circa 1903-1910 as the first major attempts to measure
      the trigonometric parallaxes of stars by photography.
      Compares instrumentation used by two projects, and their
      relative accuracies, based upon a comparison with modern
      values. 9 citations.

841a. Hoskin, Michael. "Herschel's Determination of the Solar
      Apex." *Journal for the History of Astronomy*, 11 (1980),
      153-163.

      Examines why William Herschel, a good observer but with
      no special mathematical talent, successfully detected the
      motion of the Sun in space when contemporaries, including
      Tobias Mayer and J. Lalande, considered the problem in-
      soluble. Shows that Herschel's success was fortuitous.
      20 citations.

842. Hunter, A., and E.G. Martin. "Fifty Years of Trigonometrical
      Parallaxes." *Vistas in Astronomy*, 2 (1956), 1023-1030.

      Reviews the progress of photographic stellar parallax
      work based upon techniques established primarily by Frank
      Schlesinger. 10 citations.

*   Imaeda, K., and T. Kiang. "The Japanese Record of the
      Guest-Star of 1408."

      Cited herein as item 1094.

843. Ingrao, H.C., and E. Kasparian. "Photographic Star Atlases."
      Sky & Telescope, 34 (Nov. 1967), 284-287.

      Compares major atlases from the late 19th century to
      1965. 23 citations.

844. Jackson, J. "The Distances of the Stars: A Historical
      Review." *Vistas in Astronomy*, 2 (1956), 1018-1022.

      Reviews early trigonometric parallax work, identifies
      limitations of accuracy in the late 19th century, and
      progress through international cooperation in the 20th
      century. No direct citations.

*   Jaschek, Carlos. "Data Growth in Astronomy."

      Cited herein as item 170.

*   Kaye, G. Rusby. *The Astronomical Observatories of Jai
      Singh.*

      Cited herein as item 413.

845. Knobel, E.B. "The Chronology of Star Catalogues." *Memoirs
      of the Royal Astronomical Society*, 43 (1877), 1-76.

      Provides tabular listing of 530 star catalogues includ-
      ing ancient and oriental sources, with a 20-page historical
      introduction. Includes information on date of publication,
      number of stars, description of stars, coordinates utilized
      epoch, and explanatory notes. Extensive citations.

*   McCluskey, Stephen C. "The Astronomy of the Hopi Indians."

      Cited herein as item 211.

846. Stevenson, Edward Luther. *Terrestrial and Celestial Globes.*
      2 volumes. New Haven: Yale University Press, 1921. Pp.
      xxvi + 218; xi + 291 + 230 illustrations.

      General study of the nature and construction of globes
      as aids in geography and astronomy. Volume 1 treats the
      period from antiquity through the late 16th century.
      Volume 2 covers the 17th and 18th centuries including
      discussions and descriptions of all forms of globes and
      their makers. Volume 2 includes a discussion of the
      techniques of globe construction and provides a 28-page
      bibliography and a 24-page index of globes and globe
      makers. Heavily illustrated and documented. Review in
      Isis, 4 (1922), 549-553, includes short discussion of
      early historical studies of globes and provides a bibliog-
      raphy of Stevenson's writings on cartography.

847. Strand, K. Aa. "Determination of Stellar Distances."
      Science, 144 (June 1964), 1299-1309.

Concise history of the determinations of stellar distances by trigonometric parallaxes from attempts in the mid-18th century, through first successes in the mid-19th century, to present-day techniques. Centers upon program of the U.S. Naval Observatory. 19 citations.

848. Struve, Otto. "The First Stellar Parallax Determination." Herbert M. Evans, ed., *Men and Moments in the History of Science*. Seattle: University of Washington Press, 1959. Pp. 177-206.

One of nine essays presented as part of the 25th anniversary celebration of the History of Science Dinner Club. Detailed review of early unsuccessful methods of determining stellar parallaxes followed by an exposition of successful measures in the 1830s. 32 citations.

849. Turner, H.H. *The Great Star Map*. London: John Murray, 1912. Pp. vii + 159.

Based upon a series of articles in *Science Progress* between July 1910 and April 1911 reviewing the origins, development and present status of the "Astrographic Chart" or "Carte du Ciel," an international project to photograph the entire sky. Traces the development of star charts in general from the establishment of the Royal Greenwich Observatory, and discusses the various needs for accurate and comprehensive charts including completeness for the recognition of novae, comets and new planetary bodies; as an aid to astronomical navigation; and for the improvement in the theories of motion of planetary bodies. Provides, in an Appendix, an analysis of the Eros Campaign in 1900 to measure the Solar Parallax. No direct citations.

* Van Biesbroeck, G. "Visual Binary Stars and Stellar Parallaxes."

Cited herein as item 1086.

850. Warner, Deborah Jean. "Celestial Technology." *The Smithsonian Journal of History*, 2 (1967), 35-48.

Compares constellation drawings by the Abbe Nicolas Louis de Lacaille in the middle of the 18th century found on celestial globes, maps and charts, and finds variations in the forms of three themes: the telescope, the microscope and the air pump. Discusses how the designation of these new star groups in the southern sky reflected attitudes during the French Enlightenment, notably

a preoccupation with instruments in science and art. 8
citations and 13 figures.

851. Warner, Deborah Jean. "The First Modern Sky Maps Recon-
      sidered." *Archives Internationale d'Histoire des Science*
      22 (1969), 263-266.

      General review of early geocentric star maps in the
      mid-16th century. Identifies problem areas in study of
      the history of celestial cartography. 3 citations.

852. Warner, Deborah Jean. "Johannes Bayer and His Star Atlas--
      Reconsidered." *Journal of the British Astronomical
      Association*, 86 (1975), 53-55.

      Brief statement clarifying the historical importance of
      J. Bayer's *Uranometria*, its origins, and contemporary
      celestial charts. 2 citations.

853. Warner, Deborah Jean. *The Sky Explored, Celestial Cartog-
      raphy, 1500-1800.* New York: Alan Liss, 1979. Pp. xviii +
      293.

      Comprehensive catalogue of flat star maps, heavily
      illustrated with detailed annotation. The period covered
      begins with the invention of printing and ends when
      celestial cartography became a functional tool of modern
      science. Detailed attention is given to the accuracies
      of the maps. Reviewed in: *JHA*, 11 (1980), 73-75.

854. Williams, M.E.W. "Flamsteed's Alleged Measurement of Annual
      Parallax for the Pole Star." *Journal for the History
      of Astronomy*, 10 (1979), 102-116.

      Reviews Flamsteed's attempts to detect the parallax of
      Polaris starting in the late 1670s to his announcement
      of "success" in 1699. After his announcement, and the
      initial favorable reaction by several astronomers, objec-
      tions to it were raised that were based upon discrepancies
      in the predictable behavior of parallactic motion. Williams
      analyzes in detail Flamsteed's original rationale, his
      reaction to his critics, and shows that the discrepancies
      were not readily obvious, that his critics (notably
      Cassini) were not clear in their criticism, and that what
      Flamsteed had actually detected was annual aberration,
      as Bradley noted in 1728. 43 citations.

855. Woolard, Edgar W. "The Historical Development of Celestial
      Coordinate Systems." *Publications of the Astronomical
      Society of the Pacific*, 54 (1942), 77-90.

Traces origin and development of the horizon, equator, and ecliptic systems from antiquity through the time of Tycho Brahe. Examines in detail the origins of the coordinates' right ascension and declination in the equatorial system. 11 citations.

*   Woolley, Richard. "The Stars and the Structure of the Galaxy."

    Cited herein as item 1182.

ALMANACS AND EPHEMERIDES

*   Bedini, Silvio A. *The Life of Benjamin Banneker.*

    Cited herein as item 1241.

*   Brown, Lloyd A. *The Story of Maps.*

    Cited herein as item 751.

*   Clemence, G.M. "On the System of Astronomical Constants."

    Cited herein as item 871.

*   Clemence, G.M. "The Concept of Ephemeris Time: A Case of Inadvertent Plagiarism."

    Cited herein as item 676.

*   Cotter, Charles H. *A History of Nautical Astronomy.*

    Cited herein as item 758.

*   Cotter, Charles H. *Studies in Maritime History.*

    Cited herein as item 759.

856.   Cotter, Charles H. *A History of Nautical Astronomical Tables.* London: Mansell, 1978. 10 microfiche cards, approximately 1000 pages.

    Documents the history of the production of nautical tables and ephemerides in the 18th through 20th centuries. Reviewed in: *JHA*, 11 (1980), 210-212.

857.   Forbes, Eric G. "Tobias Mayer's Lunar Tables." *Annals of Science*, 22 (1966), 105-116.

Reviews the 18th-century problem of the determination
of longitude at sea and how Mayer's lunar tables provided
"the first reliable basis for the ... calculations of
lunar distance," thus making the use of the tables of the
Moon practical for navigation. Describes how at sea the
comparison of Mayer's Tables predicting the Moon's position
(i.e., celestial longitude) were compared to observed
positions; how the differences could be converted to
longitude; and how well they worked. Includes translations
of the Royal Warrant presenting prize for a successful
longitude method to Mayer's widow. 22 citations.

858.   Forbes, Eric G. "The Bicentenary of the 'Nautical Almanac'
       (1767)." *British Journal for the History of Science*,
       3 (1967), 393-394.

       Brief review of origins of the *Nautical Almanac* and its
       original purpose to facilitate calculations of longitude
       based upon Maskelyne's application of Tobias Mayer's lunar
       tables. 8 citations.

*   Forbes, Eric G. "The Life and Work of Tobias Mayer (1723-
       62)."

       Cited herein as item 1280.

859.   Forbes, Eric G. "Tobias Mayer's Contributions to the
       Development of Lunar Theory." *Journal for the History
       of Astronomy*, 1 (1970), 144-154.

       Discusses the areas in which Mayer aided the refinement
       of lunar theory including improvement in star positions,
       reduction of instrumental errors, and a simplified method
       of computing solar eclipses circa 1750-1761. See also
       item 767.

*   Forbes, Eric G. *The Birth of Scientific Navigation*.

       Cited herein as item 768.

*   Forbes, Eric G. "Die Entwicklung der Navigationswissen-
       schaft im 18. Jahrhundert."

       Cited herein as item 769.

*   Forbes, Eric G. "Tobias Mayer's Contributions to Observa-
       tional Astronomy."

       Cited herein as item 1285.

860.   Gingerich, Owen, and Barbara Welther. "Note on Flamsteed's
       Lunar Tables." *British Journal for the History of Science*
       7 (1974), 257-258.

Examines Flamsteed's numerical tabulation of the "equations of the moon's centre" that achieved greater accuracy than equant models popular in the 1670s. Argues that the simplicity of Horrox's method, as extended by Flamsteed, indicates that he used Kepler's equal areas law. 7 citations.

860a. Hetherington, Norriss S. "Almanacs and the Extent of Knowledge of the New Astronomy in Seventeenth-Century England." *Proceedings of the American Philosophical Society*, 119 (1975), 275-279.

Examines 93 editions of English almanacs during the 17th century (or about 5% of the total volume) to conclude that they "devoted little if any space to the science of astronomy." In contrast they did, of course, devote much attention to planetary positions and configurations for astrological forecasting. Identifies several almanacs that did provide knowledge of astronomical science. 18 citations. See item 862.

* Howse, Derek. *Greenwich Time and the Discovery of the Longitude*.

Cited herein as item 778.

* Marguet, F. *Histoire générale de la navigation du XVe au XXe siècle*.

Cited herein as item 792.

861. Neugebauer, Paul Victor. *Astronomische Chronologie*. 2 volumes. Berlin: Walter de Gruyter, 1929. Pp. xii + 190 + 136 (tables).

Provides general background to the calculation of astronomical phenomena, including planetary and lunar positions, eclipses and periodic comets. Of use to scholars interested in the classical-medieval periods. Reviewed in: *Isis*, 14 (1930), 450-454.

862. Nicolson, Marjorie H. "English Almanacs and the 'New Astronomy.'" *Annals of Science*, 4 (1939), 1-33.

Presents synthesis of views expressed in some 800 popular almanacs published during the years 1600 and 1710 on the breakdown of astrology and its assimilation of modern views based upon the heliocentric hypothesis. Shows that before 1640 Copernicanism was rarely discussed and only came into its own after 1660. Shows also that Copernicanism was of little interest until Galileo's observations with the telescope. 49 citations.

863. O'Neil, W.M. *Time and the Calendars*. Manchester: University Press, 1976. Pp. x + 138.

Provides astronomical background for understanding the many calendars of history from the Babylonian, Egyptian and Chinese, to the Mayan. Based upon secondary sources. Reviewed in: *Isis*, 68 (1977), 632-633; *JHA*, 9 (1978), 221-222.

*   Sadler, D.H. "The Bicentenary of the Nautical Almanac."

Cited herein as item 386.

864. Stahlman, William D., and Owen Gingerich. *Solar and Planetary Longitudes for Years -2500 to +2000*. Madison: University of Wisconsin Press, 1963. Pp. xxix + 566.

Provides geocentric longitudes of the Sun and five visible planets at ten-day intervals with positions to integer degrees. Reviewed in: *Isis*, 55 (1964), 221-222.

865. Taylor, E.G.R. *The Haven-Finding Art. A History of Navigation from Odysseus to Captain Cook*. New York: Abelard-Schuman, 1957. Pp. xii + 295.

General account to the invention and application of the chronometer. Reviews early navigation without compass or charts; the introduction of these devices; improvements leading to reliable methods. Reviewed in: *Isis*, 49 (1958), 352-353.

865a. Tuckerman, Bryant. *Planetary, Lunar and Solar Positions*. 2 volumes. Philadelphia: American Philosophical Society, 1962-1964 (*Memoirs of the Am. Phil. Soc.*, 56; 59).

Covers period 601 B.C. to 1649 A.D. providing positions of planets, Sun and Moon at five- and ten-day intervals. Includes extensive bibliography.

866. Waters, David Watkins. "The Finding of the Longitude of Places, 1415-1767." *XIIe Congrès International d'Histoire des Sciences, Tome XA (Paris, 1968)*. Paris: Blanchard, 1971. Pp. 115-124.

General survey examines the method of lunar distances, first described by Werner in 1514. This parallactic method is developed through Maskelyne's 1763 publication "The British Mariner's Guide" based upon Tobias Mayer's *Lunar*

Tables. The determination of longitude using the parallactic displacements of Jovian satellites is also reviewed
along with use of marine chronometers from 1530 which
were finally deemed reliable with Harrison's chronometers
circa 1737-1759 and their tests in the 1760s.

ASTRONOMICAL CONSTANTS

867. Armitage, Angus. "Chappe D'Auteroche: A Pathfinder for
Astronomy." *Annals of Science*, 10 (1954), 277-293.

Reviews this astronomer's two expeditions for observations of the two transits of Venus in 1761 and 1769 from
Tobolsk in Siberia and from California, where he met his
untimely death. 46 citations.

* Badger, G.M., ed. *Captain Cook, Navigator and Scientist.*

Cited herein as item 1236.

* Ball, Robert S. *The Story of the Sun.*

Cited herein as item 997.

868. Blackwell, D.E. "The Discovery of Stellar Aberration."
Quarterly Journal of the Royal Astronomical Society,
4 (1963), 44-46.

Brief review of the circumstances leading to James
Bradley's 1729 announcement of his discovery of the
aberration of starlight due to the Earth's motion. Reproduces a fragment from Bradley's observing record
confirming the shift in position of a star.

869. Boyer, Carl B. "Early Estimates of the Velocity of Light."
Isis, 33 (1941), 24-40.

After brief commentary on pre-Roemer attempts, primarily
Galileo's experiments, Boyer concentrates upon the work
of Roemer, James Bradley and others. Reviews the great
range of values attributed to Roemer in 19th- and 20th-
century review texts and the source of the confusion. One
point of confusion was that Roemer himself did not provide
an actual value other than the fact that light required
22 minutes to traverse the Earth's orbit. Further confusion
arose over the fact that the size of the Earth's orbit
was not precisely known. Notes that Roemer's work was not

fully accepted by Cassini or other contemporaries. Bradley's work, however, "furnished an unexpected confirmation of Roemer's hypothesis," but was also dependent upon the poorly determined solar distance, and hence the general confusion over the actual value of the velocity of light continued well into the 19th century. 163 citations.

*   Brush, Stephen G. "Nineteenth-Century Debates about the Inside of the Earth: Solid, Liquid or Gas."

        Cited herein as item 607.

870.   Christie, W.H.M. "Universal Time." *Royal Institution Library of Science: Astronomy* (item 140), 252-259.

        Lecture dated March 19, 1886. Examines the social and economic factors in the adoption of a universal system of time reckoning noting various recommendations for its establishment including those of the American Metrological Society. Argues that the idea of the use of hourly meridians "can only be considered a provisional arrangement." Notes the influence of the telegraph and the question of the adoption of a prime meridian at the Washington Conference in October 1884.

871.   Clemence, G.M. "On the System of Astronomical Constants." *Astronomical Journal*, 53 (1948), 169-179.

        Describes the fundamental system of astronomical constants with the use of Ephemeris Time and an improved lunar ephemeris. 23 citations.

872.   Clemence, G.M. "Inertial Frames of Reference." *Quarterly Journal of the Royal Astronomical Society*, 7 (1966), 10-21.

        Within a contemporary discussion, examines the historical changes in the nature of inertial systems employed in astronomy starting with the stellar system and ending with the system of external galaxies. Provides commentary on the determination of fundamental constants, noting that for precession, "The observations of Bradley, embodied in the catalogue of 1755, mark so great an epoch in systematic accuracy, that it is not only useless, but detrimental, to consider any earlier observations." Examines Bessel's use of Bradley's data, Otto Struve's improved values circa 1825-1840, and later work by Nyren, Dreyer, Newcomb and J.H. Oort. 21 citations.

* Clemence, G.M. "The Concept of Ephemeris Time: A Case
of Inadvertent Plagiarism."

Cited herein as item 676.

873. Cohen, I. Bernard. "Roemer and the First Determination
of the Velocity of Light (1676)." *Isis*, 31 (1939), 327–
379.

Extensive review examining early belief in finite
velocity of light; Roemer's early years at the Paris Observatory and immediate background to his measurements;
the determination itself and its reception; Roemer's later
work in Denmark. Includes detailed bibliography, and a
facsimile with translation of his 1676 announcement. 119
citations. Reprinted by the Burndy Library, New York,
1942.

* Cortie, A.L. *Father Perry, The Jesuit Astronomer.*

Cited herein as item 1262.

874. DeBray, M.E.J. Gheury. "The Velocity of Light. History of
Its Determination from 1849 to 1933." *Isis*, 25 (1936),
437–448.

Reviews state of knowledge of velocity of light noting
the many clerical errors and misquotations of its value
in recent history. Provides a detailed listing of all
known velocity determinations during the period, compiled
from original sources beginning with Fizeau, Foucault
and Cornu, then Michelson and Newcomb, Perrotin, and
finally Pease and Pearson at Mt. Wilson in 1932. Examines
the question of the possible decrease in the velocity of
light claiming that, at "this time, the amplitude of the
variation is distinctly greater than the probable errors."
After a general analysis, he concludes that "the constancy
of the velocity of light has proved to be but a working
hypothesis which must now make way for a new theory more
in agreement with the facts." Extensive citations.

875. Dingle, Herbert. "A Re-examination of the Michelson-Morley
Experiment." *New Aspects in the History and Philosophy
of Astronomy* (item 123), 97–100.

Argues that relativistic explanations of the results
of the experiment are circular and calls for a reconsidera-
tion of the experimental results within a framework "in
which its truth is not presupposed." 6 citations.

*   Fernie, Donald. *The Whisper and the Vision.*

       Cited herein as item 282.

876.   Forbes, G. *The Transit of Venus.* London: Macmillan, 1874.
       Pp. 99.

       Descriptive review of techniques in preparation by many
       expeditions to the forthcoming 1874 transit.

*   Forbes, G. *David Gill: Man and Astronomer.*

       Cited herein as item 1286.

*   Goldstein, Bernard R. "Some Medieval Reports of Venus
       and Mercury Transits."

       Cited herein as item 649.

876a. Goldstein, S.J. "On the Secular Change in the Period of
       Io, 1668-1926." *The Astronomical Journal*, 80 (1975),
       532-539.

       Detailed technical study based upon 50 eclipse observa-
       tions of Io from 1668 to 1678 by Picard and Roemer, and
       156 observations during the modern period, 1908-1926. Ex-
       tensive citations and analysis of earlier work. Concludes
       that secular changes are extremely small, and that "Io
       is an elegant clock." 34 citations.

876b. Goldstein, S.J., J.D. Trasco, and T.J. Ogburn, III. "On
       the Velocity of Light Three Centuries Ago." *The Astro-
       nomical Journal*, 78 (1973), 122-125.

       Modern analysis of Roemer's determination of the velocity
       of light. Utilizes Roemer's and Picard's observations of
       the eclipses of Io by Jupiter's shadow from 1668 to 1678
       but bases evaluation upon modern orbital data for the
       Earth and Jupiter. Concludes that Roemer's derivation
       was flawed only by an inaccurate value for the astronomical
       unit, and that the velocity of light obtained with the
       modern value of the astronomical unit does not differ by
       more than 0.5% from presently accepted value. 6 citations.

877.   Hogg, H.S. "Out of Old Books." *Journal of the Royal Astro-
       nomical Society of Canada*, 42 (1948), 153-159; 189-193;
       45 (1951), 37-44; 89-92; 127-134; 173-178.

       Reprints, with commentary, passages from the *Journal*
       of William Wales en route to the 1769 transit of Venus
       observation post on Hudson's Bay. Later sections treat

LeGentil's long expedition to the Indian Ocean to observe both transits.

878. McCrea, W.H. "The Significance of the Discovery of Aberration." *Quarterly Journal of the Royal Astronomical Society*, 4 (1963), 41-43.

Argues that Bradley's discovery and measurement of the aberration of starlight caused by the motion of the Earth aided: (1) the velocity of light as a constant independent of reflections, source, distance of source, direction of source; (2) the belief that the Earth was indeed in motion about the Sun; (3) the recognition of a relationship between optical and mechanical (the motion of the Earth) phenomena.

878a. Meeus, J. "The Transits of Venus 3000 B.C. to A.D. 3000." *Journal of the British Astronomical Association*, 68 (1958), 98-108.

Lists and discusses all calculated transits, with some historical notes. 7 citations.

879. Newcomb, Simon. "On Hell's Alleged Falsification of His Observations of the Transit of Venus in 1769." *Monthly Notices of the Royal Astronomical Society*, 43 (1883), 371-381.

Demonstrates that Father Hell did not tamper with his observational data made during the 1769 transit observed at Wardhus, in opposition to Littrow's accusations. Direct citation to Father Hell's *Journal*. See item 885.

880. Newcomb, Simon. "Discussion of Observations of the Transits of Venus in 1761 and 1769." *Astronomical Papers of the American Ephemeris and Nautical Almanac*, 2, Pt. 5 (1890-1891), 259-405.

Exhaustive reexamination and evaluation of historical data in light of refined longitude values of original stations.

* Newcomb, Simon. *Popular Astronomy.*

Cited herein as item 1402.

881. Newcomb, Simon. "The Elements of the Four Inner Planets and the Fundamental Constants of Astronomy." *Supplement to the American Ephemeris and Nautical Almanac for 1897*. Washington, D.C.: U.S. Government Printing Office, 1895. Pp. ix + 202.

Reviews the derivation of lunar inequalities; precession
and nutation; figure of the Sun; masses of the planets;
and the solar parallax. Discussion of fundamental constants
repeated in *Astronomical Journal*, 14 (1895), 185–189.
Reviewed in: *MNRAS*, 56 (1896), 267–272; *Observatory*, 18
(1895), 202–205.

* Newcomb, Simon. *Reminiscences of an Astronomer.*

   Cited herein as item 1346.

* Nielsen, Axel V. "Ole Romer and His Meridian Circle."

   Cited herein as item 448.

* Olmsted, John W. "The Scientific Expedition of Jean Richer
   to Cayenne (1672–1673)."

   Cited herein as item 899.

882. Perry, Stephen Joseph. "The Transit of Venus." *Royal
   Institution Library of Science: Astronomy* (item 140),
   153–159.

   Lecture dated February 25, 1876. Traces history of transit
   observations of Mercury and Venus from 1631, centering
   on those of Venus in 1761 and 1769, and ends with a description of expeditions to observe the Venus transit
   of 1874.

883. Proctor, Richard A. *Transits of Venus: A Popular Account
   of Past and Coming Transits.* New York: R. Worthington,
   1875. Pp. xiv + 236.

   Reviews the results of the transits of 1761 and 1769,
   and Encke's determination that the true value of the Solar
   Parallax is far different than that derived from them.
   This placed the importance of the coming transits of
   1874 and 1882 in proper historical context. Reviewed by
   S. Newcomb in: *Nation*, 20 (April 1, 1875), 230.

884. Sarton, George. "Discovery of the Aberration of Light."
   Isis, 16 (1931), 233–265.

   Includes facsimile reproduction of James Bradley's
   letter to Halley announcing his discovery. Examines
   Bradley's early life and training, and general attempts
   by Molyneaux and others during the period to detect
   stellar parallaxes. Concludes that Bradley's success was
   due more "to his thoroughness and persistence" in contrast
   to Hooke and others. Searches for the earliest use of the

term "aberration" and provides bibliographical information
on Bradley and a name analysis of Bradley's memoir in
1729: *Philosophical Transactions*, Volume 35, pp. 637-661.
See item 888.

885. Sarton, George. "Vindication of Father Hell." *Isis*, 35
      (1944), 97-105.

      Briefly reviews the life and work of the 18th-century
      Jesuit Maximilian Hell who was interested in the use of
      magnetism as a curative power but was mainly known for
      his astronomical work, especially his observations of the
      1769 transit from within the Arctic Circle. Hell's ob-
      servations were discredited by his successors, until Simon
      Newcomb in 1883 showed that Hell's observations were
      genuine and not fake. Includes a bibliography of Hell's
      work, and secondary sources about him. 16 citations plus
      bibliography. See item 879.

886. Swenson, Loyd S. *The Ethereal Aether: A History of the
      Michelson-Morley-Miller Aether-Drift Experiments, 1880-
      1930*. Austin: U. of Texas Press, 1972. Pp. xxii + 361.

      Examines the progression of experiments by A.A. Michel-
      son, et al. from 1880 until 1930 to detect the motion of
      the Earth through a fixed aether. Shows how experiments
      were done at various stages of development of Michelson's
      interferometer, the various motivations to repeat the
      experiment (especially after the rise of relativity theory)
      and how Michelson, his colleagues, and other scientists
      reacted to their continued "null" results. Abstract paper
      published in *JHA*, 1 (1970), 56-78. Reviewed in: *JHA*, 4
      (1973), 196-199; *Isis*, 64 (1973), 431-432.

*   Tatham, W.G., and K.A. Harwood. "Astronomers and Other
      Scientists on St. Helena."

      Cited herein as item 338.

*   Turner, H.H. *The Great Star Map*.

      Cited herein as item 849.

887. Turner, H.H. *Astronomical Discovery* [1904]. Berkeley:
      University of California Press, 1963. Pp. xiii + 225.

      Collection of episodes including the discovery of Neptune;
      the determination of the Astronomical Unit; James Bradley's
      discovery of aberration and nutation; Schwabe's discovery
      of the sunspot period; and the accidental discoveries of

novae, comets and asteroids. Original edition published
in 1904. No direct citations.

888. Webb, R.S. "Aberration of Light." *Isis*, 18 (1933), 438.

Letter commenting upon Sarton paper in *Isis*, 16 (1931),
235-236, and identification of first use of term "aberration." See item 884.

889. Whatton, A.B., ed. *The Transit of Venus Across the Sun*.
London: W. Macintosh, 1859. Pp. xvi + 216.

Translation of Jeremiah Horrox's *Memoir* of the 1631
transit of Venus, with a biographical study of Horrox.

890. Wilkins, G.A. "The System of Astronomical Constants, Part
i." *Quarterly Journal of the Royal Astronomical Society*,
5 (1964), 23-31.

As part of a general contemporary discussion, provides
a brief history of international cooperation in the adoption of fundamental astronomical constants: aberration,
nutation, solar parallax, starting in 1896. 12 citations.

891. Wilson, David B. "George Gabriel Stokes on Stellar Aberration and the Luminiferous Ether." *British Journal for
the History of Science*, 6 (1972-1973), 57-72.

Studies the creation of Stokes's theory of the ether,
developed in the 1840s to explain observed phenomena such
as stellar aberration. Stokes's model embodied the properties of both fluids and solids to account for Fresnel's
demonstration of the wave-nature of light. Shows how
Stokes's "theory of aberration was a response and a chal-
lenge to Fresnel's theory" through his application of
hydro-dynamics. 65 citations.

892. Woolf, Harry. "Eighteenth-Century Observations of the
Transits of Venus." *Annals of Science*, 9 (1953), 176-
190.

Presents material later contained within his text on
the transits of Venus (item 893).

893. Woolf, Harry. *The Transits of Venus: A Study of Eighteenth
Century Science*. Princeton: Princeton University Press,
1959. Pp. xiii + 258.

Comprehensive review and analysis including an introductory chapter on the role of transits in the progress
of astronomy. Examines in detail the French and British

expeditions for the 1761 transit; the improved observations and analysis of the far more successful 1769 transit; and appendices on planetary phenomena, Bode's Law, the solar parallax, etc. Extensive citations and comprehensive bibliography. Reviewed in: *Isis*, 52 (1961), 607-608; *Centaurus*, 9 (1963-1964), 60-61.

GEODESY

*   Chapman, Sydney. "The Earth."

Cited herein as item 612.

894.   Cook, A.H. "The Contribution of Observations of Satellites to the Determination of the Earth's Gravitational Potential." *Space Science Review*, 2 (1963), 355-437.

Provides a brief historical introduction to studies of the Earth's figure and gravitational potential beginning with Newton and Cassini. Reviews the principles by which the potential may be determined by modern techniques. No historical citations.

895.   Killick, Victor W. "California's Early Astro-Geodetic Observatories." *The Griffith Observer*, 26 (1962), 18-23.

Follows George Davidson's establishment of geodetic facilities around the state for the determination of political boundaries in the mid-19th century. 9 citations.

896.   King-Hele, D.G. "The Shape of the Earth." *Journal of the Institute of Navigation*, 17 (Jan. 1964), 1-16.

Elementary review of the history of theory and determination of the Earth's shape, and the role of this work in the reception of Newtonian mechanics over Cartesian mechanics. Describes improvements in observational data provided by satellite measurements. 25 citations.

897.   Müller, Franz J. *Studien zur Geschichte der theoretischen Geodäsie*. Augsburg: 1918. Pp. viii + 203.

Contains collected essays written between 1906 and 1909 and constitutes an important history of geodesy beginning with the origins of the science at the time of Newton. Chapters arranged in somewhat chronological order whereby major figures are treated in turn including abstracts from

their writings with annotations. Major figures after Newton
and Huygens include Johann Bernoulli, Clairaut, Euler,
Delambre, Legendre, Lagrange, and then Bohnenberger,
Soldner, Oriani, Gauss and Bessel. Müller contrasts French
and German geodesy of the late 18th century, and then continues through the end of 19th century discussing the
work of K.G.J. Jacobi, P.A. Hansen and many others. Includes
detailed bibliographical citations and major new textbooks
and geodetic encyclopaedias. Reviewed in: *Isis*, 3 (1920),
438-439.

898. Munk, W.H. "Polar Wandering: A Marathon of Errors." *Nature*,
      117, No. 4508 (24 March 1956), 551-554.

      Reviews early theories of G.H. Darwin and Lord Kelvin,
      and later studies, in light of present knowledge. 15 cita-
      tions.

899. Olmsted, John W. "The Scientific Expedition of Jean Richer
      to Cayenne (1672-1673)." *Isis*, 34 (1942-1943), 117-128.

      Examines the importance of this episode to understanding
      late 17th-century astronomy as well as the institutionalism
      of science of the period. The expedition was organized to
      make observations of the variation of the period of a
      pendulum with latitude as a measure of the Earth's figure,
      and observations to increase knowledge of the solar paral-
      lax. Argues that the significance of this expedition was
      that it apparently was the first organized to examine
      specific well-defined problems, rather than to simply
      collect new, unrelated data. Notes contemporary expedi-
      tion of Picard to Uraniborg. 112 citations.

900. Payne, W.W. "Attraction and Figure of the Earth." *Popular
      Astronomy*, 9 (1901), 7-13; 117-123.

      Reviews the history of the study of the Earth's shape
      through the mid-18th century. 11 bibliographical citations.

901. Stearns, Raymond Phineas. "The Course of Capt. Edmond
      Halley in the Year 1700." *Annals of Science*, 1 (1936),
      294-301.

      Recounts Halley's voyage in 1699-1700 for the determina-
      tion of longitude, and for the establishment of the varia-
      tion in the magnetic compass. 31 citations.

*   Todhunter, I. *History of the Mathematical Theories of
      Attraction and the Figure of the Earth, from the Time
      of Newton to that of Laplace*.

      Cited herein as item 813.

902. Turner, A.J. "Hooke's Theory of the Earth's Axial Displacement: Some Contemporary Opinion." *British Journal for
the History of Science*, 7 (1974), 166-170.

Examines the reactions to Hooke's hypothesis of a non-
spherical Earth. 15 citations.

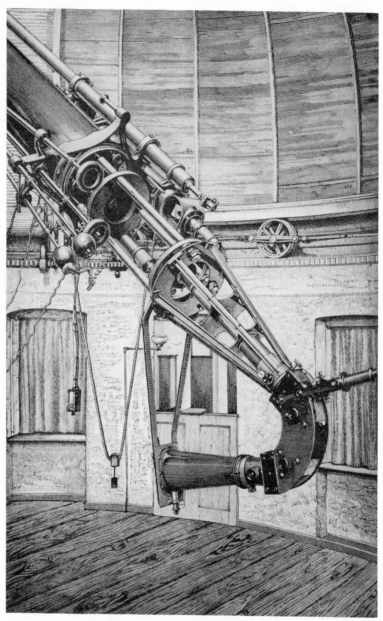

Illustration of the large Potsdam spectrograph attached to an 11-inch refractor. *(From J. Scheiner,* Astronomical Spectroscopy *(Item 930), p. 98, f. 42.)*

ASTROPHYSICS

GENERAL

903. Aller, Lawrence H. *Astrophysics: The Atmospheres of the Sun and Stars*. New York: Ronald, 1953; 1963. Pp. xi + 650.

General review of observational and theoretical studies of stellar atmospheres. Includes bibliographical annotation to historically significant primary works.

904. Aller, Lawrence H. *Astrophysics: Nuclear Transformations, Stellar Interiors, and Nebulae*. New York: Ronald, 1954. Pp. x + 291.

General review of recent developments and state of knowledge of the origin of the elements, nucleosynthesis, stellar structure and variability, gaseous nebulae and the interstellar medium. Includes annotated bibliographies at the ends of chapters noting historically significant primary sources.

905. Ambartsumian, V.A. "On Some Trends in the Development of Astrophysics." *Annual Review of Astronomy and Astrophysics*, 18 (1980), 1-13.

Short review of three areas of personal research in theoretical astrophysics by a major contemporary figure: 1. Principles of invariance applied to the theory of radiative transfer. 2. The significance of inverse problems in mathematics and quantum physics and their role in astrophysics. 3. The empirical approach to an understanding of evolutionary processes in the Universe, specifically in stars and stellar systems. Briefly examines relative merits of observation and theory. He concludes: "Nature keeps still many of its secrets. Our aim is to disclose them. It is natural to try this by observing the places where they are hidden. We can hardly reach our aim by only theorizing." No direct citations.

906. Baade, Walter. *Evolution of Stars and Galaxies*. C. Payne-
      Gaposchkin, ed. Cambridge, Mass.: Harvard University
      Press, 1963. Pp. xiii + 321.

      Based upon a series of lectures given at Harvard in
      1958 concentrating on a descriptive narrative of topics
      in stellar, galactic, and extragalactic astronomy. In-
      cludes a historical review of progress resulting from
      the discovery of the Hertzsprung-Russell Diagram; Shapley's
      determination of the scale of the Milky Way from the ob-
      served distribution of star clusters; Hubble's measure-
      ment of distances to galaxies; and Baade's many discoveries
      including the existence of stellar populations and the
      revision of the extra-galactic distance scale through a
      recalibration of the Cepheid Period-Luminosity relation-
      ship. Provides many personal anecdotes on his research.
      No direct citations.

907. Babcock, H.W. "The Zeeman Effect in Astrophysics." *Physica*,
      33 (1967), 102-121.

      Describes instrumentation and techniques for analyzing
      spectra of the Sun and stars for the detection of magnetic
      fields. 37 citations.

*   Berendzen, R., ed. *Education in and History of Modern
      Astronomy*.

      Cited herein as item 125.

908. Breene, R.G. *The Shift and Shape of Spectral Lines*.
      Oxford: Pergamon, 1961. Pp. xii + 323.

      Reviews early (mid-19th through early 20th centuries)
      research and discovery of the physical conditions in-
      fluencing the nature of spectral lines. 241 citations.

*   Campbell, William Wallace. *Stellar Motions*.

      Cited herein as item 1164.

909. Clerke, Agnes M. *The System of the Stars*. London: Long-
      mans Green and Co., 1890. Pp. xix + 424; 2nd Edition.
      London: Adam and Charles Black, 1905. Pp. xvi + 403.

      General exposition of late 19th-century stellar astronomy
      and research on stellar motions, the structure of the
      Milky Way, the nature of nebulae, and the structure of
      star clusters. Detailed discussions of stellar evolution
      based upon observed stellar characteristics and gravita-
      tional contraction. The second edition, though similar

in format, is extensively revised in light of the 15-year difference in time. Extensive citations.

910. Clerke, Agnes M. *Problems in Astrophysics*. London: Adam and Charles Black, 1903. Pp. xvi + 567.

Historical critical survey of recent astrophysics. About one-third of the text treats solar physics including: the nature of the "Reversing Layer" or the region where the Fraunhofer lines were believed to arise; the Corona; the Solar Cycle and Solar rotation. Clerke attempts to identify what is not known and what problems exist in the 1903 picture of the Sun. Part II, on "Sidereal Physics," includes problems in the analysis of stellar spectra and their interpretation for stellar evolution, especially the place of nebulae in evolution. Here Clerke suggests her own system of spectral classification and attempts estimates of the masses and surface brightnesses of stars. Chapters are also provided on variable stars, binaries and the Milky Way. Extensive citations. Detailed essay review and commentary by W.W. Campbell: *ApJ*, 18 (1903), 156-166.

911. Cowling, T.G. "Magnetic Fields in Astronomy." *Quarterly Journal of the Royal Astronomical Society*, 12 (1971), 348-351.

Reviews progress in detecting the solar magnetic field, from Hale in 1908 in his sunspot studies to the work of H.W. and H.D. Babcock in 1952 on the general field. Notes discoveries of stellar magnetic fields, magnetic fields in interstellar space, and planetary magnetism. Provides commentary on his own pioneering theoretical studies in the period 1930s-1950s on magnetohydrodynamics. See item 996.

912. DeKosky, Robert K. "Spectroscopy and the Elements in the Late Nineteenth Century: The Work of Sir William Crookes." *British Journal for the History of Science*, 6 (1972-73), 400-423.

Provides commentary on problems facing pioneer astronomical spectroscopists and the relationship of those problems to laboratory and theoretical studies. 85 citations.

913. Dingle, Herbert. *Modern Astrophysics*. New York: Macmillan, 1924; 1926. Pp. xxviii + 420.

Reviews the present state of general astrophysics circa 1923 in descriptive and comprehensive manner. Examines

in detail the techniques of spectrum analysis; characteristics of stars and the Sun as a star; stellar systems
and cosmology. Notes in Preface the omission of Eddington's
1924 paper on the mass-luminosity relationship and adds:
"The possible implications of this paper are so revolutionary in character that it is not possible to take
account of them without considerably modifying the whole
ground-work on which the book is constructed. It may be
that the book will mark an epoch in astrophysical theory—
characterised by the ascendancy of the giant and dwarf
theory of stellar evolution." Dingle's words have proven
to be correct. No direct citations.

914. Dingle, Herbert. *A Hundred Years of Spectroscopy being
the Fifty Third Robert Boyle Lecture.* Oxford: Blackwell,
1952. Pp. 23.

Surveys development of spectroscopy from Newton through
the 20th century. Includes personal reminiscences of work
with Alfred Fowler and others.

915. Dingle, Herbert. "A Hundred Years of Spectroscopy."
British Journal for the History of Science, 1 (1963),
199-216.

Based upon his 1951 Boyle Lecture and read at the
November 1962 meeting of the Society. Reviews advances
in theories of matter through spectroscopic investigation.
Provides background for appreciating astronomical applications but does not deal explicitly with astronomical
spectroscopy. Includes studies by Lockyer and Alfred
Fowler. No direct citations.

*   Douglas, A. Vibert. *The Life of Arthur Stanley Eddington.*

Cited herein as item 1269.

*   Eddington, A.S. *The Internal Constitution of the Stars.*

Cited herein as item 1140.

916. Emden, Robert. *Gaskugeln: Anwendungen der mechanischen
Warmetheorie.* Leipzig: B.G. Teubner, 1907. Pp. 498.

Comprehensive summary of convection theory and the
configurations of a gas sphere under the influence of
gravity and gas pressure. At the time, theoretical studies
of stellar structure were based solely upon convection
theory, and this technical work reviews the state of
knowledge as of 1907 together with chapters and numerical

tables designed to facilitate further research in stellar
structure, planetary atmospheres and related problems.
Written at a time when many theoreticians were not convinced that stars and the Sun behaved as perfect gasses.
See J.H. Jeans' review (item 1142) and A.S. Eddington's
commentary (item 1275) upon Emden's death in 1941.

* Forbes, Eric Gray. "A History of the Solar Red Shift
Problem."

Cited herein as item 1003.

917. Fowler, Alfred. *Report on Series in Line Spectra*. London:
Fleetway Press, 1922. Pp. 182.

Accounts for the recent development and present state
of knowledge of regularities in spectra. Includes a
historical chapter on the origins of the recognition of
series in spectra from the 1860s through 1900 including
the laboratory work of Liveing and Dewar, Kayser and
Runge; the telescopic discovery of the ultra-violet lines
in the hydrogen series by Huggins; Balmer's initial recognition of the series in 1885; and Huggins' and Vogel's
later discoveries of additional lines predicted by the
Balmer series. Later chapters review the work of Hicks,
the general development of series formulae, and the
culminating work of Bohr. Extensive citations.

918. Graff, Kasimir. *Grundriss der Astrophysik*. Leipzig: Teubner,
1928. Pp. viii + 751.

A general descriptive survey of astrophysics with detailed reference to the work of late 19th- and early
20th-century German workers. Few direct citations.

* Hale, George Ellery. *The Study of Stellar Evolution*.

Cited herein as item 1141.

* Herrmann, D.B. *Geschichte der Astronomie von Herschel bis
Hertzsprung*.

Cited herein as item 94.

* Huggins, William, ed. *The Scientific Papers of Sir
William Huggins*.

Cited herein as item 1316.

919. Hunt, T. Sterry. "Celestial Chemistry from the Time of
Newton." *American Journal of Science*, 23 (1881), 123-133.

Attempts an argument that Newton's concepts of matter and light as expressed in his "Queries" in the third book of the *Optics* "anticipated most of the recent speculations and conclusions regarding cosmic chemistry." Hunt is referring to late 19th-century speculation on the existence of "Celestial Dissociation" by N. Lockyer, B. Brodie, and Hunt. At the base of this theory of dissociation was the idea that forms of matter that exist in stars are more fundamental than any known on Earth. See also: W.H. Brock, "Lockyer and the Chemists: The First Dissociation Hypothesis." *Ambix*, 16 (1969), 81-99.

920. Hynek, J.A., ed. *Astrophysics: A Topical Symposium.* New York: McGraw-Hill, 1951. Pp. xii + 703.

Technical review of the progress of astrophysics during the period 1900-1950, in honor of the first half century of the Yerkes Observatory. The 14 chapters by leading astronomers cover virtually all topics except galactic structure and cosmology. Chapter topics include: overview of 50 years of progress in astrophysics; classification of spectra; the analysis of normal and peculiar spectra; molecular spectra; solar radiation; comets; origin of the solar system; all types of binary and variable stars; methods of stellar distance determinations; interstellar medium; and stellar structure, composition and energy. Contains items 935, 937, 970, 1025, 1048, 1065, 1075, 1079, 1086, 1100, 1107, 1114, 1122, 1136, 1154.

921. Jeans, J.H. *Astronomy and Cosmogony.* Cambridge: Cambridge University Press, 1928. Reprint of second revised edition (1929), New York: Dover, 1961. Pp. xiv + 428.

Comprehensive study, both technical and descriptive, reviewing Jeans' views on stellar structure, evolution and energy, the evolution of binary stars and the dynamical properties of nebulae. Includes detailed arguments on his views of the origin of the solar system by tidal interaction, matter annihilation as a stellar energy source, and the importance of ionization in stellar interiors. Technical reviews appear on the equilibrium of rotating fluids, and on regions of stable and unstable configurations in the Russell (now Hertzsprung-Russell) Diagram. In the light of recent work by Edwin Hubble, Jeans' treatment of spiral nebulae here contrasts greatly with his earlier discussions. Numerous footnote references.

922. Kayser, H. *Handbuch der Spektroscopie.* 5 volumes. Leipzig: S. Hirzel, 1901.

Comprehensive and standard German work. The fifth volume is devoted to astronomical spectroscopy presented in a narrative style with discussions topically arranged. The first volume includes a general history of spectroscopy.

923. Keeler, James E. "The Importance of Astrophysical Research and the Relation of Astrophysics to Other Physical Sciences." *Astrophysical Journal*, 6 (1897), 271-288.

Presented at the dedication of the Yerkes Observatory, October 21, 1897. Explores the relationships of astrophysics to physics and chemistry, as well as to astronomy and to the solution of astronomical problems. Reviews diverse reactions to the emerging discipline noting: "There may be some who view with disfavor the array of chemical, physical, and electrical appliances crowded around the modern telescope, and who look back to the observatory of the past as to a classic temple whose severe beauty had not yet been marred by modern trappings." Concludes with specific examples of the importance of astrophysics, specifically the discoveries made with "its chief instrument, the spectroscope," and its state of perfection afforded by advances in physical optics. Other studies of note include the discovery of helium in solar and terrestrial spectra, the behavior of spectra of substances under different physical conditions, the search for astronomical evidence of magnetic phenomena, the invention of the spectroheliograph, the detailed study of the solar spectrum, and the discovery of spectroscopic binary stars.

924. King, Jean I.F., ed. "Space and Atmosphere Physics." *Journal of Quantitative Spectroscopy and Radiative Transfer*, 8 (Jan. 1968), 1-16.

Proceedings of a 1966 symposium with reprinted texts of papers. Includes an historical account of the theory of radiative transfer by J.I.F. King beginning with the pioneer studies of A. Schuster and K. Schwarzschild at the turn of the century. No historical citations.

* Lockyer, J. Norman. *Contributions to Solar Physics*.

Cited herein as item 1014.

* Lockyer, J. Norman. *The Chemistry of the Sun*.

Cited herein as item 1049.

*   Lockyer, T. Mary, and Winifred L. Lockyer, eds. *Life and
        Work of Sir Norman Lockyer*.

        Cited herein as item 1332.

*   Maunder, E.W. *Sir William Huggins and Spectroscopic
        Astronomy*.

        Cited herein as item 1115.

*   Mayall, Margaret W., ed. *Centennial Symposia*.

        Cited herein as item 141.

925.   McGucken, William. *Nineteenth Century Spectroscopy.
        Development of the Understanding of Spectra 1802-1897*.
        Baltimore: Johns Hopkins Press, 1969. Pp. xii + 233.

        Heavily documented interpretive history of physical
        theories intended to explain the origin of laboratory
        and, to some extent, terrestrial and celestial spectra.
        Reviewed in: *JHA*, 3 (1972), 69; *Isis*, 65 (1974), 284-285.

*   Meadows, A.J. *Science and Controversy: A Biography of
        Sir Norman Lockyer*.

        Cited herein as item 1341.

926.   Menzel, Donald H., ed. *Selected Papers on the Transfer
        of Radiation*. New York: Dover, 1966. Pp. v + 269.

        Collection of six historical papers by A. Schuster,
        K. Schwarzschild, A.S. Eddington, S. Rosseland and E.A.
        Milne on radiative transfer. The first five papers dating
        from 1905 through 1924 form the basis for the modern study,
        while the sixth paper, by Milne, reviews the entire field
        in exhaustive detail, as it was known circa 1930.

927.   Plaskett, J.S. "Sixty Years' Progress in Astronomy."
        Journal of the Royal Astronomical Society of Canada,
        21 (1927), 295-310.

        Emphasizes the introduction of the spectroscope into
        astronomy as "probably the greatest advance in the history
        of the science, indeed marking the development of the
        sister science of astrophysics." Provides a brief over-
        view of present state of knowledge of structure and scale
        of the Universe, just after Hubble's measurement of the
        distances to galaxies.

928. Roscoe, Henry E. *Spectrum Analysis* [1869]. 4th Edition.
      Revised by Roscoe and Arthur Schuster. London: Mac-
      millan and Co., 1885. Pp. xvi + 452.

      The first (1869) edition was based upon a series of
      six lectures on all aspects of the field, delivered in
      London in 1868. Each lecture includes associated appendices
      reprinting excerpts of major works in the field. Each
      lecture provides introductory material and linking com-
      mentary for the extracts reprinted. Includes copies of
      solar spectrum maps by Kirchhoff, Angstrom and Huggins
      and extensive documentation. Throughout the various edi-
      tions (1869, 1870, 1873, 1885) some of the appended papers
      are omitted and new ones added to represent the progress
      of the field. The third edition (1873) contains an extensive
      bibliography omitted in the fourth edition. General subject
      areas include historical progress of spectrum analysis
      since Newton; the spectrum of the Sun; laboratory tech-
      niques; methods of mapping spectra; the spectra of gases;
      foundations of "solar and stellar chemistry" beginning
      with Fraunhofer; and celestial spectroscopy.

929. Saha, Megh Nad. "Dissociation Equilibrium." *Life and Work
      of Sir Norman Lockyer* (item 1332), 316-335.

      Provides analysis of present knowledge (1927) of Saha's
      theory of ionization and how it contrasts with Lockyer's
      pioneering attempt to create such a theory, based upon
      laboratory studies of spectra of gases and astronomical
      speculation. Few direct citations.

930. Scheiner, Julius. *Die Spectralanalyse der Gestirne*.
      Leipzig: Engelmann, 1890. Translated, enlarged and
      revised by Edwin Brant Frost. *Astronomical Spectros-
      copy*. Boston: Ginn, 1894; 2nd Edition, 1898. Pp. xvii
      + 484.

      Standard treatise on astronomical spectroscopy covering
      all aspects of the subject. Provides detailed spectro-
      scopic tables and an extensive bibliography listing some
      1300 sources. Concentrates upon an exposition of H.C.
      Vogel's system of stellar spectral classification as well
      as other particular research interests at Potsdam. E.B.
      Frost was in residence at Potsdam shortly after the first
      edition of Scheiner's text appeared, and in his transla-
      tion and revision remained in contact with him. Prefatory
      remarks and annotation by Scheiner in Frost's revision
      provide useful perspective on their differing opinions,
      mainly on the value of the recent Harvard spectral classi-
      fication work, and Lockyer's studies. Extensive essay

review by James Keeler in: *Astronomy & Astrophysics*, 13
(1894), 688-693.

931. Schellen, H. *Spectrum Analysis*. 2nd. Edition. Jane and
      Caroline Lassell, translators; W. Huggins, editor. New
      York: D. Appleton & Co., 1872. Pp. xviii + 455.

      Based upon a set of lectures delivered in Cologne in
      1869, this comprehensive popular introduction emphasizes
      the applications of spectrum analysis to astronomical
      work. Of great value in this otherwise important but non-
      definitive review are Huggins' frequent annotations com-
      menting upon the significance of his own contributions
      and his observations on the general state of the science.
      Includes appendices, extensive illustrations, some direct
      citations and a 15-page bibliography arranged by topic.

932. Siegel, Daniel M. "Balfour Stewart and Gustav Robert
      Kirchhoff: Two Independent Approaches to 'Kirchhoff's
      Radiation Law.'" *Isis*, 67 (1976), 565-600.

      Examines in detail the work of both experimentalists
      leading to this fundamental law: Stewart's thermodynamical
      arguments supporting the law, and Kirchhoff's arguments
      leading to the same conclusion. Shows that the ensuing
      priority dispute between British and German science arose
      not over the question of chronological priority or in-
      dependence, but "the quality, the scientific stature, of
      Stewart's earlier work *vis-a-vis* Kirchhoff's later work."
      The controversy very much involved Kirchhoff and Stewart,
      but was symptomatic of larger issues, as Siegel concludes:
      "The vehemence and longevity of the debate derived in part
      from its association with larger controversies concerning
      the history of thermodynamics and spectrum analysis, and
      behind it all stood the clash of British and German
      nationalistic scientific pretension." 170 citations.

933. Sitterly, Charlotte Moore. "Collaboration with Henry
      Norris Russell over the Years." *In Memory of Henry
      Norris Russell* (item 1356), 27-41.

      Discusses and evaluates her work with Russell on the
      photographic determination of the position of the Moon;
      atomic spectroscopy in collaboration with Russell and
      Saunders (two-electron spectra of alkaline earths in
      1924-25); the analysis of complex spectra; collaborative
      work with Mount Wilson staff in the identification and
      calibration of lines in solar and stellar spectra in-
      cluding her production of the invaluable "Multiplet Table

of Astrophysical Interest"; the production of catalogues providing dynamical parallaxes and masses for binary stars. 21 citations.

*   Smyth, Piazzi. "Practical Spectroscopy in 1880."

    Cited herein as item 482.

934.   Stratton, F.J.M. *Astronomical Physics*. New York: E.P. Dutton, 1924. Pp. xi + 213.

    General introduction to astrophysics emphasizing applications of spectroscopic techniques and instrumentation. Provides historical introductions to most topics. Extensive citations to primary literature.

935.   Strömgren, Bengt. "On The Development of Astrophysics During the Last Half Century." *Astrophysics: A Topical Symposium* (item 920), 1–11.

    Reviews theories of stellar structure, energy, and evolution circa 1900 taking detailed excerpts from the writings of Arthur Schuster and G.E. Hale. Notes that C.A. Young's picture of the reversing layer was still in vogue in largely qualitative theories of stellar atmospheres, and convective equilibrium was the only acceptable energy transport mechanism in stellar interiors. 1 citation.

936.   Strömgren, Bengt. "The Rise of Astrophysics." *Education in and History of Modern Astronomy* (item 125), 245–254.

    Brief general review and personal recollections primarily of advances in theoretical astrophysics including the structure and energy sources of stars and the theory of stellar atmospheres. 3 citations.

*   Sutton, M.A. "Sir John Herschel and the Development of Spectroscopy in Britain."

    Cited herein as item 483.

*   Sutton, M.A. "Spectroscopy and the Chemists: A Neglected Opportunity."

    Cited herein as item 484.

937.   Swings, Pol. "Molecular Spectra in Cosmic Sources." *Astrophysics: A Topical Symposium* (item 920), 145–171.

    Examines the physics of molecular spectroscopy and the identification of molecular lines in optical spectra.

Provides a historical overview of 20 years of progress
to 1950 and makes suggestions for future research. 84
citations.

*   Tuckerman, Alfred. *Index to the Literature of the Spectro-
       scope.*
       Cited herein as item 60.

938.   Unsöld, Albrecht. *Physik der Sternatmosphären.* Berlin:
       Julius Springer, 1938; 1955. Pp. ix + 866.

       Comprehensive technical review of all aspects of quan-
       titative stellar atmospheres by a major figure in the
       field. Includes chapters on radiation theory; the Sun;
       theory of Fraunhofer lines and the composition of stellar
       atmospheres. Extensive citations to primary works. 1675
       citations.

939.   Woolf, Harry. "The Beginnings of Astronomical Spectroscopy."
       Mélanges Alexandre Koyré. Volume 1. *Histoire de la
       Pensée,* XII. Paris: Hermann, 1964. Pp. 619-634.

       Traces the history from W.H. Wollaston circa 1802 through
       Fraunhofer in the 1820s. Notes William Herschel's work
       on the detection of invisible infra-red radiation. 33
       citations.

940.   Woolley, R. v.d. R., and D.W.N. Stibbs. *The Outer Layers
       of a Star.* Oxford: Clarendon Press, 1953. Pp. xi + 306.

       Technical introduction to and review of the mathematical
       and physical study of solar and stellar atmospheres circa
       1950. Numerous citations to primary literature.

                    TERRESTRIAL AND ATMOSPHERIC

941.   Akasofu, S.I. "A Search for the Interplanetary Quantity
       Controlling the Development of Geomagnetic Storms."
       Quarterly Journal of the Royal Astronomical Society,
       20 (1979), 119-139.

       A brief historical introduction appears in this other-
       wise contemporary review. 58 citations.

942.   Biermann, Kurt-R., ed. *Briefwechsel zwischen Alexander
       von Humboldt und Carl Friedrich Gauss.* Berlin: Akademie-
       Verlag, 1977. Pp. 202.

Contains 32 letters from Humboldt and 21 from Gauss between the years 1807 and 1854 on many topics including astronomy and terrestrial magnetism. Extensive annotation and documentation. Reviewed in: *Annals of Science*, 35 (1978), 433-434; *Isis*, 69 (1978), 629.

* Burchfield, Joe D. "Darwin and the Dilemma of Geological Time."

Cited herein as item 1037.

* Burchfield, Joe D. *Lord Kelvin and the Age of the Earth.*

Cited herein as item 1038.

943. Chandrasekhar, S. *Radiative Transfer*. Chicago: University of Chicago Press, 1950. Reprinted, New York: Dover, 1960. Pp. xiv + 393.

Rigorous presentation of physical and mathematical elements of the theory of stellar and planetary atmospheres; the theory of radiation, absorption, scattering, and radiative equilibrium. Annotated historical and contemporary chapter bibliographies.

* Chapman, S. "Alexander von Humboldt and Geomagnetic Science."

Cited herein as item 611.

* Chapman, S. "Aurora and Geomagnetic Storms."

Cited herein as item 610.

944. Chapman, S. *Solar Plasma, Geomagnetism and Aurora*. New York: Gordon and Breach, 1964. Pp. 141.

Comprehensive review of studies of geomagnetic storms carried out largely by the author. 132 citations.

* Chapman, S. "The Earth."

Cited herein as item 612.

945. Chapman, S., and J. Bartels. *Geomagnetism*. 2 volumes. Oxford: Oxford University Press, 1940. Pp. xxx + 1049.

Provides a historical introduction to geomagnetic studies with comprehensive exposition of geomagnetic phenomena; their analysis and physical interpretation.

*   Dorman, L.I. *Cosmic Rays, Variations and Space Explora-
      tions.*

      Cited herein as item 499.

945a. Eather, Robert H. *Majestic Lights, The Aurora in Science,
       History and the Arts.* Washington, D.C.: American Geo-
       physical Union, 1980. Pp. 323.

       General, well-illustrated popular history of auroral
       observations and speculation collected from many cultures.
       Describes modern techniques for the study of the Earth's
       magnetosphere. Very few direct citations. 536-item popular
       bibliography.

*   Emden, Robert. *Gaskugeln: Anwendungen der mechanischen
       Warmetheorie.*

       Cited herein as item 916.

946.   Ferraro, V.C.A. "The Solar-terrestrial Environment: An
       Historical Survey." *Planetary and Space Science,* 17
       (March 1969), 295-311.

       Traces the progress, since the turn of the century, of
       understanding solar-terrestrial relationships. 64 citations

947.   Fowler, W.A. "Rutherford and Nuclear Cosmochronology."
       J.B. Birks, ed. *Proceedings of the Rutherford Jubilee
       International Conference.* London: Heywood, 1961. Pp.
       640-676.

       Presented at 1961 conference in Manchester. Discusses
       Rutherford's contributions to knowledge of the cosmic
       timescale. 51 citations.

948.   Garber, Elizabeth. "Thermodynamics and Meteorology (1850-
       1900)." *Annals of Science,* 33 (1976), 51-65.

       Includes discussions of theories of convective transport
       and atmospheric phenomena that were also of some influence
       upon astronomical problems during the period, notably
       James P. Espy's theory of storms. Examines the classical
       boundaries of physics as perceived in the late 19th cen-
       tury. 41 citations.

*   Krause, Ernst H. "High Altitude Research with V-2 Rockets.

       Cited herein as item 504.

* Mahan, A.I. "Astronomical Refraction: Some History and Theories."

 Cited herein as item 616.

* Minnaert, M. *Light and Colour in the Open Air.*

 Cited herein as item 618a.

* Rossi, Bruno. *Cosmic Rays.*

 Cited herein as item 509.

949. Sharlin, Harold I. "On Being Scientific: A Critique of Evolutionary Geology and Biology in the Nineteenth Century." *Annals of Science,* 29 (1972), 271-285.

 Includes commentary on Thomson's arguments for the ages of the Earth and Sun and their influence upon geological and biological thought. 36 citations.

PLANETS AND PLANETARY ATMOSPHERES

* Alexander, A.F.O'D. *The Planet Saturn.*

 Cited herein as item 638.

950. DeVaucouleurs, Gerard. *Physics of the Planet Mars.* New York: Macmillan, 1954. Pp. 365.

 General contemporary review covering all aspects of the study of Mars from Earth-based observatories. Contains useful historical commentary in a short introduction and detailed citations in a nine-page chronological bibliography of primary works from 1905.

951. DeVorkin, David H. "W.W. Campbell's Spectroscopic Study of the Martian Atmosphere." *Quarterly Journal of the Royal Astronomical Society,* 18 (1977), 37-53.

 Reviews the active period of Campbell's work on Mars at Lick Observatory circa 1894-1911, and reactions to his conclusion, from Percival Lowell and others, that there was no definite spectroscopic evidence of water vapor and oxygen in the Martian atmosphere. Campbell set an upper limit to the oxygen and water vapor content, but was constantly criticized for his apparent rejection of the spectroscopic findings of others, notably those of Lowell and William Huggins. 45 citations.

*   Emden, Robert. *Gaskugeln: Anwendungen der mechanischen
       Warmetheorie.*

       Cited herein as item 916.

952.   Fison, A.H. *Recent Advances in Astronomy.* London: Blackie
       and Son, 1898. Pp. vi + 242.

       Collection of elementary descriptive essays on contempo-
       rary problems including stellar evolution, spectrum anal-
       ysis, the structure of the Milky Way and recent studies
       of Mars. No direct citations.

953.   Glasstone, Samuel. *The Book of Mars.* Washington, D.C.:
       NASA, 1968. Pp. vii + 315.

       General descriptive review of knowledge of Mars after
       first Mariner flights. Includes detailed introductory
       chapter on historical background. Short bibliography.

954.   Hoyt, William Graves. *Lowell and Mars.* Tucson: University
       of Arizona Press, 1976. Pp. xv + 376.

       Discusses Percival Lowell's foundation of his observa-
       tory in Arizona and his examination of the surface and
       atmosphere of Mars and the other planets. Centers upon
       Lowell's belief in life on Mars as revealed by the system
       of canals he saw on its surface, and the water vapor ap-
       parently detected in its atmosphere. Lowell's stormy re-
       lations with professional astronomers, and the controversi
       that ensued from his statements of the existence of life
       on Mars are highlighted, with documentation drawn from
       both the published and archival record. Reviewed in:
       Annals of Science, 34 (1977), 326–327; *Centaurus*, 21
       (1977), 326–327; *JHA*, 9 (1978), 224–225.

955.   Lowell, Percival. *Mars.* Boston: Houghton Mifflin, 1895.
       Pp. viii + 228. Reprinted, Greenfield, Massachusetts:
       History of Astronomy Reprints, 1978.

       Lowell's first major publication on Mars based upon his
       observations during the Martian opposition of 1894. Genera
       description of Martian surface including polar caps, canal
       oases, and the canals as evidence of advanced intelligence
       Much of what Lowell discusses here and in his other books
       is distilled from various volumes of the *Lowell Observa-
       tory Bulletin* during the period. Critically reviewed by
       W.W. Campbell in 1896: *Publications of the Astronomical
       Society of the Pacific*, 51 (1896), 207ff.

956. Lowell, Percival. *Mars and Its Canals*. New York: Macmillan, 1906. Pp. xv + 393.

Continuation and expansion of his first book, *Mars*. A re-examination of the planet through five subsequent oppositions. Finds confirmation of evidence for the present existence of advanced life there. Includes photographic as well as visual evidence.

957. Lowell, Percival. *Mars as the Abode of Life*. New York: Macmillan, 1908. Pp. xix + 288.

Based upon a series of popular lectures in 1906 published originally in *Century Magazine*. Within this work, Lowell generalizes his findings for Mars into a theory of planetary evolution, wherein Mars was seen as indicative of Earth's future history. Includes a section of 18 short notes reviewing various aspects of his study.

958. Lowell, Percival. *The Evolution of Worlds*. New York: Macmillan, 1909. Pp. xiii + 262.

General exposition of his theory for the origin of the solar system. Sees events such as novae as giving birth to spiral nebulae which in turn form solar systems. Includes continuing arguments from his earlier works showing evidence that Mars is older than Earth, and general descriptive arguments for the future evolution of the Earth.

* Maunder, E. Walter. "The 'Canals' of Mars."

Cited herein as item 654.

959. McCall, G.J.H. "The Lunar Controversy." *Journal of the British Astronomical Association*, 80 (1969), 19-29; 100-106; 190-199; 263-269; 358-360.

Historical review of the volcanic and meteoritic theories for the origins of lunar features prior to Apollo 11, and after Apollo landings. 80 citations.

960. Meadows, A.J. "The Discovery of an Atmosphere on Venus." *Annals of Science*, 22 (1966), 117-127.

Reviews the detection of a Venusian atmosphere by the Russian M.V. Lomonosov during the 1761 transit of Venus, and conflicting reports of the nature of this atmosphere by other observers at both the 1761 and 1769 events. Concludes that though Lomonosov may have been first to detect the atmosphere, its unambiguous presence was decided only by the later work of William Herschel and J.H. Schroeter,

who both noted the extension of the Venusian cusps beyond
the 180 degrees possible for a solid sphere as seen in
crescent phase. Discusses the significance of the existence
of a Venusian atmosphere for the doctrine of the plurality
of inhabited worlds. 37 citations.

*   Moore, Patrick. *The Planet Venus*.

         Cited herein as item 655.

*   Mullen, Richard D. "The Undisciplined Imagination: Edgar
         Rice Burroughs and Lowellian Mars."

         Cited herein as item 575.

961.   Pickering, W.H. *Mars*. Boston: R. Badger, 1921. Pp. 173.

         Contains his collected papers on Mars written between
         1890 and 1914. All papers have been revised and thereby
         constitute an important comparative record when used with
         original sources. Includes chapters on martian climate
         and his personal explanation for canal visibility; the
         opinions of others, including Lowell and Schiaparelli;
         and possibilities of signalling to Mars. Footnote cita-
         tions.

962.   Rosse, Lord. "On the Radiation of Heat from the Moon, the
         Law of Its Absorption by Our Atmosphere, and Its Varia-
         tion in Amount with Her Phases." *Royal Institution
         Library of Science: Astronomy* (item 140), 147-152.

         Lecture dated May 30, 1873. Briefly reviews earlier
         attempts to measure lunar heat including negative state-
         ments by J. Herschel and F. Arago, and studies by Melloni,
         Smyth, Tyndall, Joule and M. Marie-Davey, ending in his
         own researches commencing in 1868. 4 citations. See item
         497.

963.   Rosse, Lord. "The Radiant Heat from the Moon during the
         Progress of an Eclipse." *Royal Institution Library of
         Science: Astronomy* (item 140), 1-16.

         Lecture presented May 31, 1895. Reviews original studies
         of lunar heat with his 3-foot reflector, and extends his
         review to include recent studies of the variation of lunar
         heat during a lunar eclipse. Provides detailed description
         of historical instrumentation and notes similar studies
         by S.P. Langley and F.W. Very.

964.   Sagan, Carl, and J.B. Pollack. "On the Nature of the Canals
         of Mars." *Nature*, 212, No. 5058 (8 Oct. 1966), 117-121.

Compares recent Mariner IV discoveries of crater forma-
tions and tectonics on Mars to visual observations circa
1900 and later radar observations. Concludes that "the
Martian canals are ridge systems or associated mountain
chains in analogy to similar features in the terrestrial
ocean basins." 25 citations.

965.   Schorn, Ronald A. "The Spectroscopic Search for Water on
       Mars: A History." Carl Sagan, Tobias Owen and Harlan
       Smith, eds. *Planetary Atmospheres*. New York: Springer-
       Verlag New York, 1971. Pp. 223-236.

       Briefly reviews early visual studies of the martian
       surface and atmosphere, the detection of canal structure,
       seasonal changes, polar caps, and early spectroscopic
       identification of an Earth-type atmosphere containing
       water vapor. "Thus began a dreary series of observations
       which were all wrong, misleading, indecisive, or negative,"
       but with improved technique and instrumentation "the
       upper limit of ... water vapor was pushed steadily down."
       Centers upon recent work from the 1930s through the 1960s
       on the refinement of knowledge of the existence of water
       vapor, reaching its final detection by Spinrad and Münch
       in 1963. 57 citations.

966.   Slipher, V.M. "Planet Studies at the Lowell Observatory."
       Royal Institution Library of Science: Astronomy (item
       140), 341-359.

       Lecture delivered May 19, 1933. Descriptive commentary
       on the Lowell Observatory projects to photograph the sur-
       face detail and spectra of planets since the founding
       of the Observatory by Percival Lowell in the 1890s.

967.   Urey, Harold C. "The Origin and Significance of the Moon's
       Surface." *Vistas in Astronomy*, 2 (1956), 1667-1680.

       Examines the early work (1892) of the geologist G.K.
       Gilbert on the origin and character of the lunar surface
       wherein many suggestions were put forth that have since
       been "rediscovered" by modern workers. Traces the thin
       reception of Gilbert's work and then places it within the
       context of modern knowledge concerning the origin of lunar
       features: the lunar surface is ancient and is largely
       the result of collisions with meteoric material. 52
       citations.

COMETS, METEORS, AND THE INTERPLANETARY MEDIUM

968. Astapowitsch, I.S. "On the Fall of the Great Siberian
      Meteorite, June 30, 1908." *Popular Astronomy*, 46 (1938),
      310-317.

      General description of event and reconstruction of orbit.
      Includes 74 bibliographical citations to papers recounting
      event.

*   Baldet, M.F. "Liste générale des comètes de l'origine à
      1948."

      Cited herein as item 2.

969. Blackwell, D.E., D.W. Dewhirst, and M.F. Ingham. "The
      Zodiacal Light." *Advances in Astronomy and Astrophysics*,
      5 (1967), 1-69.

      Detailed review of theories of the origin of the Zodiacal
      Light from the late 17th century to date. Concentrates
      on contemporary theories and recent observations. 124
      citations.

970. Bobrovnikoff, N.T. "Comets." *Astrophysics: A Topical
      Symposium* (item 920), 302-356.

      Begins with a historical review of the study of comets
      and other diffuse matter in the Solar System. Examines
      comet catalogues, notation, historical records of cometary
      passages, the physical nature of comets, Bessel's theory
      of cometary forms and later elaborations, the variations
      in observed cometary brightness, the spectra of comets
      noting Donati's and Huggins' pioneering work, molecular
      spectra in comets, and a general theory of comets. Bobrov-
      nikoff concludes that no generally acceptable theory of
      origin as yet (1950) exists. His conclusions were made
      on the eve of Jan Oort's pioneering theory of a "comet
      cloud." 38 citations.

971. Brown, Harrison, ed. *A Bibliography on Meteorites*. Chicago:
      University of Chicago Press, 1953. Pp. viii + 686.

      Chronological listing, 1491-1950, of over 8650 citations.

972. Burke-Gaffney, Michael Walter. "Spectroscopic Observations
      of Comet 1882 II." *XIIe Congrès International d'Histoire
      des Sciences, Tome V (Paris, 1968)*. Paris: Blanchard,
      1971. Pp. 17-22.

Reviews early observations and methods of Ralph Copeland and J.G. Lohse, specifically their identification of iron in cometary spectra, an observation long disputed but only recently confirmed. 35 citations.

* Chambers, George F. *The Story of the Comets*.

Cited herein as item 644.

973. Debehogne, H. "Les Petites Planètes." *Ciel et Terre*, 82 (1966), 229-242.

Brief popular history of the discovery and studies of asteroids. 38 citations.

974. Eastman, John Robie. "The Progress of Meteoric Astronomy in America." *Philosophical Society of Washington*, 11 (1890), 275-358.

Reviews theories for the origins of meteors since the late 18th century and outlines progress in the study of meteor orbits; the relationship of meteors to comets; records of meteor falls; and catalogues of meteor phenomena. Numerous citations.

975. Farrington, Oliver Cummings. *Catalogue of the Meteorites of North America to January 1, 1909*. Washington: National Academy of Science, 1915. Pp. 511.

Includes selected reprints of major papers on American meteoritics.

976. Flight, Walter. *A Chapter in the History of Meteorites*. London: Dulau, 1887. Pp. 224.

Constitutes a review of major falls during the 19th century, with commentary on earlier falls. Extensive citations.

977. Henderson, Edward P. "American Meteorites and the National Collection." *Annual Report of the Smithsonian Institution*, (1948), 257-268.

Useful review of the American collection and listing of sighted meteorite falls in the United States.

* Hoffleit, Dorrit. "Bibliography of Meteoric Dust with Brief Abstracts."

Cited herein as item 18.

978. Krinov, E.L. *Giant Meteorites*. Oxford: Pergamon, 1966.
      Pp. xxii + 397.

      Translated by J.S. Romankiewicz and M.M. Beynon from
      an unpublished Russian text. Provides detailed historical
      commentary on observations of confirmed and suspected
      meteor impact craters in Russia and elsewhere. 163 cita-
      tions and 18-item bibliography.

979. Lancaster-Brown, P. "A Review of the Current Theories of
      the Origin and Formation of Comets." *Journal of the
      British Astronomical Association*, 75, 1 (1965), 17-29.

      Includes a short historical discussion of theories and
      general studies of comets. No citations.

*   Lange, Erwin F. "The Founders of American Meteoritics."

      Cited herein as item 209.

980. Lovell, A.C.B. *Meteor Astronomy*. Oxford: Clarendon Press,
      1954. Pp. xv + 463.

      General contemporary review covering all aspects of the
      astronomy of meteors and meteor flight. Includes some
      historical commentary on meteor observations, major meteor
      showers, and a history of velocity determinations of meteor
      orbits by different techniques including radio-echo ob-
      servations. Numerous citations.

981. Lyttleton, R.A. *The Comets and Their Origin*. Cambridge:
      Cambridge U. Press, 1953. Pp. x + 173.

      Moderately technical exposition of the author's theory
      of cometary origin, with descriptive chapters on the ob-
      servations of comets and their dynamical characteristics.
      Lyttleton created a theory of cometary origin where comets
      were caused by gravitational accretion of material in
      interstellar clouds encountered by the Sun. He also
      examined the effects of stellar perturbations but did not
      discuss J. Oort's alternative hypothesis of the existence
      of a vast comet cloud. Includes selected bibliography.

982. Marsden, Brian G. "One Hundred Periodic Comets." *Science*,
      155 (10 March 1967), 1207-1213.

      Reviews theories of cometary motion from the 19th century
      to the present including the major problems of cometary
      disintegration, orbital changes through planetary pertur-
      bations, and the stability of cometary orbits. 61 citations.

983. McKinley, D.W. *Meteor Science and Engineering*. New York: McGraw-Hill, 1961. Pp. ix + 309.

   Comprehensive review including a detailed historical survey concentrating upon the 19th and 20th centuries. Later chapters are valuable as an historical survey of the use of radio techniques in meteor reconnaissance. Twelve-page bibliography.

984. Nininger, Harvey H. *Find a Falling Star*. New York: Paul S. Eriksson, Inc., 1972. Pp. x + 254.

   Personal reminiscences of noted meteoriticist including specific recollections of major expeditions and finds, problems in maintaining support for research, and the growth and acceptance of his science. Reviewed in: *JHA*, 6 (1975), 140-141.

985. Olivier, Charles P. *Meteors*. Baltimore: Williams and Wilkins, 1925. Pp. xvi + 276.

   General monograph with brief historical introduction. Includes chapters on cometary orbits, swarms, radiants, computation of heights of meteor trails, origins of meteors (historical discussion), examination of meteoritic fragments, origin of lunar craters. Extensive citations.

*   Poynting, J.H. "Some Astronomical Consequences of the Pressure of Light."

   Cited herein as item 802.

986. Proctor, Mary, and A. Crommelin. *Comets: Their Nature, Origin, and Place in the Science of Astronomy*. London: Technical Press, 1937. Pp. xi + 204.

   Brief historical review of highlights of the recent study of comets. Concentrates upon 19th-century comet hunters. Reviewed in: *Isis*, 28 (1938), 513-514.

987. Sears, D.W. "Sketches in the History of Meteoritics 1: The Birth of the Science." *Meteoritics*, 10 (1975), 215-225.

   Examines the role of chemical analysis in promoting the extraterrestrial origins of meteorites circa 1800 and shows how chemistry remained the dominant discipline for the early study of meteorites. 31 citations.

988. Sears, D.W., and H. Sears. "Sketches in the History of Meteoritics 2: The Early Chemical and Mineralogical

Work." *Meteoritics*, 12 (1977), 27–46.

Reviews chemical and mineralogical work on meteoritics from 1800 to 1840 concentrating on advances in chemical techniques of analysis. 61 citations.

989.   Stoney, G. Johnstone. "The Story of the November Meteors." *Royal Institution Library of Science: Astronomy* (item 140), 160–171.

Lecture dated February 14, 1879. Examines recent research on the orbital motions of a swarm of meteors now referred to as the "Leonids," examined first by Humboldt in 1799 and followed by spectacular displays in 1832 in Europe and 1833 in America that made this swarm of great interest. Reviews systematic studies in Europe and America by H.A. Newton, who established their periodicity, and later evaluations by J.C. Adams, G. Schiapparelli, U. Leverrier, Stoney and Graham. 12 citations.

990.   Swings, P. "Cometary Spectra." *Quarterly Journal of the Royal Astronomical Society*, 6 (1965), 28–69.

Identifies and reviews five periods in the spectroscopic study of comets from the first observation by Donati in Florence in 1864 to contemporary studies from space vehicles. Includes visual studies of spectra (1864–1881); pioneering photographic spectra (1881–1902); the use of objective prisms (1902–1939); extension to ultra-violet and infra-red and low dispersion slit spectra (1939–1957); high resolution and space studies (1957–). 52 citations.

*   Turner, Herbert Hall. "Halley's Comet."

Cited herein as item 662.

991.   Wasson, John T. *Meteorites*. New York: Springer, 1974. Pp. x + 316.

Comprehensive study that includes sections on sources, bibliography, and history. Citations to major collections and bibliographies of meteorite literature from the 17th century to the present. Extensive citations.

992.   Whipple, Fred L. "The Incentive of a Bold Hypothesis: Hyperbolic Meteors and Comets." *Education in and History of Modern Astronomy* (item 125), 219–224.

Recalls his first interests in meteor orbits stimulated by Ernst Öpik's visit to Harvard in 1931–1932 and his suggestion that comets and meteors originate in a vast

circumsolar cloud. Reviews history of orbital analysis
and specimen studies at Harvard. 32 citations.

SOLAR PHYSICS

993. Abetti, Giorgio. "Solar Physics." *Handbuch der Astrophysik*,
    4 (item 130), 57-230.

General review with a historical introduction. Discusses
instrumentation, observations of the surface of the Sun,
and theories of the structure of the Sun. Included in
Volume 4 of the *Handbuch*. Includes short appendix on
historical study of planetary influences on solar activity.
Numerous citations.

994. Abetti, Giorgio. "Thirty Years of Solar Work at Arcetri."
    Vistas in Astronomy, 1 (1955), 624-630.

Reviews work between 1922 and 1952 on photospheric
phenomena, abundances, and the nature of the chromosphere.
No direct citations.

995. Adams, W.S. "Early Solar Research at Mount Wilson." *Vistas
    in Astronomy*, 1 (1955), 619-623.

Traces through personal knowledge the foundation of the
Observatory and its development since 1903, the various
instruments used for solar research, and the areas of
research. No direct citations.

996. Babcock, H.W. "The Sun's Magnetic Field." *Annual Review
    of Astronomy and Astrophysics*, 1 (1963), 41-58.

Reviews both the detection of magnetic fields in sunspots by Hale in 1908, and his own successful detection
of the Sun's general field in the 1950s. General review
of contemporary problems and instrumentation. 63 citations.

*   Babcock, H.W. "The Zeeman Effect in Astrophysics."

Cited herein as item 907.

997. Ball, Robert S. *The Story of the Sun* (1894). London:
    Cassell, 1901. Pp. xii + 376.

Fifth printing of 1894 work reviewing the progress of
solar physics during the latter half of the 19th century.
Includes chapters on the transits of Venus, the speed of

light, the mechanical theory of heat, as well as on all
aspects of solar physics. No direct citations.

998. Bray, R.J., and R.E. Loughead. *Sunspots*. London: Chapman
       and Hall, 1964. Pp. xi + 303. Reprinted, New York:
       Dover, 1979.

       Detailed comprehensive review with a historical intro-
       duction that outlines history of sunspot study from pre-
       telescopic observations through Hale's discovery of sun-
       spot magnetic fields. The remainder of the text, while
       contemporary, does review major 20th-century advances.
       451 citations.

*   Bushnell, David. *The Sacramento Peak Observatory*.

       Cited herein as item 268.

*   Clerke, Agnes M. *Problems in Astrophysics*.

       Cited herein as item 910.

*   Cortie, A.L. *Father Perry, The Jesuit Astronomer*.

       Cited herein as item 1262.

*   Cowling, T.G. "Magnetic Fields in Astronomy."

       Cited herein as item 911.

999. Deslandres, H. "The Progressive Disclosure of the Entire
       Atmosphere of the Sun." *Royal Institution Library of
       Science: Astronomy* (item 140), 157-176.

       Lecture dated June 10, 1910. Reviews (in French) the
       recent history of the study of the Sun's atmosphere,
       atmospheric phenomena, and photospheric phenomena center-
       ing upon knowledge derived from his own research with
       his "spectrohelioscope"--a device independently co-in-
       vented by Deslandres and G.E. Hale that allows these
       features (sunspots, faculae, filamentary structure) to
       be examined in monochromatic regions of the spectrum.

1000. Dessler, A.J. "Solar Wind and Interplanetary Magnetic
       Field." *Review of Geophysics*, 5 (February, 1967), 1-43.

       Traces theories of the solar wind from the late 19th
       century to date. Reviews present research. 121 citations.

1001. Dingle, Herbert. "The Constitution of the Sun." *Life and
       Work of Sir Norman Lockyer* (item 1332), 227-253.

Reviews the scientific background of Lockyer's early solar studies, and his development, with E. Frankland, of the first theory of the solar constitution, since that of G. Kirchhoff, to be based solely upon spectroscopic observations.

*   Earman, John, and Clark Glymour. "Relativity and Eclipses: The British Eclipse Expeditions of 1919 and Their Predecessors."

Cited herein as item 761a.

*   Eddy, J.A. "The Schaeberle 40-ft. Eclipse Camera of Lick Observatory."

Cited herein as item 431.

*   Eddy, J.A. "Thomas A. Edison and Infra-Red Astronomy."

Cited herein as item 500.

1002.   Eddy, J.A. "The Maunder Minimum." *Science*, 192 (1976), 1189-1202.

Confirms existence of a 70-year interruption in the normal solar cycle of sunspot activity that spanned the years 1645-1715.

*   Evershed, J. "Recollections of Seventy Years of Scientific Work."

Cited herein as item 1278.

*   Fison, A.H. *Recent Advances in Astronomy*.

Cited herein as item 952.

1003.   Forbes, Eric Gray. "A History of the Solar Red Shift Problem." *Annals of Science*, 17 (1961), 129-164.

Reviews the controversial problem of the interpretation of small spectral line shifts seen in the solar spectrum that apparently could not be explained by simple Doppler motion. Covers the period 1896 through the early 1960s, noting the major attempts at an explanation including the effect of pressure; comparisons of limb spectra with central disk spectra; pole-effects in arc-spectra; effects due to relativistic displacements suggested by Charles St. John and called "relativistic-radial current hypotheses," often modified through the mid-20th century. Provides background for appreciating the improvements in experimental technique and equipment,

the elimination of sources of error, and other general
aspects. Concludes that "the proposed relativity Doppler
current hypothesis" apparently provides a satisfactory
model, but problems still remain. 143 citations.

*   Friedman, Herbert. "Ultraviolet and X-rays from the Sun."

        Cited herein as item 502.

1004.   Goldberg, L., and A.K. Pierce. "The Photosphere of the
        Sun." *Encyclopaedia of Physics*, 52 (1959), 1-79.

        General review of theoretical studies beginning with
        Karl Schwarzschild's application of radiative equilibrium
        at the turn of the century. Covers theory and observa-
        tion of both the continuous and line spectrum of the
        Sun. 301 citations.

1005.   Hagihara, Yusuke. "Recent Eclipse Work in Japan." *Vistas
        in Astronomy*, 1 (1955), 708-713.

        Summarizes Japanese work at eclipses in 1936, 1941,
        1943, and 1948. 28 citations.

1006.   Hale, George Ellery. "Solar Vortices and Magnetic Fields."
        Royal Institution Library of Science: Astronomy (item
        140), 113-136.

        Lecture dated May 14, 1909. Describes his discovery of
        magnetic fields in sunspots from the observation of
        Zeeman splitting in sunspot spectra. Reviews history,
        since Galileo, of theories for the structure of sunspots,
        and the recent history of spectroscopic studies of the
        solar surface and atmosphere.

1007.   Hetherington, Norriss S. "Adriaan van Maanen's Measure-
        ments of Solar Spectra for a General Magnetic Field."
        Quarterly Journal of the Royal Astronomical Society,
        16 (1975), 235-244.

        Argues that van Maanen's spurious detection in 1914 of
        a general solar magnetic field was not due to a "personal
        bias," but is still undetermined, "and what [mystery]
        remains must contribute additional uncertainty to any
        interpretation of van Maanen's supposed measurement
        of internal motions in spiral nebulae." Provides a general
        history of the search for the general solar magnetic
        field until its detection by Harold and Horace Babcock
        circa 1955. 59 citations.

*   Hey, J.S. "Solar Radio Eclipse Observations."

         Cited herein as item 487.

1008.   Howard, Robert. "Research on Solar Magnetic Fields from
         Hale to the Present." *The Legacy of George Ellery Hale*
         (item 1391), 257-265.

         Reviews the growth of instrumentation at the Mount
         Wilson Observatory and research on the solar magnetic
         field from Hale's identification of the Zeeman Effect
         in sunspots to H.W. and H.D. Babcock's detection of the
         general magnetic field of the Sun in the 1950s. 21 cita-
         tions.

1009.   Kangro, Hans. "Ultrarotstrahlung bis zur Grenze elektrisch
         erzengter Wellen, das Lebenswerk von Heinrich Rubens."
         Annals of Science, 27 (1971), 165-200.

         Includes a short discussion of Rubens' work on the
         infrared spectrum of the Sun circa 1914. Extensive cita-
         tions and bibliography.

*   Kleczek, J.J., L. Leroy, and F.Q. Orrall. *A General
         Bibliography of Solar Prominence Research, 1880-1970.*

         Cited herein as item 29.

1010.   Langley, S.P. *The New Astronomy*. Boston: Houghton Mifflin,
         1884; 1889. Pp. xii + 260.

         Concentrates on the physical study of the Sun including
         its physical appearance, sun spots, the nature of its
         energy spectrum and the structure of its atmosphere.
         Later chapters include descriptive reviews of the physical
         characteristics of objects in the solar system, stars
         and nebulae. Includes commentary on contemporary solar
         engines, solar energy collectors, and burning glasses.
         No direct citations.

1011.   Langley, S.P. "The New Spectrum." *Popular Astronomy*, 9
         (1901), 415-425.

         Based upon a paper read at the National Academy of
         Sciences, April 18, 1901. Reviews his work with the
         "Bolograph" extending knowledge of the solar spectrum
         out to the region called "the new spectrum" in the range
         of 5 microns. 11 citations.

1012.   Lockyer, J. Norman. "On Recent Discoveries in Solar
         Physics Made by Means of the Spectroscope." *Royal*

Institution Library of Science: Astronomy (item 140), 87–102.

Lecture dated May 28, 1869. Concentrates on divergent theories of the nature of sunspots. Examines the English theory by De la Rue, Steward and Loewy that sunspots are dark due to the absorption of light by infalling cooler material, and contrasts this with the French theory of H.E.A. Faye, who argued that spots are holes in the luminous solar atmosphere revealing an interior so hot that its gases are invisible. Reviews his and Janssen's independent development of instrumentation to study chromospheric activity without an eclipse, and applies the knowledge thus gained to the question of the nature of the spots. Concludes that sunspots are regions of greater absorption.

1013. Lockyer, J. Norman. "On the Recent Solar Eclipse." *Royal Institution Library of Science: Astronomy* (item 140), 113–129.

Lecture dated March 17, 1871. Reviews pre-spectroscopic and post-spectroscopic observations that established the existence and general character of the Solar Corona. Begins with naked-eye and early telescopic observations since 1722. No direct citations.

1014. Lockyer, J. Norman. *Contributions to Solar Physics*. London: Macmillan, 1874. Pp. xxi + 676.

Constitutes a collection of Lockyer's early papers on solar spectroscopy, laboratory spectrum analysis, and solar physics in general. Organized in two parts. Part I is a "Popular Account of Ancient and Modern Sun-Work" wherein 26 papers have been organized and to some extent altered to provide continuity. These include descriptions of eclipse expeditions, reviews of the history of spectrum analysis as applied to solar studies, and critiques of the work of Faye and Carrington. Part II includes direct unaltered texts of 13 technical contributions and 16 short notes on solar physics and laboratory spectroscopy.

*   Lockyer, J. Norman. *The Chemistry of the Sun*.

Cited herein as item 1049.

1015. Lockyer, J. Norman. *The Sun's Place in Nature*. London: Macmillan and Co., 1897. Pp. xvi + 360.

Based upon a series of lectures in 1894 extending
arguments set forth in his 1890 text *The Meteoritic
Hypothesis* (item 1143). Chief among the advances during
the interval was the identification of helium in terrestrial spectra thus providing a new set of spectral
criteria for the hottest stars; next was the 1895 out-
burst of Nova Aurigae which Lockyer interpreted as support for his "Meteoritic Hypothesis." His 1897 text
reviews the history of both the celestial and terrestrial
discoveries of helium. Detailed annotations and citations.

*   Lockyer, T. Mary, and Winifred L. Lockyer, eds. *Life
        and Work of Sir Norman Lockyer*.

        Cited herein as item 1332.

*   McCrea, W.H. "The Constitution and Evolution of Stars."

        Cited herein as item 1144.

1016.   Meadows, A.J. *Early Solar Physics*. Oxford: Pergamon,
        1970. Pp. viii + 312.

        Solar physics between 1850 and 1900. A collection of
        27 original papers published during this period (some
        translated into English), introduced in a detailed
        historical essay by Meadows. Solar spectroscopy and
        solar activity are emphasized, but pioneer papers on the
        structure of the Sun are also included. Reviewed in:
        JHA, 2 (1971), 126.

1017.   Menzel, D.H. "The History of Astronomical Spectroscopy."
        Education in and History of Modern Astronomy (item
        125), 225-244.

        Brief narrative history of qualitative and quantitative
        methods in astronomical spectroscopy including review
        of Menzel's own work on the solar atmosphere, starting
        in 1921.

1018.   Milne, Edward A. "The Sun's Outer Atmosphere." *Royal
        Institution Library of Science: Astronomy* (item 140),
        296-304.

        Lecture delivered March 9, 1928. Descriptive analysis
        of the structure of the solar chromosphere based upon
        the assumption that it is wholly gaseous. Argues for the
        importance of the role of "monochromatic radiative
        equilibrium" in the chromosphere to produce its observed
        spectrum rather than the generally received condition
        of "local thermodynamic equilibrium." In other words

the chromospheric energy balance is achieved through
radiative rather than convective processes. Milne notes
difficulty in determining just where the radiative
chromosphere meets the convective reversing layer and
muses: "Somehow or other the sun knows how to arrange
its layers in this region in accordance with the laws
of physics, and in so doing it propounds an attractive
puzzle for the mathematician and the solar physicist."
No direct citations.

1019.   Minnaert, M. "Forty Years of Solar Spectroscopy." C. De
        Jager, ed. *The Solar Spectrum: Proceedings of the
        Symposium Held at the University of Utrecht, August
        1963*. Dordrecht: Reidel, 1965. Pp. 3-25.

        Recollections and personal reminiscences including a
        general historical survey. No direct citations.

1020.   Mitchell, S.A. "Eclipses of the Sun." *Handbuch der
        Astrophysik*, 4 (item 130), 231-357.

        Chronological exposition of studies of solar eclipse
        phenomena from the mid-19th century through 1925. Provides
        topical sections on phenomena. Included in Volume 4 of
        the *Handbuch*. Numerous citations.

1021.   Mitchell, S.A. *Eclipses of the Sun*. New York: Columbia
        University Press, 1923; 4th Edition. 1935. Pp. xvii +
        520.

        General popular review covering eclipse events since
        historical antiquity, personal experiences at eclipse
        events from 1900 through 1930, and topical discussions
        of knowledge gained about the Sun during eclipses. Few
        direct citations.

1022.   Mohler, O. "Solar Spectroscopy." *Science*, 128 (Sept. 5,
        1958), 505-510.

        Reviews study of the Sun concentrating upon history
        of Michigan's McMath-Hulbert Observatory and its solar
        instrumentation. 9 citations.

1023.   Newall, Hugh Frank. "Eclipse Problems and Observations."
        Royal Institution Library of Science: Astronomy (item
        140), 61-74.

        Lecture presented February 9, 1906. Reviews recent
        advances in physics and in instrumentation for solar
        research that might increase knowledge of the solar

atmosphere and Corona, the behavior of comets as they travel around the Sun, and the origin of the Earth's magnetism and its relation to sunspot activity. Discusses Poynting's studies of the pressure of radiation, Arrhenius' use of the concept for explaining cometary tails, and K. Schwarzschild's refined analysis. Presents observational results from the 1905 Eclipse.

1024. Newton, H.W. "The Lineage of the Great Sunspots." *Vistas in Astronomy*, 1 (1955), 666-674.

Reviews the progress of observational solar physics since 1874 against the background of progress in observing the behavior of major sunspot events. 23 citations.

1024a. Oppolzer, Theodor. *Canon of Eclipses*. [1887]. Owen Gingerich, Tr. New York: Dover, 1962. Pp. lxx + 376.

Translation of the classic *Canon der Finsternisse* originally published in 1887. Provides data for observations of 8000 solar and 5200 lunar eclipses for all years from antiquity to the 22nd century, 1200 B.C.-2100 A.D. Includes descriptions of paths of totality and partiality.

1025. Pettit, Edison. "The Sun and Stellar Radiation." *Astrophysics: A Topical Symposium* (item 920), 259-301.

Reviews present (1950) knowledge of the character of general solar radiation and the Sun as a star. Examines dynamical aspects including the angular diameter and distance of the Sun, its mass and rotation, its spectral energy curve and the Solar Constant. Reviews methods for determining photospheric temperatures, variable star light curves, planetary heat and lunar heat. Sunspots and chromospheric phenomena are discussed, and Pettit states at the outset that "What sunspots really are, and what their origin is, remains nearly as much of a mystery as it was in 1900." Ends with discussions of prominences, the Corona, and solar-terrestrial relationships. No direct citations.

1026. Plotkin, Howard. "Henry Draper, the Discovery of Oxygen in the Sun, and the Dilemma of Interpreting the Solar Spectrum." *Journal for the History of Astronomy*, 8 (1977), 44-71.

Examines Draper's spurious discovery in 1877 of a bright-line spectrum of oxygen in the solar atmosphere and the reactions of spectroscopists to this controversial observation. Reactions were first positive and

stimulated a revised theory of the solar spectrum by
Draper, Arthur Schuster and others, but then opinion
turned against it, notably from Lockyer, Christie, and
several American physicists. 38 citations.

*   Redman, R.O. "The JEPC."

        Cited herein as item 373.

1027.   Richardson, Robert S. "Sunspot Problems Old and New."
        Popular Astronomy, 55 (1947), 120-133.

        Reviews studies of solar rotation, the solar cycle of
        sunspot activity, and observed characteristics of sun-
        spots. Provides an overview of theories of the structure
        and origins of sunspots from the 18th-century speculations
        of Alexander Wilson and William Herschel to the magnetic
        theory of Hans Alfven. Notes studies of solar terrestrial
        relationships, and pioneer radio studies of the Sun.
        40 citations.

*   Saha, Megh Nad. "Dissociation Equilibrium."

        Cited herein as item 929.

1028.   Secchi, Angelo. *Le Soleil*. Paris: Gauthier-Villars,
        1870. Pp. xvi + 422.

        First major review of general knowledge of the Sun and
        solar constitution after the application of spectrum
        analysis. Includes discussions of spectroscopic instrumen-
        tation, the structure and behavior of sun spots, the
        solar atmosphere and visible surface, solar radiation,
        gravitation, and the Sun as a star compared with the
        characteristics of other stars. Translated into German
        in 1872 and enlarged somewhat by H. Schellen as: *Die
        Sonne*. Braunschweig: Westermann, 1872. Reviewed in:
        Nature, 4 (May 18, 1871), 41-43.

*   Spitzer, Lyman. "Russell and Theoretical Astrophysics."

        Cited herein as item 1084.

1029.   St. John, C.E. "The Constitution of the Sun: A Modern
        View." *Life and Work of Sir Norman Lockyer* (item 1332),
        254-265.

        Reviews the present state of knowledge of the structure
        of sunspots circa 1927 and identifies Lockyer's influence,
        notably his original argument that sunspot spectra must
        be absorption phenomena and not emission phenomena, as
        dictated by early solar theories.

1030. Strömgren, Bengt. "On the Chemical Composition of the Solar Atmosphere." *Festschrift für Elis Strömgren.* Copenhagen: Eina Munksgaard, 1940. Pp. 218-257.

Reviews early phases of modern study of the solar atmosphere beginning with the work of Albrecht Unsöld and Henry Norris Russell in the late 1920s. 37 citations.

1031. Tousey, R. "The Spectrum of the Sun in the Extreme Ultraviolet." *Quarterly Journal of the Royal Astronomical Society,* 5 (1964), 123-144.

Provides a brief historical introduction to the study of the ultra-violet spectrum of the Sun and its connection with the Earth's ionosphere by reviewing the pioneering work of Edlén and others and the first V-2 rocket soundings organized by the Naval Research Laboratory. Concentrates on the growth of contemporary knowlege of the subject. Concludes that "extreme ultraviolet and x-ray spectroscopy of the Sun are still in the observational stage." 51 citations.

* Woolley, R. v.d. R., and D.W.N. Stibbs. *The Outer Layers of a Star.*

Cited herein as item 940.

* Wright, Helen, J.N. Warnow, and Charles Weiner, eds. *The Legacy of George Ellery Hale.*

Cited herein as item 1391.

1032. Young, Charles A. *The Sun.* New York: D. Appleton and Co., 1881; 1883 (2nd Edition). Pp. 331.

Reviews general knowledge about the Sun including its distance and dimensions; instrumentation for examining its surface and spectrum; the use of the spectroscope and spectrum analysis; the solar chromosphere, prominences and corona; the Sun's light and heat. The majority of the discussion reviews work accomplished during the period 1860-1880, and centers on Young's own work, and that of Secchi, Lockyer, Proctor, Vogel, and Langley. This famous monograph went through many editions, including a general revision in 1895. No direct citations.

MAINTENANCE OF SOLAR HEAT AND ORIGIN OF SOLAR SYSTEM

1033.   Atkinson, R. d'E. "The Energy of the Stars." *The Observatory*, 69 (1949), 161–184.

Constitutes the Halley Lecture for 1949, delivered at Oxford on May 12. Summary of nuclear theory of stellar energy by one of the pioneers in the field. Reviews the many aspects of observational astronomy and theoretical physics combined in the study over the previous 30 years. 28 citations.

1034.   Ball, Robert S. "The Nebular Theory." *Royal Institution Library of Science: Astronomy* (item 140), 31–35.

Lecture presented May 10, 1902. Presents contemporary arguments in favor of the nebular theory of the origin of the Solar System in a brief popular review. Notes as favorable arguments: similarity of composition of Earth and Sun; recent photographs of nebulae by G.E. Hale, W.W. Campbell, Isaac Roberts and W.E. Wilson; motions observed in the recent Nova (1901) in Perseus.

1035.   Brush, Stephen G. "A Geologist Among Astronomers: The Rise and Fall of the Chamberlin–Moulton Cosmogony." *Journal for the History of Astronomy*, 9 (1978), 1–41; 77–104.

Examines the state of the study of cosmogony at the turn of the century, primarily the dynamical problems associated with the Nebular Hypothesis. Details T.C. Chamberlin's growing interest in modifying the Nebular Hypothesis with his planetesimal hypothesis starting in 1892; his association with the celestial mechanician Forest Ray Moulton, and their collaboration. Emphasizes the geological bases of Chamberlin's speculations, the importance of the interpretation of spiral nebulae as proto-solar systems, and the reception of the planetesimal hypothesis by astronomers through the 1930s. 319 citations.

1036.   Brush, Stephen G. "Poincaré and Cosmic Evolution." *Physics Today*, 33 (March 1980), 42–49.

Traces Poincaré's growing interests in, and study of the origin and stability of the Solar System; the ultimate fate of the Universe as dictated by a rigid interpretation of the Second Law of Thermodynamics; the problems of reconciling this law with the established stability

of the Solar System by Laplace; and the applicability
of studies of rotating fluid masses to the origin and
evolution of the Earth-Moon system and the Solar System
in general. 20 citations.

1037.   Burchfield, Joe D. "Darwin and the Dilemma of Geological
        Time." *Isis*, 65 (1974), 301-321.

        Traces Darwin's need for the vast time scale available
        from uniformitarian geology and the severe limitations
        of the highly restricted scale derived from Kelvin's
        physical arguments for the age of the Earth. Examines
        objections to both time scales and Darwin's study of the
        denudation of the Weald that required a long scale. Notes
        arguments by Kelvin, James Croll and others for a time
        scale limited by the age of the Sun's heat. 85 citations.

1038.   Burchfield, Joe D. *Lord Kelvin and the Age of the Earth.*
        New York: Science History Publications, 1975. Pp. xii +
        260.

        Traces the development of William Thomson's (Lord
        Kelvin) use of his newly-produced laws of thermodynamics
        to show that the age of the Earth, as well as of the
        Sun, is far less than time scales demanded by uniformi-
        tarian geology. Details Kelvin's meteoric theory of the
        Sun's heat, and then his refined gravitational time
        scales through the 1860s similar to those of Helmholtz
        in the mid-1850s. Shows that Kelvin's arguments caused
        those involved in geochronology to come to erroneous
        conclusions based upon Kelvin's time limitations, which
        persisted until geologists began to use radioactive dating
        techniques and also came to the realization, via Ruther-
        ford's work, that the Earth could not be a simple cooling
        body. Reviewed in: *Annals of Science*, 33 (1976), 485-
        488; *Isis*, 67 (1976), 492-494.

1039.   Chapman, S. "The Source of the Sun's Energy." *Monthly
        Notices of the Royal Astronomical Society*, 102 (1942),
        110-130.

        Review of the problem contrasting what was known in
        Kelvin's time to knowledge in the 1920s and 1930s result-
        ing from nuclear physics and the first successful me-
        chanisms suggested in 1938. Numerous citations.

*   Clarke, Henry L. "The Life-History of Star Systems."

        Cited herein as item 1069.

1040. Clerke, Agnes M. *Modern Cosmogonies*. London: Adam and
       Charles Black, 1905. Pp. vii + 287.

       Brief and popular review of theories of the origin of
       the Solar System from ancient speculation through Kant,
       Laplace and 19th-century variations of the Nebular
       Hypothesis, including the meteoritic theories of Lockyer
       and G.H. Darwin and the encounter theories of Moulton
       in the 1890s. Clerke returns at various points in her
       history to the question of the form of primordial matter,
       or "Protyle," noting "The notion of a primordial form
       of matter meets us at every stage of cosmogonical specu-
       lation." Some citations.

1041. Croll, James. *Stellar Evolution and Its Relations to
       Geological Time*. New York: D. Appleton, 1889. Pp. xi +
       118.

       To account for the requirements of geological time,
       Croll argues that the source of solar energy was the con-
       version of kinetic energy in a collision of two bodies
       that produced the Sun. This "impact theory" of stellar
       evolution was a recurrent theme through the latter half of
       the 19th century, and is well-represented in this work.
       Constitutes an extension of an earlier work: *Climate
       and Cosmology*. London: Edward Stanford, 1889. Includes
       numerous citations.

1042. Faye, H. *Sur l'origine du monde, théories cosmogoniques
       des anciens et des modernes*. Paris: Gauthier-Villars,
       1896. Pp. xi + 313.

       Exposition of theories of cosmogony in three main
       periods: Greece; Descartes and Newton through Kant and
       Laplace; 19th-century theories, centering on Faye's
       personal speculation based upon knowledge of the struc-
       ture and composition of nebulae, the nature of the Sun,
       and evidence from geology.

1043. Herczeg, Tibor. "Planetary Cosmogonies." *Vistas in
       Astronomy*, 10 (1968), 175-206.

       Reviews theories of the origin of the Solar System
       since the 1930s and attempts a general synthesis. 67
       citations.

*   Hetherington, Norriss S. "Adriaan van Maanen on the
       Significance of Internal Motions in Spiral Nebulae."

       Cited herein as item 1189.

*   Irons, James Campbell. *Autobiographical Sketch of James Croll with [a] Memoir of his Life and Work.*

Cited herein as item 1319.

1044.   Jaki, Stanley L. "The Five Forms of Laplace's Cosmogony." *American Journal of Physics*, 44 (1976), 4-11.

Reviews the development of Laplace's theory of the origin of the Solar System between the years 1796 and 1824, arguing that throughout these revisions Laplace was reluctant to produce a quantitative test of his model in light of available astronomical data. 81 citations.

1044a.   Jaki, Stanley L. *Planets and Planetarians: A History of Theories of the Origins of Planetary Systems.* New York: John Wiley, 1978. Pp. vi + 266.

A general reconnaissance of theories since the Greeks, with emphasis upon post-Newtonian theories. Complains that the tendency to search for modes of origin of planetary systems as commonly occurring results of star formation actually may hide the full complexity of the process. Numerous citations. Reviewed in: *Isis*, 72 (1981), 118.

*   Jeans, J.H. *Problems of Cosmogony and Stellar Dynamics.*

Cited herein as item 1171.

1045.   Jeans, J.H. *The Nebular Hypothesis and Modern Cosmogony.* Oxford: Clarendon Press, 1923. Pp. 31.

The Halley lecture delivered May 23, 1922. Reviews the basic tenets of Laplace's Nebular Hypothesis and how they have fared in light of subsequent theoretical and observational studies, the former by Jacobi, Kelvin, Poincaré, G.H. Darwin, and the latter by the nebular observations of G.W. Ritchey, F.G. Pease, and especially Adriaan van Maanen's measurements of rotational proper motions in spiral nebulae. Argues that Laplace's view of the formation of planets from rings of matter thrown off the contracting nebular stars is not possible, and that "the small bodies of the solar system can never have existed in the gaseous state." Discusses fission and tidal hypotheses, reconciling them with Laplace's theory, but concludes that no comprehensive theory of cosmogony is yet at hand. No direct citations.

1046. Jeans, J.H. "The Origin of the Solar System." *Royal
       Institution Library of Science: Astronomy* (item 140),
       271-295.

       Lecture delivered February 15, 1924. Popular exposition
       that links spiral nebulae to the recently modified evolu-
       tionary cosmogonies of Kant and Laplace based partly upon
       the then-popular acceptance of van Maanen's spurious
       detection of rotational proper motions in spiral nebulae.
       Notes results from his own dynamical studies that appar-
       ently confirmed the hoped-for similarity between masses
       of the central condensations of nebulae and the average
       masses of the stars. Written at the high point of the
       acceptance of Russell's theory of stellar evolution
       shortly before its demise, and when the demise of the
       credibility of van Maanen's work was on the horizon.
       Bulk of the text concerns Jeans' own tidal interaction
       theory for the origin of the Solar System.

*   Jeans, J.H. *Astronomy and Cosmogony*.

       Cited herein as item 921.

1047. Jeffreys, H. "The Origin of the Solar System." *Monthly
       Notices of the Royal Astronomical Society*, 108 (1948),
       94-103.

       Critically examines possible theories of the origin
       of the Solar System suggested since his 1929 edition of
       The Earth. 26 citations.

1048. Kuiper, Gerard P. "On the Origin of the Solar System."
       Astrophysics: A Topical Symposium (item 920), 357-424.

       General technical review of work since the 1920s
       emphasizing the intercomparison of observation and theory.
       Includes the origin of comets as "the condensation prod-
       ucts of the outer parts of the solar nebula which formed
       the planets," and a general study of the origins of
       planets presented from three points of view. Concludes
       with satisfaction that "present developments indicate
       that the process of planetary formation is but a special
       case of the almost universal process of binary star
       formation." 39 citations.

*   Lindsay, Robert Bruce. *Julius Robert Mayer, Prophet of
       Energy*.

       Cited herein as item 1330.

1049.   Lockyer, J. Norman. *The Chemistry of the Sun*. London:
        Macmillan and Co., 1887. Pp. xix + 457.

        General exposition of the author's spectroscopic study
        of the atmosphere of the Sun with a detailed historical
        introduction to the origins of solar spectroscopy includ-
        ing the contributions of Kepler, Newton, Wollaston,
        Fraunhofer, Brewster, John Herschel, etc., through
        the discoveries of Kirchhoff and subsequent work on the
        solar spectrum by Angstrom and Thalen. Lockyer then
        describes his important early laboratory techniques for
        establishing temperature criteria for spectra that led
        to his "Dissociation Hypothesis": the idea that with an
        increase in temperature, atoms could be broken down into
        fundamental particles whose spectra were revealed as
        "basic" lines supposed to be common to many atomic lines
        in the spectra of different elements. Lockyer's theory,
        though controversial, held the germ of truth in the
        ionization theory of Saha in 1920. Numerous citations.

*   Lockyer, J. Norman. *The Sun's Place in Nature*.

        Cited herein as item 1015.

1050.   Lockyer, J. Norman. *Inorganic Evolution as Studied by
        Spectrum Analysis*. London: Macmillan and Co., 1900.
        Pp. x + 198.

        Lockyer's ultimate revision of his theory of dissocia-
        tion as described in detail in *The Chemistry of the Sun*
        (item 1049). Acts as a résumé of much of his earlier
        work including his *Meteoritic Hypothesis* (item 1143).
        Lockyer here attempts to answer the many criticisms of
        his controversial theories, and ends with a general dis-
        cussion of evolution in the inorganic world. Numerous
        citations.

*   Lowell, Percival. *The Evolution of Worlds*.

        Cited herein as item 958.

*   Nieto, Michael Martin. *The Titius-Bode Law of Planetary
        Distances: Its History and Theory*.

        Cited herein as item 797.

*   Numbers, Ronald L. *The American Kepler: Daniel Kirkwood
        and his Analogy*.

        Cited herein as item 798.

*   Numbers, Ronald L. *Creation by Natural Law: Laplace's Nebular Hypothesis in American Thought.*

        Cited herein as item 579.

1051.   Poincaré, H. *Leçons sur les hypothèses cosmogoniques.* Paris: A. Hermann, 1911; 1913. Pp. lxx + 294.

        General review of theories of the origin of the Solar System since Kant and Laplace. Includes the contributions of Roche in aiding the understanding of the formation of Laplacian rings of matter from a rotating gaseous mass; the further researches of G.H. Darwin, Poincaré and Liapounoff extending the fundamental work of Maclaurin Jacobi and others on the rotation of fluids; general problems associated with tidal interaction and tidal disruption; 19th-century thermodynamical theories of stellar evolution based upon studies of the Sun by Helmholtz, Kelvin, and J. Homer Lane; general problems in the statistical mechanics of gaseous, particulate masses, and of stellar systems. 1913 edition, edited by E. Lebon, contains a biography of Poincaré.

1052.   Russell, Henry Norris. *The Solar System and Its Origin.* New York: Macmillan, 1935. Pp. 144.

        Popular general review based upon lectures given at the University of Virginia in 1934 and divided into three parts: the dynamical properties of the system; physical and chemical properties; and theories of origin. In the third part the classical Nebular Hypothesis is reviewed along with the many criticisms of it, which resulted in numerous modifications. Alternative close encounter theories of Chamberlin, Moulton, Jeans and Jeffreys, and Jeffreys' 1929 revival of a collision hypothesis are also examined, but Russell draws no final conclusions. These lectures were given several years before Lyman Spitzer, a Russell student, demonstrated that the close encounter and collisional mechanisms could not produce planets. No direct citations.

*   Sarton, George. "The Discovery of the Law of Conservation of Energy."

        Cited herein as item 1147.

*   See, T.J.J. *Researches on the Evolution of the Stellar Systems.*

        Cited herein as item 1082.

*   Sharlin, Harold I. "On Being Scientific: A Critique of
        Evolutionary Geology and Biology in the Nineteenth
        Century."

        Cited herein as item 949.

*   Sticker, Bernhard. *Bau und Bildung des Weltalls*.

        Cited herein as item 1222.

1053.   Ter Haar, D. "Cosmogonical Problems and Stellar Energy."
        Reviews of Modern Physics, 22, 2 (1950), 119-152.

        Surveys theories of the origin of the Solar System and
        planetary systems in general including the associated
        problems of nucleosynthesis and stellar energy. Written
        before nuclear processes in giant stars were understood.
        239 citations.

1054.   Ter Haar, D., and A.G.W. Cameron. "Historical Review of
        Theories of the Origin of the Solar System." Robert
        Jastrow and A.G.W. Cameron, eds. *Origin of the Solar
        System*. New York: Academic Press, 1963. Pp. 1-37.

        Detailed review by Ter Haar adapted by Cameron. Begins
        with the Cartesian Vortex Theory, early tidal theories,
        the Nebular Hypothesis as delineated by Kant and later
        by Laplace, and continues with alterations to the Nebular
        Hypothesis, the role of magnetism, turbulence and angular
        momentum. Concludes with exposition of modern theories
        extending the work of von Weizsäker. 59 citations.

*   Whitrow, G.J. "The Nebular Hypothesis of Kant and Laplace."

        Cited herein as item 596.

1055.   Williams, I.P., and A.W. Cremin. "A Survey of Theories
        Relating to the Origin of the Solar System." *Quarterly
        Journal of the Royal Astronomical Society*, 9 (1968),
        40-62.

        Broad technical survey of 20th-century theories examined
        in three topical categories: concurrent formation of
        planets and Sun; capture of interstellar material; tidal
        disruption. After a critical examination of major theories,
        concludes that "The amount of evidence we have available
        at present does not permit us to distinguish between
        these theories but it is likely that the correct theory ...
        will be somewhat similar to one or other of these." 103
        citations.

1056. Woolfson, M.M. "Cosmogony Today." *Quarterly Journal of the Royal Astronomical Society*, 20 (1979), 97-114.

Argues that no "completely accepted theory of the origin of planetary systems exists." Reviews range of theories from Laplace's Nebular Hypothesis to the "collision" theories of Chamberlin, Moulton, Jeans and Jeffreys. Examines in detail four recent theories and discusses their strengths and weaknesses. 39 citations.

THE STARS

1057. DeVorkin, David H. "Michelson and the Problem of Stellar Diameters." *Journal for the History of Astronomy*, 6 (1975), 1-18.

Reviews A.A. Michelson's development of a stellar interferometer from tests in 1890 and the measurement of the angular diameters of Jovian satellites in 1891 to the measurement of the angular diameter of Betelgeuse at Mount Wilson in 1920. Argues that 30-year delay was as much due to the lack of a suitable prediction for a stellar angular diameter as it was due to technological limitations. 59 citations.

1058. Herrmann, D.B. "Aus der Entwicklung der Grössenklassen-Definition im 19. Jahrhundert." *Die Sterne*, 48 (1972), 20-30; 113-120.

Detailed review of systems for the determination of stellar brightnesses. After a brief introduction to brightness scales from Hipparchus and Ptolemy to Argelander, Herrmann examines early attempts at physiologically-based systems by Weber and Fechner, C.A. Steinheil, and finally that of Norman Pogson. Pogson's successful though initially unappreciated system is reviewed in detail in the second part of this work, which reviews the establishment of his relationship between the intensity of a star's radiation and its measured "magnitude"; the calibration and determination of the scale of this visual system; and subsequent work. 38 citations.

*   Herrmann, D.B. "Karl Friedrich Zöllner und die 'Potsdamer Durchmusterung,' Versuch einer Rekonstruktion."

Cited herein as item 436.

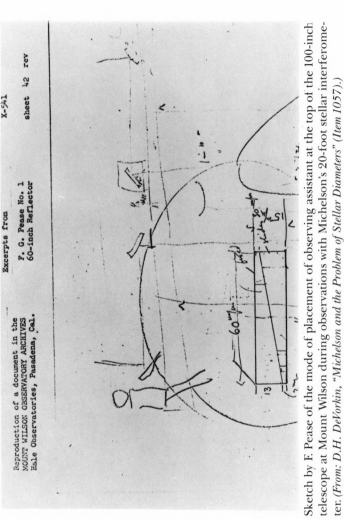

Sketch by F. Pease of the mode of placement of observing assistant at the top of the 100-inch telescope at Mount Wilson during observations with Michelson's 20-foot stellar interferometer. (*From: D.H. DeVorkin, "Michelson and the Problem of Stellar Diameters" (Item 1057).*)

*   Herrmann, D.B., and D. Hoffmann. "Astrofotometrie und
        Lichttechnik in der 2. Hälfte des 19. Jahrhunderts."

        Cited herein as item 469.

1059.   Hetherington, Norriss S. "Sirius B and the Gravitational
        Redshift: An Historical Review." *Quarterly Journal of
        the Royal Astronomical Society*, 21 (1980), 246-252.

        Of interest and value for its detailed review of
        knowledge about this star since its discovery in the
        1860s. The analysis and conclusion, notably an accusa-
        tion that astronomers are guilty of ignoring earlier
        observations shown later to be spurious, should be read
        with caution. 43 citations.

1060.   Hoffleit, Dorrit. "The Discovery and Exploitation of
        Spectroscopic Parallaxes." *Popular Astronomy*, 58
        (1950), 428-438; 483-501; 59 (1951), 4-19.

        Examines the history of solar and stellar spectroscopy
        to trace the development of the recognition of luminosity
        sensitive criteria in stars. Shows that several early
        classifiers, including A. Secchi and N. Lockyer, detected
        and utilized luminosity sensitive criteria, but that the
        first really successful detection was the application
        of A. Maury's detailed system by E. Hertzsprung in 1905-
        1909. Reviews the origins of the HR Diagram, the consen-
        sus on a notation for classification in 1910, and the
        discovery of a viable spectroscopic technique for lumi-
        nosity measurement by W.S. Adams and A. Kohlschütter
        between 1913-1916 at Mt. Wilson. Ends with continued
        refinement of technique and refinement of luminosity
        classes on the HR Diagram. Extensive citations.

1061.   Jones, Derek. "Norman Pogson and the Definition of Stellar
        Magnitude." *Astronomical Society of the Pacific Leaf-
        let No. 469* (July, 1968). Pp. 8.

        Reviews attempts from the Greeks to the Herschels to
        measure the apparent brightnesses of stars on a con-
        sistent and reliable scale, ending in Pogson's work in
        the mid-19th century which formed the basis of the
        present system. Brief sketch of Pogson's life. No direct
        citations.

1062.   Lundmark, Knut. "Luminosities, Colours, Diameters, Den-
        sities, Masses of the Stars." *Handbuch der Astrophysik*,
        5 (item 130), 210-574; 575-697.

Comprehensive technical review, in Volume 5, of the
state of knowledge of basic data for stars. Involves all
aspects of observational astrophysics presented in the
form of a history. Covers, for all years, knowledge of
apparent stellar magnitudes and colors; visual and
photographic photometry; the problems of absolute magnitudes including methods of distance determinations
for stars; stellar spectra and the HR Diagram; stellar
diameters, densities and masses and methods for the
reduction of binary star observations. Exhaustive citations.

1063.   Newcomb, Simon. *The Stars. A Study of the Universe.*
New York, Putnam, 1901. Pp. xi + 333.

Popular descriptive review, reprinted many times, of
late 19th-century knowledge about the stars; their
distances, positions and motions as derived by the work
of Argelander, Schonfeld, Thome, Gill and Kapteyn;
photometric surveys; general results of stellar spectrum
analysis; variable stars and novae; history of stellar
parallaxes; binary and multiple stars; stellar systems;
stellar structure and evolution; cosmology. No direct
citations. Appendices include lists of stars with determined parallaxes and stars with large proper motions,
and a short list of identified spectroscopic binaries.
Reviewed in: *PASP*, 14 (Apr. 1902), 51-53; *The Observatory*,
25 (Feb 1902), 96-98.

*   Payne, Cecilia H. *The Stars of High Luminosity.*

Cited herein as item 1117.

1064.   Pickering, E.C. "The Light of the Stars." *International
University Lectures at the Congress of Arts and Science
Universal Exposition, Saint Louis (1904).* Volume VI.
New York: University Alliance, 1909. Pp. 67-82.

Reviews late 19th-century advances in the classification
of stellar spectra and the production of stellar photometric data. Centers upon Harvard projects in these
areas, a detailed description of the development of the
meridian photometer, and general developments in photometry. No direct citations. Brief bibliography of secondary
sources.

*   Russell, Henry Norris. *The Composition of the Stars.*

Cited herein as item 1119.

1065. Strömgren, Bengt. "The Growth of our Knowledge of the
       Physics of the Stars." *Astrophysics: A Topical Symposium*
       (item 920), 172-258.

       Traces in technical detail "the main stream of astro-
       physical thought and effort during the past half century:
       the study of the physics of the stars." Divides review
       into three areas: the continuous spectra and radiation
       of the stars, stellar absorption lines and the interior
       structure of the stars.

*   Thiele, Joachim. "Zur Wirkungsgeschichte des Doppler-
       prinzips im Neunzehnten Jahrhundert."

       Cited herein as item 485.

1066. Wesselink, A.J., K. Paranya, and K. DeVorkin. "Catalogue
       of Stellar Dimensions." *Astronomy and Astrophysics
       Supplement*, 7 (1972), 257-289.

       Introduction to a catalogue of 2392 calculated angular
       and linear diameters includes a historical review of
       knowledge of stellar diameters from the work of A.A.
       Michelson through Hanbury Brown's intensity interferometer
       measures in 1967. Adds a discussion of predictions of
       stellar diameters by E.C. Pickering, H.N. Russell, E.
       Hertzsprung and others. 22 citations.

                    BINARY AND MULTIPLE STARS

1067. Aitken, Robert Grant. *The Binary Stars*. New York: D.C.
       McMurrie, 1918. 2nd Edition: New York: McGraw-Hill,
       1935. Reprinted by Dover, 1964. Pp. x + 309.

       Standard classic introduction to techniques of study
       of all types of binary stars. Includes two chapters on
       the history of the study starting with the discovery
       of the first "double star" by J.B. Riccioli in 1650.
       Reviews the work of William Herschel, F.G.W. Struve,
       John Herschel and later S.W. Burnham, E.E. Barnard and
       Hussey, all visual binary observers. Attention is then
       turned to early spectroscopic and eclipsing binary ob-
       servers and the production of double star catalogues.
       Numerous citations. Reviewed in: *ApJ*, 82 (1935), 368.

1068. Batten, Alan H. "The History of U Coronae Borealis."
Quarterly Journal of the Royal Astronomical Society,
5 (1964), 145-157.

General review of observations of this Algol-type
eclipsing binary since its discovery by Winnecke in
1869. Examines difficulties over the interpretation of
the derived spectroscopic orbit and photometric light
curve, the early techniques of observation, and suggestions for future observations. 28 citations.

* Beer, Arthur, ed. "The Henry Norris Russell Memorial
Volume."

Cited herein as item 1242.

* Brush, Stephen G. "Poincaré and Cosmic Evolution."

Cited herein as item 1036.

1069. Clarke, Henry L. "The Life-History of Star Systems."
Popular Astronomy, 3 (1896), 489-519.

Descriptive review of Darwin's and See's methodology
in examining the evolution of multiple star systems.
Examines Poincaré's similar studies and the general investigation of the role of tidal friction. No direct
citations.

* Eggen, Olin J. "Sherburne Wesley Burnham and his Double
Star Catalogue."

Cited herein as item 1276.

1070. Eggen, Olin J. "The Astrophysics of Visual Binaries."
Quarterly Journal of the Royal Astronomical Society,
3 (1962), 259-287.

Examines observational aspects, orbit calculations,
and theoretical calculations of the structure of stars
in visual binary systems. Compares and contrasts historical
and contemporary techniques and accuracy of data, noting
that "The modus operandi in double star astronomy has
been little changed for 150 years." 24 citations.

1071. Fernie, J.D. "The Period-Luminosity Relation: A Historical
Review." *Publications of the Astronomical Society of
the Pacific*, 81 (1969), 707-731.

Examines problems associated with the calibration of
the Cepheid Period-Luminosity relation since its first

detection by H.S. Leavitt in 1908 and first attempts at
a calibration by E. Hertzsprung, H.N. Russell and H.
Shapley. Tentatively identifies a strong "herd instinct"
among astronomers to follow established leads in problems
of calibrating empirical relationships. Carries discussion
through to recent work. Extensive citations.

*   Hardin, Clyde L. "The Scientific Work of the Reverend
           John Michell."

           Cited herein as item 1298.

1072.   Henroteau, F.C. "Double and Multiple Stars." *Handbuch*
           der Astrophysik, 6 (item 130), 299–468.

           Provides, in Volume 6, a detailed historical review
of the work of double star observers from pre-telescopic
through modern times as an introduction to a comprehensive
exposition of the classification of all types of binary
systems, the determination of orbits, and the statistics
of binary and multiple systems. Extensive citations.

1073.   Huffer, C.M., and G.W. Collins. "Computation of Elements
           of Eclipsing Binary Stars by High-Speed Computing
           Machines." *Astrophysical Journal Supplement*, 7, 71
           (1962), 351–410.

           Provides a short historical introduction to a discussion
of computational methods.

1074.   Hussey, William J. "Notes on the Progress of Double
           Star Astronomy." *Publications of the Astronomical*
           Society of the Pacific, 12 (1900), 91–103.

           Concentrates on history of observations of visual binary
stars beginning with the Herschels, James South, the
Struves at Dorpat and Pulkova, Mädler, Dawes, Secchi and
Dembowski. Divides the history into two sections, before
and after 1870, and provides detailed analysis of the
number of systems detected by post-1870 observers,
showing that S.W. Burnham discovered by far the most:
fully one-third of the total number discovered in the
30-year period.

1075.   Hynek, J.A. "Spectroscopic Binaries and Stars with
           Composite Spectra." *Astrophysics: A Topical Symposium*
           (item 920), 448–478.

           Examines recent growth (to 1950) of knowledge of
spectroscopic binary stars starting in 1889 but con-

centrating on 20th-century advances in observation and
reduction techniques. Includes both dynamical and physical
studies, and theories of the origin and evolution of
binary stars. 17 citations.

1076.   Johnson, Martin. "The Interpretation of Some Changes in
        Orientation of a Binary Star Orbit." *Quarterly Journal
        of the Royal Astronomical Society*, 2 (1961), 9-23.

        Reviews both observational and theoretical techniques
        for probing stellar interiors during the period 1928-
        1960. Centers on H.N. Russell's 1928 formulation of the
        "Apsidal Motion Test" and theoretical discussions derived
        from it and from observations of suitable binary star
        systems. 13 citations.

1077.   Kopal, Zdenek. *Close Binary Systems*. London: Chapman and
        Hall, 1959. Pp. xiv + 558.

        General technical account of the theory of close binary
        systems including exposition of reduction procedures.
        Provides brief historical introduction and bibliographical
        remarks at the ends of each chapter including annotated
        historical citations.

*   Lundmark, Knut. "Luminosities, Colours, Diameters, Den-
        sities, Masses of the Stars."

        Cited herein as item 1062.

1078.   Olivier, Charles P. "Double Stars." *Popular Astronomy*,
        52 (1944), 417-428.

        Brief review of work since 1800 but concentrates upon
        past 50 years. 23 citations.

1079.   Pierce, Newton L. "Eclipsing Binaries." *Astrophysics:
        A Topical Symposium* (item 920), 479-494.

        Describes the early history of photometric analyses
        of the light curves of eclipsing binaries in the late
        19th century and efforts to interpret them. Notes that
        proof of the eclipse character of the light curve came
        only in 1889 with H.C. Vogel's radial velocity observa-
        tions of Algol. Examines later methods by H.N. Russell
        and Harlow Shapley for the interpretation of eclipsing
        binary light curves, elaborations and extentions in the
        20s and 30s, and variations to account for changing
        periods, apsidal motions, etc. 39 citations.

1080. Russell, Henry Norris, and Charlotte E. Moore. *The Masses
       of the Stars*. Chicago: University of Chicago Press,
       1940. Pp. ix + 236.

       Outgrowth of technical lectures given by Russell at
       the Harvard Tercentenary in 1938. Provides general in-
       troduction to astrometric, statistical, and dynamical
       means for determining the masses of stars, with detailed
       references to recent astronomical literature. Includes
       general catalogue of stellar masses.

1081. Russell, Henry Norris. "The Royal Road of Eclipses."
       Centennial Symposia (item 141), 181-209.

       Reviews knowledge obtainable by observations of eclipsing
       binary systems. Briefly traces history of study concen-
       trating on late 19th- and early 20th-century work. 29
       citations.

1082. See, T.J.J. *Researches on the Evolution of the Stellar
       Systems*. 2 volumes. Lynn, Mass.: Nichols, 1896; 1910.
       Pp. vii + 252; vii + 734.

       Volume 1 includes a 64-page historical section on the
       development of double star astronomy and on the mathe-
       matical theories of binary star motion. Volume II is
       the author's exposition of his "Capture Theory of Cosmical
       Evolution." Volume 1 reviewed by S.W. Burnham in: *Popular
       Astronomy*, 4 (1897), 471-474.

1083. Sidgreaves, Walter. "Spectroscopic Studies of Astrophysical
       Problems at Stonyhurst College Observatory." *Royal In-
       stitution Library of Science: Astronomy* (item 140),
       47-60.

       Lecture presented January 22, 1904. Concentrates on
       pioneer work there on the spectroscopic orbit of Beta
       Lyrae, dubbed by Agnes Clerke the "Problem Star."

*   Sitterly, Charlotte Moore. "Collaboration with Henry
       Norris Russell over the Years."

       Cited herein as item 933.

1084. Spitzer, Lyman. "Russell and Theoretical Astrophysics."
       In Memory of Henry Norris Russell (item 1356), 3-8.

       Brief review of scientific highlights of Russell's
       career centering upon a series of landmark papers between
       1928-1929 when Russell refined his derivative technique
       of dynamical parallaxes (the use of binary star orbits

as distance indicators); provided a method for the analysis
of binary star systems with eccentric orbits to provide
data on the internal densities of the member stars; and
published defining work establishing the quantitative
relative abundances of 56 elements in the solar atmosphere
showing the predominance of hydrogen and helium.

*   Struve, Otto. "The Analysis of Peculiar Stellar Spectra."

   Cited herein as item 1122.

1085.   Struve, Otto. "Milestones in Double Star Astronomy." *Sky
   & Telescope*, 24 (July 1962), 17-19.

   Popular sketch of highlights from John Goodricke's
   contributions to the present. No historical citations.

1086.   Van Biesbroeck, G. "Visual Binary Stars and Stellar
   Parallaxes." *Astrophysics: A Topical Symposium* (item
   920), 425-447.

   Examines historical relationship between studies of
   visual binary orbits and parallax measures. Bulk of
   paper is a contemporary (1950) review of techniques for
   the observation and reduction of visual binary orbits,
   and the geometric and indirect methods of stellar distance
   determination. 19 citations.

1087.   Wood, Frank Bradshaw. "The Contributions of Henry Norris
   Russell to the Study of Close Binary Stars." *In Memory
   of Henry Norris Russell* (item 1356), 47-49.

   Reviews Russell's papers on the analysis of orbits of
   eclipsing binary stars for masses and interior density
   distributions.

VARIABLE STARS, NOVAE AND SUPERNOVAE

1088.   Argelander, Fr. "The Variable Stars [1844]." *Popular
   Astronomy* , 20 (1912), 91-99; 148-156; 207-217.

   Translation by Annie J. Cannon of an 1844 paper out-
   lining the development and state of variable star work.
   No direct citations.

1089.   Ashworth, William B., Jr. "A Probable Flamsteed Observation of the Cassiopeia A Supernova." *Journal for the
   History of Astronomy*, 11 (1980), 1-9.

Argues that 3 Cas, which appeared in Flamsteed's
original *Historia Coelestis* (1725), and which was
excised by Francis Baily for his corrected edition, the
British Catalogue of 1835, was indeed an observation of
the supernova now identified as Cassiopeia A. Note added
in proof identifies simultaneous work of others that
differ in conclusion of identity. 31 citations.

*   Baade, Walter. *Evolution of Stars and Galaxies*.

        Cited herein as item 906.

*   Bobrovnikoff, N.T. "The Discovery of Variable Stars."

        Cited herein as item 666.

1090.   Clark, David H., and F. Richard Stephenson. *The Historical
        Supernovae*. Oxford: Pergamon, 1977. Pp. x + 233.

        Use of historical records of novae and the recovery
        and evaluation of these records. General compilation and
        discussion of all known events, together with a compre-
        hensive listing of 120 remnants known in our galaxy to
        have resulted from supernovae. Reviewed by C. Payne-
        Gaposchkin: *JHA*, 10 (1979), 47-50.

*   Dorman, L.I. *Cosmic Rays, Variations and Space Explora-
        tions*.

        Cited herein as item 499.

1091.   Dreyer, J.L.E. "William Herschel's Observations of
        Variable Stars and Stars Suspected of Variability."
        Monthly Notices of the Royal Astronomical Society,
        78 (1918), 554-568.

        Exposition of unpublished observations recently un-
        covered among Herschel's papers.

1092.   Duyvendak, J.J.L., N.U. Mayall, and J.H. Oort. "Further
        Data Bearing on the Identification of the Crab Nebula
        with the Supernova of 1054 A.D. Part I: The Ancient
        Oriental Chronicles; Part II: The Astronomical Aspects."
        Publications of the Astronomical Society of the Pacific,
        54 (1942), 91-104.

        Examines the historical data and early attempts to
        link the Crab with the "guest star" observed in the Far
        East in 1054 A.D. Argues for new evidence showing that
        event was indeed a supernova based upon its daylight
        visibility. 38 citations.

*   Fernie, J.D. "The Period–Luminosity Relation: A Historical
        Review."

        Cited herein as item 1071.

1092a. Goldstein, Bernard R. "Evidence for a Supernova of A.D.
        1006." *The Astronomical Journal*, 70 (1965), 105–111.

        Important historiographical case study of a very bright
        stationary object observed in 1006 and recorded in records
        from Spain, Europe, Egypt, Iraq and the Far East. Ex-
        tensive bibliographical citations. Additional commentary
        provided by Goldstein and Ho Peng Yoke in: *The Astro-
        nomical Journal*, 70 (1965), 748–753, adding sightings
        from China and Japan. The radio remnant of this event
        is reported by F.F. Gardner and D.K. Milne in: *The Astro-
        nomical Journal*, 70 (1965), 754. The optical remnant
        is discussed by S. van den Bergh in: *The Astrophysical
        Journal (Letters)*, 208 (1976), L17.

*   Hoskin, Michael A. "Ritchey, Curtis and the Discovery of
        Novae in Spiral Nebulae."

        Cited herein as item 1193.

1093.   Hoskin, Michael A. "Goodricke, Pigott and the Quest for
        Variable Stars." *Journal for the History of Astronomy*,
        10 (1979), 23–41.

        Portrays the astronomical alliance between Edward
        Pigott (1753-1825) and John Goodricke (1765-1786), gentry
        in York, and their mutual interests in deriving the
        physical cause for the variability of stars, especially
        Algol. Their belief in the cause of variability as due
        to eclipses by a dark body in orbit around the star was
        contested by William Herschel. Highlights their contact
        and relations with Herschel, and their changing views on
        the causes of stellar variability. Establishes Pigott's
        leading role in the intellectual partnership. 109 citations.

1094.   Imaeda, K., and T. Kiang. "The Japanese Record of the
        Guest-Star of 1408." *Journal for the History of Astronomy*,
        11 (1980), 77–80.

        Concludes that the event was probably a supernova,
        similar to Tycho's Nova, and that it possibly can be
        identified with the x-ray source Cygnus X-1. 13 citations.

*   Jaschek, Carlos. "Data Growth in Astronomy."

        Cited herein as item 170.

1095. Jones, Kenneth Glyn. "S Andromedae, 1885: An Analysis
       of Contemporary Reports and a Reconstruction." *Journal
       for the History of Astronomy*, 7 (1976), 27-40.

       Examines record of the first, and the brightest, super-
       nova ever observed in an external galaxy and argues that
       its contemporary interpretation "remained an obstacle
       in the way of progress to a firmer grasp of the scale
       of the universe for nearly forty years." The great
       apparent brightness of the object suggested that its
       distance could not be very great. 52 citations.

1096. Ledoux, P., and T. Walraven. "Variable Stars." *Encyclo-
       paedia of Physics*, 51 (1958), 353-604.

       Provides historical introduction within a general con-
       temporary review. 498 citations.

*   Mayall, R. Newton. "The Story of the AAVSO."

       Cited herein as item 368.

1097. McLaughlin, Dean B. "The Study of Stellar Variation."
       Popular Astronomy, 53 (1945), 323-340; 369-388.

       Reviews 50 years of variable star study noting progress
       in catalogue data; techniques of observation and reduction;
       studies of eclipsing binaries as well as intrinsic variable
       of all major types. 97 citations.

1098. Milne, E.A. "Theory of Pulsating Variables." *Handbuch der
       Astrophysik*, 3, pt. 2 (item 920), 804-821.

       Detailed mathematical treatment contained within part
       2 of Volume 3 of the *Handbuch*. Includes brief historical
       commentary on origins of both the theory of pulsation and
       theories of fluid motion in a radiation field. Numerous
       citations.

1099. Payne-Gaposchkin, Cecilia. "The Intrinsic Variable Stars."
       Astrophysics: A Topical Symposium (item 920), 495-525.

       Contemporary review (1950) of knowledge of stars that
       vary in brightness, color, and spectrum due to changes
       in their physical structure. These include five main
       types: W Virginis (cluster variables), Cepheids, RV Tauri
       types, semiregular variables and long period variables.
       Examines the variations in physical characteristics of
       these types in period, motion, and spectrum as a con-
       tinuous sequence. Provides a concise bibliography of
       catalogues of intrinsic variable stars.

*   Payne-Gaposchkin, Cecilia. *Variable Stars & Galactic Structure*

      Cited herein as item 1175.

1100. Payne-Gaposchkin, Cecilia, and Sergei Gaposchkin. *Variable Stars*. Cambridge, Mass.: Harvard College Observatory, 1938. Pp. xiv + 382.

      Comprehensive descriptive review and analysis of recent observational research on all types of variable stars including eclipsing binaries, all forms of intrinsic periodic and a-periodic variables, and extrinsic variables (associated with nebulosity). Includes a chapter on techniques. Tables, numerous footnote references.

1101. Rosseland, Svein. *The Pulsation Theory of Variable Stars*. Oxford: Clarendon Press, 1949. Pp. viii + 152. Reprinted, New York: Dover, 1964.

      Provides an introductory chapter on the origin and growth of the pulsation theory since John Goodricke's detection of Delta Cephei in 1784, and the first pulsation hypothesis advanced by August Ritter in 1879. Monograph concentrates on the mathematical theory of pulsation. Numerous citations.

*   Rossi, Bruno. *Cosmic Rays*.

      Cited herein as item 509.

*   Shapley, Harlow. "A Half Century of Globular Clusters."

      Cited herein as item 1161.

1102. Stratton, F.J.M. "Novae." *Handbuch der Astrophysik*, 6 (item 130), 251-298.

      Reviews, in Volume 6, data on nova events (including supernovae, which were not identified as a separate class at the time) including historical records. Identifies the observed physical characteristics of novae and their relationship with nebulae. Provides a historical review of theories explaining their cause. Numerous citations.

*   Struve, Otto. "The Analysis of Peculiar Stellar Spectra."

      Cited herein as item 1122.

1103. Turner, H.H. "The New Star in Gemini." *Royal Institution Library of Science: Astronomy* (item 140), 36-46.

Lecture presented June 5, 1903. Reviews the discoveries
of novae listing 18 since Tycho Brahe's discovery of the
1572 "new star." Emphasizes role of photographic patrols,
and the discovery of Nova Geminorum found at the University
Observatory at Oxford on March 24, 1903. Examines speculation on nature of novae noting that passing from a discussion of their discovery to a discussion of their nature
"is like leaving firm ground for quicksand."

1104.   Walborn, Nolan R., and Martha H. Liller. "The Earliest
        Spectroscopic Observations of Eta Carinae and its
        Interaction with the Carina Nebula." *Astrophysical
        Journal*, 211 (1977), 181-183.

        Shows that spectral changes are analogous to normal
        novae, and that the star is indeed associated with the
        nebula. Argues from this work "that the discrepancy
        between [John] Herschel's 1847 description of the nebula
        and its present appearance is real." 27 citations.

1105.   Wesselink, A.J. "Modern Views on Cepheid Variables."
        *Monthly Notices of the Astronomical Society of South
        Africa*, 22 (1963), 91-98.

        Reviews 20th-century history of the determination of
        stellar luminosities, peculiarities in spectra, the cali-
        bration of the Cepheid Period/Luminosity relationship
        and the growth and development of the pulsation theory.
        Describes his precise method for the determination of
        Cepheid characteristics. 36 citations.

1106.   Ze-zong, Xi, and Bo Shu-ren, "Ancient Novae and Super-
        novae Recorded in the Annals of China, Korea, and
        Japan and Their Significance in Radio Astronomy."
        Science, 154 No. 3749 (4 Nov. 1966), 597-603.

        Translation, with notes and references, of a catalogue
        of 90 novae events observed in the Far East to 1700. 40
        citations.

STELLAR SPECTROSCOPY AND SPECTRAL CLASSIFICATION

1107.   Aller, Lawrence H. "Interpretation of Normal Stellar
        Spectra." *Astrophysics: A Topical Symposium* (item 920),
        29-84.

Reviews late 19th-century qualitative analyses of stellar and solar spectra from the work of Young and Lockyer and the then-present limitations upon quantitative studies. The bulk of the chapter examines the state of the art (1950) of the measurement of stellar spectra; spectral line formation; line intensities; and the dependence of the strength of a spectral line upon the number of atoms required to produce it. 61 citations.

*   Aller, Lawrence H. *Astrophysics: The Atmospheres of the Sun and Stars.*

Cited herein as item 903.

*   Cowling, T.G. "Magnetic Fields in Astronomy."

Cited herein as item 911.

1108.   Curtiss, R.H. "Classification and Description of Stellar Spectra." *Handbuch der Astrophysik* 5 (item 130), 1-108.

Comprehensive technical and descriptive review, in Volume 5, of all known systems of spectral classification; their criteria, observations, and instrumental techniques required. Detailed analysis of each system and comparisons of principal systems. Centers upon the development of the Draper System at Harvard. Includes a bibliography of catalogs of stellar spectra and extensive citations.

1109.   Dingle, Herbert. "The Dissociation Hypothesis." *Life and Work of Sir Norman Lockyer* (item 1332), 292-315.

Review of Norman Lockyer's controversial theory of the constitution of matter and its change of physical state on the atomic level as a function of changes in temperature, as revealed by spectrum analysis. No direct citations.

*   Dingle, Herbert. "The Meteoritic Hypothesis and Stellar Evolution."

Cited herein as item 1138.

1110.   Dunham, Theodore. "Methods in Stellar Spectroscopy." *Vistas in Astronomy*, 2 (1956), 1223-1283.

Reviews in technical detail the development of Coudé, or fixed focus, spectrographs at Mount Wilson beginning with W.S. Adams' work in 1911, and continuing through the design of the 100-inch Hooker telescope Coudé focus, of special importance in the early years of modern high-

dispersion stellar spectroscopy. Traces the advance of
technique and design including the introduction of
gratings in 1932, the use of Schmidt cameras in 1934
and the introduction of photo—electric scanning devices
in later years. 69 citations.

1111. Gaposchkin, Cecilia Payne. "Russell and the Composition
of Stellar Atmospheres." *In Memory of Henry Norris
Russell* (item 1356), 15-18.

Emphasizes Russell's application of Catalán's theory of
multiplet spectra to the analysis of solar and stellar
atmospheres through the use of laboratory spectra. Discusses how for the first time intensities of spectral
features could be properly calibrated for the effective
study of composition. 6 citations.

1111a. Herbig, G.H., ed. *Spectroscopic Astrophysics: An Assessment of the Contributions of Otto Struve*. Berkeley:
University of California Press, 1970. Pp. ix + 462.

Reprints 10 slightly edited selections from Struve's
published works that represent the areas he was active
in. Each reprint is followed by detailed commentary by
a contemporary astronomer on Struve's contribution and
the general state of the problem area discussed. Problem
areas include both quantitative and qualitative observational stellar astrophysics. Prepared as a memorial to
Struve's career, and includes a brief appreciation of
Struve by Herbig. Extensive citations to primary literature.

*   Hoffleit, Dorrit. "The Discovery and Exploitation of
Spectroscopic Parallaxes."

Cited herein as item 1060.

1112. Huggins, William. "Celestial Spectroscopy." *Report of
the British Association*, 61 (1891), 3-37.

Presidential address to the British Association reviewing state of spectroscopic astronomy in all its aspects
from observational techniques to theoretical deductions
on the nature of matter in stars and nebulae. Of special
interest is his attention to the role of thermodynamics
in the study of celestial evolution, and the identity
of the chief nebular line.

1113. Huggins, William. *An Atlas of Representative Stellar
Spectra*. London: William Wesley, 1899. Pp. ix + 165 +
12 plates.

Provides detailed commentary on the methods employed
in his observatory, with chapters on instrumentation
and observing technique as well as his interpretation
of the character of his stellar spectra in terms of an
evolutionary classification. The spectra included in
this text represent his classification scheme.

* Huggins, William, ed. *The Scientific Papers of Sir William
  Huggins.*

  Cited herein as item 1316.

* Jaschek, Carlos. "Data Growth in Astronomy."

  Cited herein as item 170.

* Jones, Bessie Zaban, and Lyle Gifford Boyd. *The Harvard
  College Observatory: The First Four Directorships,
  1839-1919.*

  Cited herein as item 305.

* Kayser, H. *Handbuch der Spektroscopie.*

  Cited herein as item 922.

1114. Keenan, P.C., and W.W. Morgan. "Classification of Stellar
  Spectra." *Astrophysics: A Topical Symposium* (item
  920), 12-28.

  After a brief historical introduction, outlines present
  (1950) state of spectral classification centering upon
  the authors' established "MK System" in 1943 that accounts
  for the temperature and absolute magnitudes of stars--
  the basic system in use today. 18 citations.

* Lockyer, J. Norman. *The Chemistry of the Sun.*

  Cited herein as item 1049.

* Lockyer, J. Norman. *Inorganic Evolution as Studied by
  Spectrum Analysis.*

  Cited herein as item 1050.

1115. Maunder, E.W. *Sir William Huggins and Spectroscopic
  Astronomy.* London: T.C. & E.C. Jack, 1913. Facsimile
  reprint: London: P.M.E. Erwood, 1979. Pp. 94.

  Provides a brief overview of fifty years of progress
  in solar, stellar and planetary spectroscopy centering
  upon the contributions of William Huggins.

*   Meadows, A.J. *Science and Controversy: A Biography of Sir Norman Lockyer.*

        Cited herein as item 1341.

*   Menzel, D.H. "The History of Astronomical Spectroscopy."

        Cited herein as item 1017.

1116.   Newall, H.F. *The Spectroscope and Its Work*. London: Society for Promoting Christian Knowledge, 1910. Pp. iv + 163.

        Popular introduction covering all aspects of celestial spectroscopy with emphasis upon its historical development. No direct citations.

*   Nielsen, Axel V. "The History of the Hertzsprung-Russell Diagram."

        Cited herein as item 1127.

1117.   Payne, Cecilia H. *The Stars of High Luminosity*. New York: McGraw-Hill, 1930. Pp. xiii + 320.

        Comprehensive technical monograph reviewing early quantitative estimates of the composition of stellar atmospheres. Reviews the development of spectrophotometry, establishes the contemporary data base, and discusses observational results of studies of stars of all spectral classes as well as variable stars, and their incorporation into theories of stellar atmospheres. Intended to follow up her classic 1925 Ph.D. thesis: *Stellar Atmospheres* (Cambridge, Mass.: Harvard, 1925). Extensive citations to primary literature.

*   Pickering, E.C. "The Light of the Stars."

        Cited herein as item 1064.

1118.   Plotkin, Howard. "Edward C. Pickering, the Henry Draper Memorial, and the Beginnings of Astrophysics in America." *Annals of Science*, 35 (1978), 365-377.

        Reviews the pioneering work of Henry Draper in the photography of stellar spectra and his legacy of support for similar research at Harvard College Observatory. The development of the Henry Draper Memorial at Harvard by E.C. Pickering and the generation and evolution of the Harvard system of spectral classification through 1912 are presented. 63 citations.

1119. Russell, Henry Norris. *The Composition of the Stars*.
       Oxford: Clarendon Press, 1933. Pp. 31.

       The Halley Lecture for 1933 reviewing recent advances,
       primarily the applications of quantum physics, in the
       analysis of stellar atmospheres. Includes growth of
       observational knowledge of the spectra of the Sun and
       stars, advances in the theory of complex spectra, the
       theory of ionization and two-electron spectra, and the
       calibration of line strengths.

*   Scheiner, Julius. *Die Spectralanalyse der Gestirne*.

       Cited herein as item 930.

*   Sitterly, Charlotte Moore. "Collaboration with Henry
       Norris Russell over the Years."

       Cited herein as item 933.

1120. Struve, Otto. "Some New Trends in Stellar Spectroscopy."
       Popular Astronomy, 43 (1935), 483-496; 559-568; 628-639.

       General review of the quantitative analysis of stellar
       atmospheres. No direct citations.

1121. Struve, Otto. "The Observation and Interpretation of
       Stellar Absorption Lines." *Popular Astronomy*, 46 (1938),
       431-451; 497-509.

       Technical review of recent progress in the quantitative
       analysis of stellar spectra. No direct citations.

1122. Struve, Otto. "The Analysis of Peculiar Stellar Spectra."
       Astrophysics: A Topical Symposium (item 920), 85-144.

       Reviews the state of knowledge, circa 1950, of eight
       important types of objects that produce unusual spectra:
       hot blue stars with extended atmospheric shells; close
       binaries with gaseous rings; spectra with peculiar line
       intensities; Wolf-Rayet stars; T-Tauri variables; general
       spectra of stars with extended atmospheres; novae; emis-
       sion spectra in long-period and Cepheid-type stars. 3
       citations.

*   Unsöld, Albrecht. *Physik Der Sternatmosphären*.

       Cited herein as item 938.

*   Woolley, R. v.d. R., and D.W.N. Stibbs. *The Outer Layers
       of a Star*.

       Cited herein as item 940.

THE HERTZSPRUNG-RUSSELL DIAGRAM

*   Baade, Walter. *Evolution of Stars and Galaxies*.

      Cited herein as item 906.

*   DeVorkin, David H. "Michelson and the Problem of Stellar
      Diameters."

      Cited herein as item 1057.

1123.   DeVorkin, David H. "The Origins of the Hertzsprung-
      Russell Diagram." *In Memory of Henry Norris Russell*
      (item 1356), 61-77.

      Examines Russell's early career circa 1898-1910 and
the influences that led him to the independent co-discovery of the existence of two separate luminosity
classes of stars. 33 citations.

1124.   DeVorkin, David H. "Steps Toward the Hertzsprung-Russell
      Diagram." *Physics Today*, 31 (1978), 32-39.

      Traces the origins of the diagram through the independent
researches of Russell and Hertzsprung, the motives that
brought both to the relationship, and the factors that
both aided and prevented its initial acceptance—notably
E.C. Pickering's scepticism over the reality of Hertzsprung's findings during the very same period Pickering
was aiding Russell in research that was to confirm
Hertzsprung's work. 12 citations.

1125.   Herrmann, D.B., ed. *Zur Strahlung der Sterne* (Ostwalds
      Klassiker Band 255). Leipzig: Geest & Portig, 1976.
      Pp. 100.

      After a 28-page introduction (in German) to the life
and work of Ejnar Hertzsprung, Herrmann reprints, with
extensive annotation, Hertzsprung's three famous but
quite inaccessible papers (except for the third, which
appeared in the *Astronomische Nachrichten*) which established the bases for the present Hertzsprung-Russell
Diagram. Reviewed in: *Annals of Science*, 34 (1977), 327.

1126.   Herrmann, D.B., ed. *Das Hertzsprung-Russell Diagram*.
      Berlin-Treptow: Archenhold-Sternwarte Vorträge und
      Schriften No. 56, 1978. Pp. 39.

      Six articles (in German) from a 1977 Symposium held at
Archenhold-Sternwarte in honor of Russell's 100th birth-

day and the 10th anniversary of Hertzsprung's death.
Papers include "Die historischen Wurzeln des HRD" (D.B.
Herrmann); "Die Veränderlichen im HRD-einst und Jetzt"
(P. Enskonatus)--a discussion of the changing theories
of Cepheid variables; "Der Briefwechsel zwischen Hertz-
sprung und Russell über das HRD" (D.B. Herrmann and H.
Zach); and "Konstruktion eines Hertzsprung-Diagrams
nach Originaldaten" (F. Jansen).

1126a. Herrmann, D.B. "W.H.S. Monck über Eigenbewegungen und
Spektraltypen der Sterne." *Die Sterne*, 56 (1980),
170-174.

Reviews the work of this obscure but brilliant Dublin
amateur who, in the 1890s, came very close to detecting
the existence of giant and dwarf stars from a statistical
study of proper motions and spectra. 22 citations.

* Hoffleit, Dorrit. "The Discovery and Exploitation of
Spectroscopic Parallaxes."

Cited herein as item 1060.

* Jeans, J.H. *Astronomy and Cosmogony*.

Cited herein as item 921.

1127. Nielsen, Axel V. "The History of the Hertzsprung-Russell
Diagram." *Centaurus*, 9 (1963), 219-253.

Detailed examination of the discovery that there is a
well-defined double-valued correlation of the colors (or
spectra) of stars with their luminosities. Centers upon
the life and work of Hertzsprung, his early years as a
"private astronomer" in Copenhagen, his defining papers
between 1905 and 1909, and the eventual acceptance of
his discoveries only after their independent co-discovery
and announcement by H.N. Russell in the period 1910-1913.
Provides historical background on the growth of spectral
classification and stellar statistics leading to the
diagram. 64 citations.

* Philip, A.G. Davis, and David H. DeVorkin, eds. *In Memory
of Henry Norris Russell*.

Cited herein as item 1356.

1128. Sitterly, Bancroft W. "Changing Interpretations of the
Hertzsprung-Russell Diagram, 1910-1940: A Historical
Note." *Vistas in Astronomy*, 12 (1970), 357-366.

Describes and comments upon three major changes that took place. The first was Russell's theory of stellar evolution based upon the diagram where stars began as red giants, contracted and heated to become Main Sequence stars, and then cooled and moved down the Main Sequence to extinction. The second occurred in the period 1923-1925 when Russell, Adams and Joy, and independently Hertzsprung, all on the observational side, and Eddington on the theoretical side, derived the mass/luminosity relationship for stars which indicated that the Main Sequence could not be a cooling sequence in evolution. The third major change came in the late 1930s when Gerard Kuiper showed that curves of constant composition on the diagram, produced theoretically by Strömgren, bore a marked similarity to curves and patterns produced by stars in open clusters. Kuiper suggested that this indicated differing internal compositions for stars in different clusters, which then stimulated Strömgren to examine the evolution of a star that was transmuting hydrogen into helium, and to find that as this occurred, a star passed from being a Main Sequence star to being a giant, exactly opposite to what Russell envisioned in 1913. 43 citations.

*   Smith, Robert W. "Russell and Stellar Evolution--His 'Relations Between the Spectra and Other Characteristics of the Stars.'"

          Cited herein as item 1148.

1129.   Strand, K.Aa. "Hertzsprung's Contributions to the HR Diagram." *In Memory of Henry Norris Russell* (item 1356), 55-59.

          Brief review of Hertzsprung's three papers between 1905 and 1909 that resulted in his discovery of two separate luminosity classes of stars, today called "giant" and "dwarf" stars. 5 citations.

*   Strömgren, Bengt. "Evolution of Stars."

          Cited herein as item 1149.

1130.   Turner, H.H. "Giant Suns." *Royal Institution Library of Science: Astronomy* (item 140), 229-238.

          Lecture delivered January 31, 1919. Reviews the various lines of work that led to the recognition of the existence of giant stars. Includes parallax determinations, the origin and significance of the Hertzsprung-Russell Diagram studies in stellar structure and evolution, and the new technique of spectroscopic parallaxes of W.S. Adams and Anton Kohlschütter that greatly increased knowledge of stellar luminosities.

1131. Waterfield, R.L. "The Story of the Hertzsprung-Russell Diagram." *Journal of the British Astronomical Association*, 67 (1956), 1-24.

General review of independent development of diagram by Hertzsprung and Russell during the period 1905-1913. Continues with a discussion of Russell's theory of evolution based upon the diagram, events that led to the abandonment of the theory, and contemporary advances, including the discoveries of the technique of spectroscopic parallaxes, the mass/luminosity relationship, and the source of energy of the Sun and stars. Examines the developing knowledge of stellar structure and evolution within the context of modern stellar and galactic astrophysics.

STELLAR STRUCTURE AND EVOLUTION

*   Aller, Lawrence H. *Astrophysics: Nuclear Transformations, Stellar Interiors, and Nebulae.*

Cited herein as item 904.

1132. Atkinson, Robert D'E. "Reminiscences of Henry Norris Russell." *In Memory of Henry Norris Russell* (item 1356), 19-25.

Recalls association with Russell starting in 1929 and their mutual interests in discovering the source of energy of the Sun.

*   Baade, Walter. *Evolution of Stars and Galaxies.*

Cited herein as item 906.

1133. Bishop, Jeanne E. "The Golden Era of Theoretical Physics and the Black Box of Stellar Energy." *The Griffith Observer*, 42 (1978), 3-17.

Examines the introduction of nuclear physics into the problem of the source and maintenance of solar and stellar energy, centering upon events in the 1920s and 30s; changing theories of evolutionary tracks for stars on

the HR Diagram; the discovery of a workable fusion
mechanism by Bethe and Critchfield; and the important
meeting where many of these problems first came together--
the Washington Conference of 1938. Selected bibliography
lists 10 citations.

1134. Burbidge, E.M., G.R. Burbidge, W.A. Fowler, and F. Hoyle.
"Synthesis of the Elements in Stars." *Reviews of Modern Physics*, 29 (1957), 547-650.

General review of processes of nucleosynthesis including
extensive citations to historical literature.

1135. Chandrasekhar, S. *An Introduction to the Study of Stellar
Structure*. Chicago: University of Chicago Press, 1939.
Reprinted, New York: Dover, 1975. Pp. ix + 509.

Rigorous mathematical treatment of all aspects of
problems of stellar structure. Extensive annotated
bibliographical essays at the end of each chapter mark
this work, with others by Chandrasekhar, as a fundamental historical reference. A chapter on stellar energy
was written just before the classic work by Bethe in
1938.

1136. Chandrasekhar, S. "The Structure, the Composition, and
the Source of Energy of the Stars." *Astrophysics: A
Topical Symposium* (item 920), 598-674.

Provides a detailed technical introduction to the theory
of stellar interiors at a time (1950) when it was just
becoming evident that the proton-proton cycle might be
as important as the CNO cycle in solar type stars. Examines
structure and constitution of white dwarf stars, high-
lighting his work on the existence of a limiting mass
for white dwarfs (completely degenerate configurations)
which "must have an important bearing on the final stages
of evolution of massive stars." This last comment bears
note since at the time late stages of stellar evolution
were not at all clear.

1137. Cowling, T.G. "The Development of the Theory of Stellar
Structure." *Quarterly Journal of the Royal Astronomical
Society*, 7 (1966), 121-137.

Presidential address to the RAS reviewing theories of
stellar structure from Lane, Ritter and Emden's polytropic
models in the period 1869-1907; Eddington's introduction
of radiative equilibrium circa 1917-1924; differing views
of stellar structure by Eddington, Jeans, Milne and others;

advances in the 1930s toward models possessing regions
of convection and radiation (largely through Cowling's
efforts); Chandrasekhar's white dwarf theory; the work
of Strömgren, Bethe, and von Weizäcker in the late 30s;
the development of stratified composition models by Öpik,
Hoyle and Lyttleton; to Schwarzschild's and Sandage's
detailed studies in 1952 of the evolution of a star that
has exhausted its supply of hydrogen. Emphasizes the
need to retain a balance between mathematics, physics,
and observational astronomy showing that "excessive
reliance" on any one or the exclusion of any one "can
lead one astray."

1138.   Dingle, Herbert. "The Meteoritic Hypothesis and Stellar
        Evolution." *Life and Work of Sir Norman Lockyer* (item
        1332), 363-381.

        Reviews Norman Lockyer's theory of the origins of stars
        out of nebulae composed of meteoritic material, and his
        system of spectral classification based upon his theory
        of stellar evolution. No direct citations.

1139.   Eddington, A.S. "The Interior of a Star." *Royal Institu-
        tion Library of Science: Astronomy* (item 140), 243-
        257.

        Lecture delivered February 23, 1923. Reviews theoretical
        knowledge of the interiors of gaseous giant stars center-
        ing upon his own contributions to the study of the role
        of radiative equilibrium. Examines, in popular descrip-
        tive fashion, the giant and dwarf theory of stellar evolu-
        tion; M.N. Saha's study of ionization equilibrium; the
        recognition and determination of absorption coefficients
        for different elements; and the source of solar and
        stellar heat.

1140.   Eddington, A.S. *The Internal Constitution of the Stars*.
        Cambridge: Cambridge University Press, 1926. Pp. xvi +
        407. Reprinted: New York: Dover, 1959.

        Classic review and exposition of stellar structure and
        evolution by the leading pioneer in the modern period.
        Provides detailed discussions and evaluations of the
        progress of theory and observation in stellar astrophysics.
        Examines the thermodynamics of radiation, applications
        of quantum theory, stellar models, the sources of stellar
        energy, the question of the existence of a general inter-
        stellar medium. Written when the author had just concluded
        that H.N. Russell's theory of stellar evolution was un-
        tenable. Extensive bibliographical citations at the end
        of each chapter. Reviewed in: *Observatory*, 50 (1927), 56-60.

*   Eggen, Olin. "Stellar Kinematics and Evolution."
         Cited herein as item 1170.

*   Emden, Robert. *Gaskugeln: Anwendungen der mechanischen
         Warmetheorie.*
         Cited herein as item 916.

*   Fison, A.H. *Recent Advances in Astronomy.*
         Cited herein as item 952.

1141.   Hale, George Ellery. *The Study of Stellar Evolution.*
         Chicago: University of Chicago Press, 1908. Pp. xi +
         252 + 104 plates.

         Reviews contemporary theories of stellar evolution
         as a theme for the general exposition of the techniques
         and progress of modern astrophysics as practiced at the
         Yerkes Observatory and at the Mount Wilson Solar Observa-
         tory. Includes discussions of astronomical photography;
         spectroscopy; the study of the Sun; the spectroheliograph;
         astronomical advantages of mountain observatories; stellar
         temperatures and sunspots; the Nebular, Meteoritic and
         Planetesimal Hypotheses; stellar evolution; possibili-
         ties of new instrumentation and the need for large tele-
         scopes. Relies upon theories of Arthur Schuster for dis-
         cussions of stellar atmospheres and evolution. Prefers
         the conventional linear theories of evolution with hot
         stars emerging from nebulae through a process of gravi-
         tational condensation where changes in spectra with
         condensation were thought to be due to distillation.
         No direct citations.

*   Huggins, William. *An Atlas of Representative Stellar
         Spectra.*
         Cited herein as item 1113.

1142.   Jeans, J.H. "Review of Emden's *Gaskugeln*." *Astrophysical
         Journal*, 30 (1909), 72-74.

         Reviews *Gaskugeln* (item 916) with a skepticism typical
         of the period, noting that it is far from certain that
         stars behave according to the perfect gas laws. Feels
         "that many of the problems will only progress when they
         are treated from a more physical point of view ..." but
         welcomes Emden's mathematical contribution.

*   Johnson, Martin. "The Interpretation of Some Changes in Orientation of a Binary Star Orbit."

Cited herein as item 1076.

*   Kuchowicz, B. *Nuclear Astrophysics—A Bibliographical Guide* (4 parts).

Cited herein as item 32.

1143.   Lockyer, J. Norman. *The Meteoritic Hypothesis.* London: Macmillan, 1890. Pp. xvi + 560.

First major exposition of observations, laboratory experimentation, and speculative analysis of Lockyer's theory that stars are formed out of vast swarms of meteoritic material. Reviewed in: *Nation*, 52 (Jan. 1, 1891), 14-15.

*   Lockyer, J. Norman. *Inorganic Evolution as Studied by Spectrum Analysis.*

Cited herein as item 1050.

*   Lockyer, T. Mary, and Winifred L. Lockyer, eds. *Life and Work of Sir Norman Lockyer.*

Cited herein as item 1332.

1144.   McCrea, W.H. "The Constitution and Evolution of Stars." *A Century of Science, 1851-1951* (item 129), 136-152.

Reviews knowledge of the physics and chemistry of the Sun comparing it to state of knowledge in 1851, and provides a brief historical overview of the applications of photography and spectroscopy to problems in solar and stellar physics. Discusses advances in theoretical astrophysics that provided modern picture of stellar structure, energy, and evolution. No direct citations.

1145.   McCrea, W.H. "Evidences of Evolution in Astronomy." *Quarterly Journal of the Royal Astronomical Society*, 3 (1962), 63-78.

Reviews recent (1930s-1960) progress in knowledge of the course of evolution of stars, from both theoretical and observational standpoints. Concentrates on post-Main Sequence stellar evolution and the work of Sandage, Schwarzschild, and other contemporary workers. 12 citations.

*   Meadows, A.J. *Early Solar Physics*.

        Cited herein as item 1016.

*   Milne, E.A. "Theory of Pulsating Variables."

        Cited herein as item 1098.

*   Milne, E.A. *Sir James Jeans, A Biography*.

        Cited herein as item 1343.

*   Russell, Henry Norris, Raymond Smith Dugan, and John Q.
        Stewart. *Astronomy*.

        Cited herein as item 1416.

1146.   Russell, Henry Norris. "Sir Norman Lockyer's Work in
        the Light of Present Astrophysical Knowledge." *Life
        and Work of Sir Norman Lockyer* (item 1332), 382-394.

        Centers upon Lockyer's theory of stellar evolution,
        as revised by Russell during the period 1910-1913, and
        discusses how it has fared in following years. 2 cita-
        tions.

1147.   Sarton, George. "The Discovery of the Law of Conserva-
        tion of Energy." *Isis*, 13 (1929), 18-44.

        Compares contributions of Robert Mayer and James
        Prescott Joule in the discovery of this law and the
        development of thermodynamics. Adds commentary on Carnot,
        Seguin, and Colding and includes facsimile reproductions
        of Mayer's and Joule's earliest papers on the subject
        with a manuscript fragment by Carnot.

1148.   Smith, Robert W. "Russell and Stellar Evolution--His
        'Relations Between the Spectra and Other Characteristics
        of the Stars.'" *In Memory of Henry Norris Russell*
        (item 1356), 9-13.

        Brief study of Russell's first complete exposition of
        his "Russell" Diagram (now the Hertzsprung-Russell
        Diagram). 10 citations.

1149.   Strömgren, Bengt. "Evolution of Stars." *Astronomical
        Journal*, 57 (1952), 65-83.

        Technical review of the problem with emphasis on the
        construction of interior models for the Sun and stars;
        the evolutionary changes in composition, structure, mass
        and rotation; the formation of stars; and the use of the

Hertzsprung-Russell Diagram for interpreting problems in stellar evolution. 65 citations.

1150. Struve, Otto. "Charles Darwin and the Problem of Stellar Evolution." *Sky & Telescope*, 18 (1959), 240-242.

Argues that present state of knowledge of stellar evolution is similar to knowledge of biological evolution just prior to Darwin's publication of *The Origin of Species*. No great synthesis has yet emerged in astronomy, but "its raw materials ... may be said to exist in the magnificent book *Stellar Structure*, edited by S. Flügge." Struve then reviews the contents of this work, volume 51 of the *Handbuch der Physik*.

*   Ter Haar, D. "Cosmogonical Problems and Stellar Energy."

Cited herein as item 1053.

TRUE NEBULAE AND THE INTERSTELLAR MEDIUM

*   Aller, Lawrence H. *Astrophysics: Nuclear Transformations, Stellar Interiors, and Nebulae.*

Cited herein as item 904.

1151. Beals, C.S. "The Material of Interstellar Space." *Popular Astronomy*, 52 (1944), 209-228.

After a brief review of late 19th- and early 20th-century studies, this review concentrates upon advances in the 1930s after the work of R.J. Trumpler. 88 citations.

*   Berendzen, Richard, Richard Hart, and Daniel Seeley. *Man Discovers the Galaxies.*

Cited herein as item 1183.

*   Cederblad, S. "Studies of Bright Diffuse Galactic Nebulae, with Special Regard to Their Spatial Distribution."

Cited herein as item 1165.

*   Cowling, T.G. "Magnetic Fields in Astronomy."

Cited herein as item 911.

1152. Curtis, Heber D. "The Nebulae." *Handbuch der Astrophysik*,
       5 (item 130), 774-936.

       Reviews, in Volume 5, the growth of knowledge of all
       classes of true nebulae, providing historical commentary
       on their cosmological roles. Centers upon the nature of
       spiral nebulae and their characteristics as true
       galaxies. Includes finding lists for the nebular classi-
       fications in historical literature. Extensive citations.

1153. Fernie, J.D. "The Historical Quest for the Nature of
       the Spiral Nebulae." *Publications of the Astronomical
       Society of the Pacific*, 82 (1970), 1189-1230.

       Traces the history of observational studies of nebulae
       from first records circa 1500 to 1925. The early period
       centers upon the observations of W. Herschel and specu-
       lations by Swedenborg, Wright, Kant, and Lambert. 19th-
       century studies include John Herschel's extension of
       his father's work and the detection of spirals by Lord
       Rosse, along with the spectroscopic verification of
       nebulae by Huggins and later photographic studies. 20th-
       century work identifying the existence of external
       galaxies includes that of V.M. Slipher, A. van Maanen,
       and E. Hubble. Extensive citations.

1154. Greenstein, Jesse L. "Interstellar Matter." *Astrophysics:
       A Topical Symposium* (item 920), 526-597.

       Provides a contemporary (1950) survey of knowledge of
       all optical aspects of the interstellar medium beginning
       with discrete absorption and general absorption and
       reddening by dust and gas. Examines various types of
       discrete nebulae, their spectra and derivation of their
       physical characteristics. Reviews the theory of the
       structure of interstellar grains, and their galactic
       distribution. The origin of spectra in emission nebulae
       is detailed along with the optical detection in 1937 of
       interstellar hydrogen by Struve and Elvey, and Strömgren's
       theoretical discussion of the properties of hydrogen
       gas in space. Attention is paid to the early detection
       of galactic radio noise providing a brief review of
       observations and theory of emission. Other topics include
       the evolutionary significance of interstellar matter
       and the detection of interstellar polarization. 60 cita-
       tions and brief bibliography.

1155. Hirsh, Richard F. "The Riddle of the Gaseous Nebulae."
       Isis, 70 (1979), 197-212.

Traces the search for the chemical identification of
the major lines seen in emission in gaseous nebulae since
their discovery by William Huggins in 1864. Examines
progress in physics that allowed for the eventual successful identification of the lines by I.S. Bowen in 1927
based upon the quantum theory of the behavior of gases
under extreme conditions. Notes the work of the physicist
J.A. Nicholson circa 1913-1914 and his attempts to identify
the nebular lines. 88 citations.

1156. Nichol, J.P. *The Architecture of the Heavens*. Ninth
       Edition. London: H. Bailliere, 1851. Pp. xiv + 300.

       Important speculative review that first appeared in
       1838 and went through many extensive revisions. This
       edition, dedicated to the wife of Lord Rosse, recounts
       many of the discoveries made with Rosse's telescopes
       and argues for the resolvability of all nebulae. Also
       of specific interest is Nichol's conviction regarding
       the existence of external galaxies and the structure of
       our own galaxy. Other titles produced by this important
       commentator on mid-19th-century cosmology include: *The
       Stellar Universe: Thoughts on Points Connected with the
       System of the World*; *The Planetary System* (which includes
       chapters on the discovery of Neptune and the Nebular
       Hypothesis); and *Contemplations on the Solar System*
       (which includes a chapter on the constitution of the
       Sun). No direct citations or indexes.

1157. Struve, Otto. "Recent Progress in the Study of Reflection
       Nebulae." *Popular Astronomy*, 45 (1937), 9-23.

       After a brief historical introduction, concentrates
       on recent studies at Yerkes Observatory on the detection
       and analysis of the nature of reflection nebulae and
       surrounding interstellar media. No direct citations.

1158. Van de Hulst, H.C. "The Solid Particles in Interstellar
       Space." *Recherches astronomiques de l'Observatoire
       d'Utrecht*, 11, No. 2 (1949), 1-10.

       Provides introduction to theories of composition of
       interstellar dust before the exposition of the author's
       "dirty snow ball" theory.

*   Walborn, Nolan R., and Martha H. Liller. "The Earliest
       Spectroscopic Observations of Eta Carinae and Its
       Interaction with the Carina Nebula."

       Cited herein as item 1104.

1159. Zanstra, H. "The Gaseous Nebula as a Quantum Counter."
       Quarterly Journal of the Royal Astronomical Society,
       2 (1961), 137-148.

       Recounts early work by himself and independently by
       D. Menzel in 1926 that the source of nebular emission
       was ultra-violet radiation. Reviews observational evi-
       dence and then examines present state of research. 18
       citations.

                          STAR CLUSTERS

*   Berendzen, Richard, Richard Hart, and Daniel Seeley.
       Man Discovers the Galaxies.

       Cited herein as item 1183.

*   Milne, E.A. *Sir James Jeans, A Biography*.

       Cited herein as item 1343.

*   Plaskett, J.S. "Modern Conceptions of the Stellar System."

       Cited herein as item 1176.

1160. Shapley, Harlow. *Star Clusters*. Cambridge, Mass.: Harvard
       College Observatory, 1930. Pp. xi + 276.

       Reviews all recent work on star clusters, both open
       and globular, from the late 19th century to date. Similar
       work appears in the *Handbuch der Astrophysik*, 5, pt. 2,
       pp. 698-773. Numerous citations.

1161. Shapley, Harlow. "A Half Century of Globular Clusters."
       Popular Astronomy, 57 (1949), 203-229.

       A retrospective of his own contributions and those of
       Harvard College Observatory examined against the back-
       ground of general progress in the area. Includes the
       identification of variables; their distribution, distances
       representative stellar populations, classifications,
       structure and motions. 34 citations.

1162. Vasilevskis, S. "On Proper Motions of Open Clusters."
       Astronomical Journal, 67 (1962), 699-706.

       Reviews work since the 19th century on the identifica-
       tion of star clusters by group proper motions.

GALACTIC STRUCTURE

*   Berendzen, Richard, Richard Hart, and Daniel Seeley.
         Man Discovers the Galaxies.

         Cited herein as item 1183.

1163.   Bok, Bart J. *The Distribution of the Stars in Space*.
         Chicago: University of Chicago Press, 1937. Pp. xvi +
         124.

         General contemporary review of methods of stellar
         statistics and problems in galactic structure. Provides
         a historical introduction and cites primary studies post-
         1900 throughout text. Written as a general review of
         progress since Shapley established the eccentric loca-
         tion of the Sun in the galactic system and Oort and
         Lindblad demonstrated that the galaxy is in differential
         rotation. 230 citations.

1164.   Campbell, William Wallace. *Stellar Motions*. New Haven:
         Yale University Press, 1913. Pp. xi + 328.

         Based upon the Silliman Lectures of 1910 given at Yale
         University. Comprehensive technical review, with detailed
         historical development, of knowledge of stellar motions
         including technique, methodology and statistical results
         circa 1910. Emphasizes the employment of spectroscopic
         radial velocities in determining the solar motion;
         kinematic properties of the stellar system; spectroscopic
         binary star orbits; and comparison of spectroscopic results
         with astrometric and photometric results. Examines in
         some detail Kapteyn's detection of star streams and the
         observed increase in radial velocity with advancing
         spectral type for which Campbell claims priority. Extensive
         citations to primary sources.

1165.   Cederblad, S. "Studies of Bright Diffuse Galactic Nebulae,
         with Special Regard to Their Spatial Distribution."
         Meddelanden fran Lunds Observatorium, II, No. 119
         (1946), 1-166.

         Includes a historical introduction to the statistical
         analysis of the spatial distribution of stars of different
         spectra and the distribution of various nebular types.

1166.   Chandrasekhar, S. *Principles of Stellar Dynamics*. Chicago:
         University of Chicago Press, 1942. Reprinted, New York:
         Dover, 1960. Pp. viii + 313.

Rigorous mathematical exposition of the dynamics of stellar systems. Relaxation times are determined through the effects of stellar encounters treated as two-body encounters in classical mechanics. Star clusters are also examined in this way. The differential rotation of the galaxy is examined through Liouville's Theorem and the solutions to the equations of continuity. Extensive annotated historical and contemporary bibliography. The Dover reprint includes articles on dynamical friction and "New Methods in Stellar Dynamics" by Chandrasekhar.

1167.   de Sitter, W. *Kosmos*. Cambridge, Mass.: Harvard University Press, 1932. Pp. xii + 138.

Based upon a series of six lectures on the history of knowledge of the structure of the Universe delivered at the Lowell Institute, Boston, in November, 1931. Traces history of astronomy from the Greeks, through Copernicus and the scientific revolution, and the growth of "sidereal astronomy" (William Herschel and his successors). Then he concentrates upon the life-work of J.C. Kapteyn and the growth of modern knowledge about the structure of the galaxy, relativity, and modern theories of the Universe. No direct citations or index. Reviewed in: *Isis*, 20 (1932), 316-318.

1168.   Eddington, A.S. *Stellar Movements and the Structure of the Universe*. London: Macmillan, 1914. Pp. xii + 266.

Comprehensive review of galactic structure circa 1914 that provides an excellent discussion of the solar motion, the star streaming models of Kapteyn, and alternatives by K. Schwarzschild, Halm and Eddington. Chapters are also devoted to the observational data base, star counts, statistical investigations, the nature of nebulae and star clusters, and stellar dynamics. Numerous references and annotated bibliography at the end of each chapter.

1169.   Eddington, A.S. *The Rotation of the Galaxy*. Oxford: Clarendon Press, 1931. Pp. 30.

The Halley Lecture delivered on May 30, 1930. Reviews recent advances in galactic structure and dynamics, notably the work of Oort and Lindblad.

1170.   Eggen, Olin. "Stellar Kinematics and Evolution." *Vistas in Astronomy*, 12 (1970), 367-414.

Reviews the historical relationship between the two studies beginning circa 1900-1914 with the detection of increasing velocity dispersion with advancing (i.e., believed to be an ageing sequence) spectral type and the influence this discovery had on theories of evolution of the stars. Contrasts the scientific styles of Eddington and Russell, and provides comparisons of present knowledge with the problems of earlier periods.

*   Gill, D., and J.C. Kapteyn. "Cape Photographic Durchmusterung Volume 1."

Cited herein as item 836.

*   Herrmann, Dieter B., ed. *Zur Strahlung der Sterne* (Ostwalds Klassiker Band 255).

Cited herein as item 1125.

1171.   Jeans, J.H. *Problems of Cosmogony and Stellar Dynamics*. Cambridge: Cambridge University Press, 1919. Pp. viii + 293.

Classic technical exposition of Jeans' studies of the motions and stability of stellar systems, nebulae, etc. Numerous citations.

*   Jones, Harold Spencer. "The Structure of the Universe."

Cited herein as item 1210.

1172.   Kapteyn, J.C. "Recent Researches in the Structure of the Universe." *Royal Institution Library of Science: Astronomy* (item 140), 78–96.

Lecture delivered May 22, 1908. Reviews his classic studies in galactic structure which led to the "Kapteyn Universe." Includes studies of stellar positions, motions, and statistical estimates of stellar distances and distribution in space. Presents his cooperative scheme for the detailed study of stellar data in selected regions of the sky, his "Plan of Selected Areas," and argues for its adoption.

1173.   Kapteyn, J.C. *First and Second Report on the Progress of the Plan of Selected Areas*. Groningen: Hoitsema Brothers, 1911. Pp. 34.

Review of major international project in statistical astronomy to provide a comprehensive data base for studies of galactic structure. Third and fourth reports published by Kapteyn's successor, P.J. van Rhijn: *Third Report*. Groningen: Hoitsema Brothers, 1923. Pp. 96;

"Fourth Report." *Bulletin of the Astronomical Institutes of the Netherlands*, 6 (1930), 75–81. Extensive citations to primary sources.

1174.   Oort, Jan H. "The Development of our Insight into the Structure of the Galaxy between 1920 and 1940." *Education in and History of Modern Astronomy* (item 125), 255–266.

Personal account from era when conflicting models of the galaxy by Kapteyn and Shapley were proposed. Examines statistical work by van Rhijn and by Oort on the detection of differential rotation, Lindblad's theoretical work and studies of interstellar absorption by Trumpler. 41 citations and discussion.

1175.   Payne-Gaposchkin, Cecilia. *Variable Stars & Galactic Structure*. London: Athlone Press, 1954. Pp. xii + 116.

Monograph based upon a series of lectures given at the University of London in 1952. Reviews knowledge of pulsating and eruptive variable stars, and their role in mapping out the structure of the galaxy during the first half of the 20th century. 149 citations.

1176.   Plaskett, J.S. "Modern Conceptions of the Stellar System." *Popular Astronomy*, 47 (1939), 239–256.

Reviews knowledge of the structure of the Milky Way including a historical introduction from W. Herschel through J.C. Kapteyn. No direct citations.

1177.   Proctor, R.A. "On Star-grouping, Star-drift, and Star-mist." *The Royal Institution Library of Science: Astronomy* (item 140), 103–112.

Briefly reviews the detection of the solar motion by William Herschel and the extension of the study by John Herschel as an introduction to the author's recent suggestion that groups of stars have common motions in space (star-drifts) and that through proper statistical analysis of these motions, the distance to the drifting group can be determined. Argues in addition that all nebulae are within the Milky Way. Lecture dated May 6, 1870.

*   Szanser, Adam J. "Marian Kowalski: A Little Known Pioneer in Stellar Statistics."

Cited herein as item 812.

1178. Thoren, Victor, Charles Gow, and Kent Honeycutt. "An
      Early View of Galactic Rotation." *Centaurus*, 18 (1973–
      1974), 301–314.

      Examines Hugo Gyldén's discovery, in 1871, of evidence
      indicating that the galaxy is in differential rotation,
      a discovery anticipating the recognized priority of
      Lindblad and Oort by some 50 years. Provides translated
      excerpts of Gyldén's "Suggestions of Regularity in the
      Motions of the Stars," and notes that part of Gyldén's
      obscurity may be in the fact that there was little faith
      in the existence of rotational order in the galactic
      system at the time. 15 citations.

1179. Trumpler, R.J., and H. Weaver. *Statistical Astronomy*.
      Berkeley: University of California Press, 1953. Pp.
      xx + 644. Reprinted, New York: Dover, 1962.

      Comprehensive technical review of statistical methods
      of galactic research. Includes annotated bibliographies
      that contain historical information for each chapter.

1180. Van de Kamp, P. "The Galactocentric Revolution, a
      Reminiscent Narrative." *Publications of the Astronomical
      Society of the Pacific*, 77 (1965), 325–335.

      Describes 20th-century observational evidence and
      arguments for the present view of the structure of the
      galaxy and the Sun's place within it. 33 citations.

1180a. van Woerden, Hugo, Willem N. Brouw, and Henk C. van de
      Hulst, eds. *Oort and the Universe: A Sketch of Oort's
      Research and Person*. Dordrecht: D. Reidel, 1980. Pp.
      viii + 210.

      A collection of 22 essays by Jan Oort's students and
      colleagues written in honor of his 80th birthday. Most
      essays are topical reviews of the progress of research
      in galactic and extra-galactic astronomy, international
      cooperation in astronomy, and the many other areas touched
      by Oort during 60 years of research. Several essays con-
      tain valuable general observations of the character of
      astronomical research. Numerous citations.

1181. Woolley, Richard. "The Kinematical and Chemical History
      of the Galaxy." *Journal of the London Mathematical
      Society*, 41 (January 1966), 29–48.

      Describes the recent history of the subject including
      research techniques and major results. No direct citations.

1182. Woolley, Richard. "The Stars and the Structure of the
       Galaxy." *Quarterly Journal of the Royal Astronomical
       Society*, 11 (1970), 403-428.

       Narrative account of galactic structure studies from
       Halley's detection of stellar proper motion in 1717;
       the detection of solar motion by William Herschel in
       1783; the increase in accuracy in stellar parallaxes
       since their first detection in the 1830s; John Herschel's
       accounts of nebulae; Huggins' application of the spectro-
       scope; stellar radial velocities and Eddington's studies
       in galactic structure.

                SPIRAL NEBULAE AND THE EXISTENCE OF GALAXIES

1183. Berendzen, Richard, Richard Hart, and Daniel Seeley.
       Man Discovers the Galaxies. New York: Science History
       Publications, 1976. Pp. x + 228.

       Exposition of development of modern cosmology from
       approximately 1900 through the 1930s. Extensive discussions
       of the Shapley-Curtis "Debate" of April 1920 and the
       influence of van Maanen's apparent detection of rota-
       tional proper motions in spiral nebulae. Demonstrates
       that modern cosmology grew out of statistical studies
       of the positions and motions of stars and the long
       debate over the nature of spiral nebulae. Developed as
       a historical textbook with extensive documentation, re-
       view questions and summaries. Expansion of a series of
       articles in the *Journal for the History of Astronomy*,
       2 (1971), 109-119; 3 (1972), 52-64, 75-86; 4 (1973),
       46-56, 73-98. Reviewed in: *Isis*, 70 (1979), 285-286;
       JHA, 9 (1978), 222-224; *Annals of Science*, 35 (1978),
       91-94.

*   Curtis, Heber D. "The Nebulae."

       Cited herein as item 1152.

*   Fernie, J.D. "The Historical Quest for the Nature of
       the Spiral Nebulae."

       Cited herein as item 1153.

1184. Gordon, Kurtiss J. "History of our Understanding of a
       Spiral Galaxy: Messier 33." *Quarterly Journal of the
       Royal Astronomical Society*, 10 (1969), 293-307.

Comprehensive history of observations and speculation on the nature of M33 from the early 18th-century through contemporary thought. Includes brief discussion of Swedenborg's, Kant's, Wright's and Herschel's contributions. Reviews Rosse's detection of its spiral structure in 1850, spectroscopic and photographic studies by Huggins, Roberts, Keeler, Slipher and modern work. 128 citations.

1185. Hetherington, Norriss S. "The Measurement of Radial Velocities of Spiral Nebulae." *Isis*, 62 (1971), 309-313.

Briefly examines observations and speculations on the motions of spiral nebulae by W.W. Campbell, who suggested a link between motion and stage of evolution; V.M. Slipher, who argued that these motions were a reflection of solar (and hence our galaxy's) motion in space; and Willem de Sitter, who utilized the observations of Slipher and others to predict the expansion of the Universe as an elaboration of Einstein's general relativity. Concludes that "de Sitter's hypothesis" was also the only one of the three to lead to observations significantly directed by theory. 25 citations.

1186. Hetherington, Norriss S. "Adriaan van Maanen and Internal Motions in Spiral Nebulae: A Historical Review." *Quarterly Journal of the Royal Astronomical Society*, 13 (1972), 25-39.

General review of van Maanen's spurious detection of rotation in spiral nebulae through an analysis of proper-motions of discrete objects within the spiral arms. Examines the "initial plausibility" of van Maanen's work; the acceptance of it since it was the result of a standard and well-tested astronomical technique; the problems the measurement raised with other examinations of nebulae for rotation; the "incompatibility" of van Maanen's findings with the island universe hypothesis; the conclusive observations of Edwin Hubble in conflict with van Maanen's results and lingering favor of the latter's work; and finally van Maanen's "retraction" in 1935. Suggests several possible explanations for van Maanen's observations. 79 citations.

1187. Hetherington, Norriss S. "Edwin Hubble on Adriaan van Maanen's Internal Motions in Spiral Nebulae." *Isis*, 65 (1974), 390-393.

Describes Hubble's dilemma in attempting to reconcile
or explain van Maanen's spurious detection of rotational
proper motions in spiral nebulae. Concludes that "no
completely satisfactory explanation of the cause of van
Maanen's specious measurements" exists. 13 citations.

1188.   Hetherington, Norriss S. "Edwin Hubble's Examination of
         Internal Motions of Spiral Nebulae." *Quarterly Journal
         of the Royal Astronomical Society*, 15 (1974), 392-418.

Recounts Hubble's re-examination of Adriaan van Maanen's
conflicting results on the distances to spiral nebulae
after Hubble's successful use of Cepheid variables in
spiral nebulae to show that these objects are extra-
galactic. Provides an extensive analysis of Hubble's
efforts to understand van Maanen's work, chiefly from
an examination of unpublished material. 12 citations.

1189.   Hetherington, Norriss S. "Adriaan van Maanen on the
         Significance of Internal Motions in Spiral Nebulae."
         Journal for the History of Astronomy, 5 (1974), 52-53.

Reprints, with commentary, a section of van Maanen's
original paper in 1916 that was omitted from publication.
Shows that van Maanen's original intent was to examine
the Chamberlin-Moulton theory of cosmogony, but that he
was also aware of the significance of his results for
the distances to spiral nebulae. 5 citations.

1190.   Hetherington, Norriss S. "The Simultaneous 'Discovery'
         of Internal Motions in Spiral Nebulae." *Journal for
         the History of Astronomy*, 6 (1975), 115-125.

Demonstrates that interest in detecting rotational
proper motions in spiral nebulae was stimulated by the
measurement by V.M. Slipher of rotational velocities by
spectroscopic means between 1912 and 1914. Reviews
photographic proper motion studies of van Maanen, Lampland
and Curtis, as well as Isaac Roberts' earlier spurious
results. Examines why van Maanen and others would expect
to find rotational motion, including expectations from
Slipher's work noted above, and from the planetesimal
hypothesis of Moulton and Chamberlin. 59 citations.

*   Hetherington, Norriss S. "Adriaan van Maanen's Measurements
         of Solar Spectra for a General Magnetic Field."

         Cited herein as item 1007.

* Hetherington, Norriss S. "Observational Cosmology in the Twentieth Century."

Cited herein as item 1208.

1191. Hoskin, M.A. "Edwin Hubble and the Existence of External Galaxies." *XIIᵉ Congrès International d'Histoire des Sciences, Tome V (Paris, 1968)*. Paris: Blanchard, 1971. Pp. 49-53.

Recounts Hubble's original announcements of his determination of the distances to galaxies and the events that ensued. 23 citations.

1192. Hoskin, M.A. "The 'Great Debate': What Really Happened." *Journal for the History of Astronomy*, 7 (1976), 169-182.

Detailed account of origin, organization, and execution of a meeting of the National Academy of Sciences in Washington in April, 1920 when Harlow Shapley and Heber D. Curtis gave opposing views over the dimensions of the Milky Way and the nature of spiral nebulae. Shows that heretofore published papers by Shapley and Curtis are not the verbatim transcripts of their arguments at the meeting. Reproduces Shapley's transcript, and Curtis' notes and sequence of slides. 50 citations.

1193. Hoskin, M.A. "Ritchey, Curtis and the Discovery of Novae in Spiral Nebulae." *Journal for the History of Astronomy*, 7 (1976), 47-53.

Reviews the contributions of Ritchey and Curtis to the establishment of the use of novae as distance indicators for spiral nebulae circa 1917. Emphasizes Curtis' careful analysis of faint novae, as distinguished from the anomalously bright event in Andromeda in 1885 and another event in 1895. 35 citations.

1194. Hubble, E.P. *The Realm of the Nebulae*. New Haven: Yale University Press, 1936; New York: Dover, 1958. Pp. xiv + 207.

Consists of Hubble's Silliman Lectures delivered at Yale in 1935 covering general topics in extra-galactic astronomy which were largely the results of his own research. Includes the recognition of extragalactic "spiral nebulae" or galaxies; the characteristics of galaxies; distribution of galaxies in space and their tendency to cluster; development of criteria for determining the distances to galaxies; the recognition of the velocity-distance relationship for galaxies from Slipher's pioneering work to its establishment by Hubble and Humason general cosmological theories of Einstein, Lemaitre and Milne. Identifies Hubble's conservative philosophy wherein he was reluctant to interpret the red shifts of galaxy spectra as indicative of a velocity of recession. Numerous citations to primary literature.

*   Jaki, Stanley L. *The Milky Way: An Elusive Road for Science*.

Cited herein as item 671.

*   Jones, Kenneth Glyn. "S Andromedae, 1885: An Analysis of Contemporary Reports and a Reconstruction."

Cited herein as item 1095.

1195. MacPherson, Hector. "Herschel's World-View in the Light of Modern Astronomy." *The Observatory*, 45 (1922), 254-261.

Identifies the apparent rehabilitation of William Herschel's conception of the morphology and extent of the visible Universe, based upon H. Shapley's then-popular conception of the structure and extent of the Milky Way. Numerous citations.

*   Proctor, R.A. "On Star-grouping, Star-drift, and Star-mist."

Cited herein as item 1177.

*   Whitney, C.A. *The Discovery of Our Galaxy*.

Cited herein as item 116.

*   Whitrow, G.J. "Kant and the Extragalactic Nebulae."

Cited herein as item 595.

1196. Whitrow, G.J. "Theoretical Cosmology in the Twentieth Century." *Human Implications of Scientific Advance* (item 362), 576-593.

Reviews the mathematical foundations of 20th-century cosmological theory beginning with Carl Neumann's 1874 identification of cosmological limitations of Newtonian theory which were later followed up by Hugo von Seeliger, C.V.L. Charlier and others. Concentrates on relativistic theories of Einstein, Willem de Sitter, G. Lemaitre, H.P. Robertson, Alexander Friedmann, E.A. Milne, A.G.

Walker, P.A.M. Dirac, and A.S. Eddington. The steady
state theories of Fred Hoyle, H. Bondi and T. Gold are
mentioned briefly. 42 citations.

EXTRA-GALACTIC ASTRONOMY AND COSMOLOGY

1197. Adam, M.G. "The Observational Tests of the Gravitation
Theory." *Proceedings of the Royal Society Series A*,
270, No. 1342 (1962), 297-305.

Opening paper at Society session, held 22 February
1962, on present state of relativity. Reviews the various
astronomical tests of relativity including the advance
of the perihelion of Mercury, the deflection of light,
the red shift of spectral lines in white dwarf stars.
Discusses laboratory measurements of the red shift. 15
citations and group discussion.

* Baade, Walter. *Evolution of Stars and Galaxies*.

Cited herein as item 906.

* Berendzen, Richard, Richard Hart, and Daniel Seeley.
Man Discovers the Galaxies.

Cited herein as item 1183.

* Brill, Dieter R., and Robert C. Perisho. "Resource
Letter GR-1 on General Relativity."

Cited herein as item 749.

1198. Dickson, F.P. *The Bowl of Night*. Cambridge, Mass.: M.I.T.
Press, 1968. Pp. ix + 228.

Descriptive introduction to cosmology including
historical aspects. Olbers' Paradox, early relativity
theory, and the development of modern cosmological
theories are presented. Reviewed in: *Isis*, 61 (1970),
125.

1199. Eddington, A.S. "The Expanding Universe." *Royal Institution Library of Science: Astronomy* (item 140), 305-
313.

Popular lecture dated January 22, 1932 on the cosmological theory of Willem de Sitter recently developed
by Friedmann and Lemaitre. Emphasizes how it agrees with

the recent observations of the recessions of galaxies
by Hubble. Examines the presence of a "cosmical constant"
in both the realm of general relativity and in quantum
mechanics. Concludes in general that "We can feel little
doubt therefore that the observed motions of the nebulae
are genuine and represent the expansion effect predicted
by relativity. We must reconcile ourselves to this
alarming rate of expansion, which plays havoc with older
ideas as to the time-scale." This lecture was expanded
into a general review for the September 1932 meetings
of the IAU at Harvard and ultimately became the basis
for his popular text *The Expanding Universe*.

1200. Eddington, A.S. *The Expanding Universe*. New York:
       Macmillan, 1933. Pp. viii + 182.

       Expansion of public lecture providing a popular exposi-
       tion of the search for the cosmical constant, the "most
       elusive constant of nature." Links astronomy, relativity
       and wave mechanics in an attempt to develop a unified
       theory, based upon the existence of a cosmical constant
       within a generalized theory of universal gravitation,
       where an association is sought between the size of the
       electron and the curvature of space. Reviewed in: *Isis*,
       21 (1934), 322–326.

1201. Field, George B., Halton Arp, and John H. Bahcall. *The
       Redshift Controversy*. Massachusetts: W.A. Benjamin,
       1973. Pp. xv + 324.

       Contains extended statements by Arp and Bahcall arguing
       for the various interpretations of redshifts of galaxies
       and quasars, and reprints important papers supporting
       these views.

*   Forbes, Eric Gray, ed. *Human Implications of Scientific
       Advance*.

       Cited herein as item 362.

1202. Greenstein, Jesse L. "Quasi-stellar Radio Sources."
       Scientific American, 209 (Dec. 1963), 54–62.

       Reviews, based partly upon direct participation, the
       discovery of QSOs at the Hale Observatories and Caltech.
       No direct citations.

1203. Harder, Allen J. "The Copernican Character of Einstein's
       Cosmology." *Annals of Science*, 29 (1972), 339–347.

Demonstrates that Einstein's role in the development
of modern relativistic cosmology was similar to that of
Copernicus in initiating the revolution that bears his
name. 30 citations.

1204.   Harder, Allen J. "E.A. Milne, Scientific Revolutions
and the Growth of Knowledge." *Annals of Science*, 31
(1974), 351-363.

Reviews Milne's alternative theory of kinematic rela-
tivity and cosmology as expounded in his 1932 work "World
Structure and the Expansion of the Universe." Contrasts
Milne's theory to that of general relativity and examines
Milne's methodology within the context of the rejection
of his theory. 69 citations.

1205.   Harre, R., ed. *Scientific Thought, 1900-1960: A Selective
Survey*. Oxford: Clarendon Press, 1969. Pp. viii + 277.

Contains a chapter by C.W. Kilmister on relativity and
cosmology which includes an annotated bibliography.

1206.   Harrison, E.R. "Olbers' Paradox and the Background Radiation Density in an Isotropic Homogeneous Universe."
Monthly Notices of the Royal Astronomical Society,
131 (1965), 1-12.

Mathematical demonstration of how Olbers' assumptions
account for the present radiation level.

1207.   Heckmann, Otto. "150 Jahre Kosmologie." *Sterne und
Weltraum*, 5 (1966), 276-285.

Descriptive survey of changing cosmological views
since the 19th century.

*   Hetherington, Norriss S. "The Measurement of Radial
Velocities of Spiral Nebulae."

Cited herein as item 1185.

1208.   Hetherington, Norriss S. "Observational Cosmology in
the Twentieth Century." *Human Implications of Scientific
Advance* (item 362), 567-575.

Examines two main themes: the problem of the structure
of the visible Universe as a monolithic stellar system
or a system of island universes; and the recognition
of the expansion of the Universe. No direct citations.

*   Holton, Gerald. "Resource Letter SRT-1 on Special
      Relativity Theory."

      Cited herein as item 19.

*   Hoskin, M.A. "Edwin Hubble and the Existence of External
      Galaxies."

      Cited herein as item 1191.

*   Hoskin, M.A. "The 'Great Debate': What Really Happened."

      Cited herein as item 1192.

*   Hoskin, M.A. "Cosmology in the Eighteenth and Nineteenth
      Centuries."

      Cited herein as item 553.

*   Hubble, E.P. *The Realm of the Nebulae*.

      Cited herein as item 1194.

*   Jaki, Stanley L. "Johann Georg Soldner and the Gravita-
      tional Bending of Light, with an English Translation
      of His Essay on It Published in 1801."

      Cited herein as item 782.

1209. Jeans, James. "The Size and Age of the Universe." *Royal
       Institution Library of Science: Astronomy* (item 140),
       360-382.

       Lecture delivered November 29, 1935. After a quick
       descriptive review of the history of cosmological specu-
       lation, Jeans turns to the contemporary problem of
       reconciling the many different "ages" of the Universe
       as derived from the observed expansion rate of the
       Universe, and his own studies based upon the dynamics
       of star systems.

1210. Jones, Harold Spencer. "The Structure of the Universe."
       A Century of Science, 1851-1951 (item 129), 153-168.

       Briefly reviews the work of William Herschel and Lord
       Rosse, and then discusses progress toward understanding
       the structure of the Milky Way galaxy and its place in
       the Universe, including growth of knowledge of stellar
       distances, Shapley's studies of globular clusters, the
       detection of differential galactic rotation by J. Oort,
       and Hubble's establishment of the red shift of galaxies.
       No direct citations.

1211. Knox-Shaw, H. "The Observational Evidence for the Expansion of the Universe." *Royal Institution Library of Science: Astronomy* (item 140), 314-340.

Lecture delivered November 25, 1932. Presents observational aspects of the problem, with the caveat that: "Many of us, who are not able to follow the mathematicians in the lofty realms where they would lead us, are inclined to examine the observed facts a little more closely, and to judge the theoretical structures not so much by their beauty as pieces of pure mathematics as by their accordance with the natural universe as it is revealed to us by our telescopes." Criticizes relevant astronomical data and techniques, including the development of the period/luminosity relationship for Cepheids and how faint nebular spectra are procured, and concludes that "it is somewhat rash to claim as support for theory the virtual coincidence of its [expansion] predicted rate with the present observational value." Concludes also that it must be left to the physicist to determine if the observed red-shift is actually due to expansion.

1212. Layzer, D. "The Formation of Stars and Galaxies: Unified Hypotheses." *Annual Reviews of Astronomy and Astrophysics*, 2 (1964), 341-362.

Reviews the cosmogonies of James Jeans, C.F. von Weizsäcker, Fred Hoyle, V. Ambartsumian and his own that attempt a "unified approach" in that they invoke a single process to account for manifold phenomena. 31 citations.

* Lockyer, J. Norman. *The Sun's Place in Nature*.

Cited herein as item 1015.

1213. Lundmark, Knut. "On Metagalactic Distance Indicators." *Vistas in Astronomy*, 2 (1956), 1607-1619.

Traces the "Island-Universe" concept from the time of Galileo. Surveys general cosmological observations and thought since the 17th century including the search for true nebulae. Emphasizes differences in speculative style during the Restoration (1660-1700); the Augustan Age (1700-1745); and the era of Rationalism and Neo-Romanticism (1750-1820). The modern era comprises the majority of the discussion. 45 citations.

1214. Merleau-Ponty, Jacques. *Cosmologie du XXᵉ Siècle*. Paris: Gallimard, 1965. Pp. 533.

Examines the development of both theoretical and observational cosmology emphasizing philosophical implications without detailed recourse to mathematics. Essay
review in: *British Journal for the History of Science*,
3 (1966), 184-187.

1215. Merleau-Ponty, Jacques. "Thèmes Cosmologiques chez les
       Fondateurs de la Thermodynamique Classique." *Human
       Implications of Scientific Advance* (item 362), 559-566.

       Examines perceived implications of classical thermo-
       dynamics for theories of cosmology in the 19th and 20th
       centuries. No direct citations.

1216. Milne, E.A. *Relativity, Gravitation and World Structure.*
       Oxford: Oxford University Press, 1935. Pp. viii + 365.

       Based upon Milne's 1933 monograph chapter in *Zeitschrift
       für Astrophysik*, 6, and constitutes a general review of
       theoretical relativistic cosmology. Describes cosmological
       models put forth by de Sitter, Einstein, Lemaitre, and
       Milne and compares each, but primarily his own, to recent
       observations of the motions and distribution of spiral
       galaxies. Predicts future tests of the various world
       models with improved telescopic equipment. Reviewed in:
       Isis, 26 (1936), 215-218.

1217. North, J.D. *The Measure of the Universe.* Oxford: Clarendon
       Press, 1965. Pp. xxi + 436.

       Comprehensive historical review of all areas of modern
       cosmology including both observational and theoretical
       aspects as well as an extensive section on philosophical
       thought. Heavily documented with extensive annotated
       bibliography. Essay review in: *British Journal for the
       History of Science*, 3 (1966), 184-187. See also: *Centaurus*,
       13 (1968-1969), 293.

1218. Robertson, H.P. "Relativistic Cosmology." *Reviews of
       Modern Physics*, 5 (1933), 62-90.

       Written by a prominent mathematician active during the
       early modern period of work. Reviews Einstein and de Sitter
       universes and the later dynamic universe models of
       Friedmann, Lemaitre and Minkowski.

1219. Russell, Henry Norris. "The Time Scale of the Universe."
       Time and Its Mysteries Series III. New York: New York
       University Press, 1949. Pp. 3-30.

Based upon a lecture delivered in New York on May 14, 1940. Popular review of various determinations of the time scale from radioactivity and the half lives of elements in the Earth's crust; dynamical studies of binary stars; the expansion rate of the galaxies as determined by Hubble; the rate of dissipation of star clusters; and the ages of stars based upon the recently discovered sources of nuclear energy through hydrogen fusion. Concludes that all lines of evidence show a time scale on the order of several billions of years, but problems remain, notably the persistence of giant stars and the existence of white dwarfs. No direct citations.

1220. Ryan, M.P., Jr., and L.C. Shepley. *Homogeneous Relativistic Cosmologies*. Princeton: Princeton University Press, 1975. Pp. xv + 320.

Traces progress of theoretical cosmology from the 1930s through the 1960s concentrating upon post-1960 developments. Provides introductory historical commentary from the 17th century to date and an extensive 26-page bibliography of primarily 20th-century works. Reviewed in: *Annals of Science*, 32 (1975), 410.

1221. Smith, Robert W. "The Origins of the Velocity-Distance Relation." *Journal for the History of Astronomy*, 10 (1979), 133-165.

Considers "the establishment in the late 1920s and early 1930s of the linear relation between a galaxy's distance and the redshift of its spectral lines: the 'velocity-distance' relation." Begins with a detailed review of V.M. Slipher's first detection of the relation between 1913 and 1915 and shows why Slipher and others were unable to put the relation on a definite footing for the lack of a reliable distance to a galaxy (then considered as spiral nebulae and not definitely extra-galactic). Provides a lucid interlude on cosmology and general relativity circa 1916-1920 and how the confirmed velocity-distance relation, as calibrated by Hubble in 1929, convinced Einstein and others of the reality of the expanding Universe. 143 citations.

1222. Sticker, Bernhard. *Bau und Bildung des Weltalls*. Freiburg: Herder, 1967. Pp. 272.

Traces development of cosmological theories from antiquity to the present through selections from the writings of Plato, Aristotle, Ptolemy, and Copernicus through Kant,

Laplace, W. Herschel and into the modern period with
Einstein, Hubble and Otto Heckmann. Includes short introduction and biographical information, with bibliographical citations. Reviewed in: *British Journal for
the History of Science*, 4 (1969), 407–408.

*   Whitney, C.A. *The Discovery of our Galaxy.*

        Cited herein as item 116.

*   Whitrow, G.J. "Theoretical Cosmology in the Twentieth
    Century."

        Cited herein as item 1196.

*   Whittaker, Sir Edmund. *From Euclid to Eddington, A Study
    of Conceptions of the External World.*

        Cited herein as item 597.

BIOGRAPHICAL, AUTOBIOGRAPHICAL
AND COLLECTED WORKS

1223. Abbot, Charles G. *Adventures in the World of Science.*
Washington, D.C.: Public Affairs Press, 1958. Pp. x +
150.

Autobiographical narrative by this past Secretary of
the Smithsonian who was a specialist in the measurement
of solar radiation. Provides description of nature of
government science before World War II. Reviewed in:
Isis, 50 (1959), 516-517.

*   Adams, C.W. "William Allen Miller and William Hallowes
Miller."

Cited herein as item 478.

1224. Adams, John Couch. *The Scientific Papers of John Couch
Adams.* 2 volumes. Cambridge: Cambridge University
Press, 1896-1900. Pp. liv + 502; xxxii + 646.

Collected works, including his papers credited with
the discovery of a planet beyond Uranus.

1225. Adams, W.S. "The Contributions of Newton to Observational
Astronomy." *The Royal Society: Newton Tercentenary
Celebrations.* Cambridge: Cambridge University Press,
1947. Pp. xvi + 92.

Address given as part of the Tercentenary of Newton's
birth. Reviews his contributions to spectroscopy and
telescope design. Reviewed in: *Isis*, 41 (1950), 114-116.

1226. Airy, Wilfrid, ed. *Autobiography of Sir George Biddell
Airy.* Cambridge: Cambridge University Press, 1896.
Pp. ix + 414.

Chronological accounting of his life and work centering
on his years at Cambridge, his career at the Greenwich
Observatory, 1836-1881, and his retirement there until
his death in 1892. The earlier period of this autobiography was taken directly from notes, but the latter part

of his life was compiled from his journals and archival letters, and written in the third person. Includes transcripts of correspondence and bibliography of Airy's works.

1227.   Aitken, Robert Grant. "Comments from the Side Lines." *Popular Astronomy*, 48 (1940), 457-465.

Brief reminiscence by a noted astronomer and past director of Lick Observatory. Recalls growth of instrumentation and of the astronomical data base during his career.

*   Ambartsumian, V.A. "On Some Trends in the Development of Astrophysics."

Cited herein as item 905.

*   Andoyer, Henri. *L'oeuvre scientifique de Laplace*.

Cited herein as item 746.

1228.   Arago, F. *Biographies of Distinguished Scientific Men*. W.H. Smyth, Rev. Baden Powell and Robert Grant, translators. London: Longman, Brown, Green, et al., 1857. Pp. viii + 607.

Provides an autobiography, and sketches of J.S. Bailly, William Herschel, S. Laplace, and many others.

1229.   Arago, F. "Laplace." *Annual Report of the Smithsonian for 1874*. Washington, D.C.: U.S. Government Printing Office, 1875. Pp. 129-168.

A eulogy presented in 1842 before the French Academy, translated by Baden Powell. Provides a general review of 18th- and early 19th-century French contributions to celestial mechanics, and Laplace's dominant role. Detailed annotation by Baden Powell provides clarifications and criticism of Arago's general representations.

1230.   Armitage, Angus. "Alexander Wilson, M.D.--A University Astronomer of Eighteenth Century Scotland." *Popular Astronomy*, 58 (1950), 388-395.

Brief biographical sketch noting Wilson's influential theory of the structure of sunspots and the solar atmosphere. 17 citations.

1231.   Armitage, Angus. "Robert Hooke as an Astronomer." *Popular Astronomy*, 59 (1951), 287-299.

Reviews Hooke's life and contributions to astronomy in the areas of instrumentation, observations and lectures on the search for stellar parallax, and the theory of gravitation. 28 citations.

1232. Armitage, Angus. "The Pilgrimage of Pingré." *Annals of Science*, 9 (1953), 47-63.

Describes the life and work of Alexandre-Gui Pingré, a private astronomer active in France during the second half of the 18th century who participated in many aspects of observational astronomy including studies of comets, the problem of longitude, and the reconstruction of 17th-century astronomical observations extracted from both printed and manuscript sources. 48 citations.

* Armitage, Angus. "The Astronomical Work of Nicolas-Louis De Lacaille."

Cited herein as item 825.

1233. Armitage, Angus. *William Herschel*. Garden City, N.Y.: Doubleday & Co., 1963. Pp. xii + 158.

Brief descriptive study of Herschel's personal life followed by an exposition of his contributions to astronomy. Includes descriptive background material on the state of astronomy in the mid-18th century. Provides a final chapter on modern state of knowledge in areas of Herschel's work. Short bibliography. Essay review in: *History of Science*, 2 (1963), 145-148. See also: *Isis*, 55 (1964), 452.

1234. Armitage, Angus. *Edmond Halley*. London: Nelson, 1966. Pp. xii + 220.

Scientific biography evaluating Halley's work within the context of his times emphasizing those researches communicated to the *Philosophical Transactions*. Reviewed in: *Annals of Science*, 22 (1966), 299-300.

1235. Armitage, Angus. *John Kepler*. London: Faber and Faber, 1966. Pp. 194.

Popular introduction to Kepler's life and science based upon secondary sources. Reviewed in: *Annals of Science*, 23 (1967), 80.

1236. Badger, G.M., ed. *Captain Cook, Navigator and Scientist*. New York: Humanities Press, 1970. Pp. ix + 143.

Summarizes James Cook's life, voyages, and scientific
accomplishments including the application of new navigational techniques and the observation of the transit of
Venus on Tahiti. This series of papers, presented at a
May, 1969 Symposium honoring Cook's bicentenary held in
Australia, includes one by R. v.d. R. Woolley on the
significance of transits of Venus to the scale of the
Solar System and the scientific value of the observations made in 1761 and 1769. Reviewed in: *Isis*, 63 (1972),
121.

1237. Baily, Francis. *An Account of the Revd. J. Flamsteed
compiled from his own manuscripts.* London: Lords
Commissioners of the Admiralty, 1835; 1837. Pp. lxxiii +
672. Reprinted, London: Dawson, 1966. Pp. lxxiii + 759.

Provides, in the first 100 pages, a biography of
Flamsteed based upon his own writings, and then reprints
281 letters as the bulk of this work. Includes Flamsteed's
Star Catalogue.

1238. Ball, R.S. *Great Astronomers.* London: George Philip &
Son, 1895. Pp. xii + 372.

Biographical chapters on eighteen astronomers from
Ptolemy through John Couch Adams. Includes, in addition
to the popular standards, sketches of Flamsteed, Bradley,
Halley, John Brinkley, the Earl of Rosse, Airy, William
Rowan Hamilton and Urbain Leverrier.

1239. Ball, W. Valentine, ed. *Reminiscences and Letters of Sir
Robert Ball.* Boston: Little, Brown, 1915. Pp. xiv +
408.

Chronological narrative of Ball's life including his
position as tutor to Lord Rosse's sons in Parsonstown
during the years 1865-1867 and his contact with Rosse's
famous observatory. Ball's subsequent positions as
Professor of applied mathematics at the Royal College
of Science (1867-1874), Astronomer Royal of Ireland
(1874-1892), and Lowndean Professor at Cambridge (1893-
1913) are outlined in detail, as are his many professional,
social and political activities. An appendix evaluating
Ball's mathematical work is provided by E.T. Whittaker.
Numerous citations from Ball's correspondence are re-
produced.

1240. Baumgardt, Carola. *Johannes Kepler: Life and Letters.*
London: V. Gollancz, 1952. Pp. 209.

Brief scholarly study of Kepler's personal life, as
revealed through his letters. A brief forward by Albert

Einstein comments on Kepler's scientific work. Short
bibliography.

1241. Bedini, Silvio A. *The Life of Benjamin Banneker*. New
    York: Charles Scribner's Sons, 1972. Pp. xviii + 434.

    Case study of Banneker's calculation and preparation
    of popular almanacs in late 18th-century America. In-
    cludes technical aspects of almanac preparation as well
    as a detailed account of Banneker's life as the first
    black in science. Reviewed in: *JHA*, V (1974), 140-141;
    Isis, 64 (1973), 126-127.

1242. Beer, Arthur, ed. "The Henry Norris Russell Memorial
    Volume." *Vistas in Astronomy*, 12 (1970), 1-426.

    Twenty-two papers written in honor of Russell center-
    ing upon eclipsing binary stars and the chemical composi-
    tion of the Sun and stars. Historical papers by K.G.
    Kron ("Some Recollections"); Rose Szafraniec ("Henry
    Norris Russell's Contribution to the Study of Eclipsing
    Variables"); and Bancroft Sitterly ("Changing Interpreta-
    tions of the Hertzsprung-Russell Diagram, 1910-1940: A
    Historical Note") are included.

1243. Beer, Arthur, and Peter Beer, eds. "Kepler, Four Hundred
    Years." *Vistas in Astronomy*, 18 (1975), 1-1033.

    Includes the full texts in English of symposia held
    in honor of Kepler's 400th birthday held at Philadelphia,
    Leningrad, Graz, Linz, London, Paris, Berlin and Evanston.
    Organized into topical chapters covering all aspects
    of Kepler's science. Includes a chronology, Kepler
    iconography, manuscript sources, and a "Bibliographia
    Kepleriana" 1967-1975, covering many previous years and
    continued in later volumes of *Vistas*. Reviewed in:
    Centaurus, 20 (1976), 179.

1244. Beer, Arthur, and K. Aa. Strand, eds. "Copernicus Yester-
    day and Today." *Vistas in Astronomy*, 17 (1975), 1-225.

    Seventeen papers presented at a commemorative confer-
    ence held in Washington, D.C., in December, 1972. In-
    cludes discussion on topics ranging from Copernicus'
    science, the nature of scientific revolutions, and the
    impact of Copernican thought. Reviewed in: *Centaurus*, 20
    (1976), 179.

*   Beer, Peter, ed. "Newton and the Enlightenment."

    Cited herein as item 721.

1245.   Bell, A.E. *Christian Huygens and the Development of
        Science in the Seventeenth Century*. New York: Longmans,
        Green and Co., 1947. Pp. 220.

        Biographical study of Huygens' life and scientific
        work, including his contributions to mechanics, optics,
        and astronomy. Mentions his construction of telescopes
        and pendulum clocks and his astronomical observations.
        Reviewed in: *Annals of Science*, 6 (1948), 103-104; *Isis*,
        40 (1949), 272-273.

1246.   Bell, Trudy E. "A Labor of Love: Biography of Edward
        Singleton Holden." *The Griffith Observer*, 37 (1973),
        2-10.

        Brief biography set against the early years of the
        Lick Observatory and the Astronomical Society of the
        Pacific. Several indirect citations.

1247.   Berry, Robert Elton. *Yankee Stargazer*. New York: McGraw-
        Hill, 1941. Pp. xi + 234.

        A popular biography of Nathaniel Bowditch.

1247a. Bessel, F. *Briefwechsel zwischen Bessel und Steinheil*.
        Leipzig: W. Engelmann, 1913.

        Transcription, with abstracts, of 85 letters between
        Bessel and the noted scientific instrument maker during
        period 1826-1844.

*   Biermann, Kurt-R., ed. *Briefwechsel zwischen Alexander
        von Humboldt und Carl Friedrich Gauss*.

        Cited herein as item 942.

1248.   Biermann, Kurt-R., ed. *Briefwechsel zwischen Alexander
        von Humboldt und Heinrich Christian Schumacher*. Berlin:
        Akademie-Verlag, 1980. Pp. 196.

        Transcribes, with annotation, 106 letters between
        Humboldt and Schumacher, the director of the Altoona
        Observatory and founder of the *Astronomische Nachrichten*.

1249.   Born, Max. "Astronomical Recollections." *Vistas in
        Astronomy*, 1 (1955), 41-44.

        Reminisces about student life at the Universities of
        Breslau and Göttingen and his courses in astronomy from
        Franz and Karl Schwarzschild. No direct citations.

1250.   Bradley, James. *Miscellaneous Works and Correspondence
        of the Rev. James Bradley*. S.P. Rigaud, ed. Oxford:
        Oxford University Press, 1832. Pp. cviii + 528.

The 108-page preface, entitled "Memoirs of Bradley" contains a general biography by Rigaud of his life and work, based upon his letters and notes. Among the collected works and correspondence are Bradley's letter to Halley on his discovery of aberration, from the *Philosophical Transactions*, Volume 35, and a previously unpublished manuscript on the "Rules for Aberration."

1251. Brasch, F.E. "John Winthrop [1714-1779], America's First Astronomer and the Science of his Period." *Publications of the Astronomical Society of the Pacific*, 28 (1916), 153-170.

Biographical exposition of Winthrop's science, and general science during the Colonial Period centering upon Harvard and the apparatus Winthrop collected for use there. 25 citations.

1252. Brasch, F.E., ed. *Sir Isaac Newton, A Bicentenary Evaluation of His Work*. Baltimore: Williams and Wilkins, 1928. Pp. ix + 351.

Collection of papers written on the bicentenary of Newton's death, and presented in New York in 1927. Includes discussions by W.W. Campbell on "Newton's influence upon the development of astrophysics" and E.W. Brown on "Developments following from Newton's work," which centers upon the history of Newtonian celestial mechanics. Another chapter, by Brasch, reviews John Winthrop's scientific life as a Newtonian in early American science. Reviewed in: *Isis*, 11 (1928), 387-93.

1252a. Brewster, Sir David. *Memoirs of the Life, Writings, and Discoveries of Sir Isaac Newton*. 2 volumes. Edinburgh: Thomas Constable, 1855. Pp. xv + 478; xi + 564.

Comprehensive classic that reprints, within the text and in an appendix in Volume 2, numerous letters between Newton and major contemporaries. Detailed footnote annotations and citations, many of great interest as representative of knowledge at mid-century.

1253. Browne, Charles Albert. "Scientific Notes and Letters of John Winthrop, Jr. (1600-1676), First Governor of Connecticut." *Isis*, 11 (1928), 325-342.

Describes the letters and library of John Winthrop, including those materials surviving from his astronomical interests, within a short biographical study. 1 citation.

1254. Brugmans, H.L. *Le Séjour de Christian Huygens à Paris et ses relations avec les milieux scientifiques français suivi de son journal de voyage à Paris et à Londres.* Paris: E. Droz, 1935. Pp. 200.

Based upon first 17 volumes of Huygens' collected works, this text provides background to the scientific life in Paris experienced by Huygens. Heavily documented. Reviewed in: *Annals of Science*, 1 (1936), 235-237.

*   Bruhns, C., ed. *Vierteljahrsschrift der astronomischen Gesellschaft.*

Cited herein as item 356.

1255.   Bruhns, C. *Johann Franz Encke.* Leipzig: E.J. Gunther, 1869. Pp. viii + 350.

Scientific biography reviewing Encke's training in Göttingen and his rise to prominence as director of the Berlin Observatory. Examines Encke's research in orbital theory and his contacts with other early 19th-century astronomers including Bessel, Gauss, Olbers, W. Struve and Hansen. Some citations and an Encke bibliography.

1256.   Burnham, S.W. "Early Life of E.E. Barnard." *Popular Astronomy*, 1 (1894), 193-195; 341-345; 441-447.

Brief popular narrative, by a famous colleague, of Barnard's life in Tennessee, early researches, and move to Lick Observatory. No direct citations.

1257.   Buttmann, Günther. *The Shadow of the Telescope: A Biography of John Herschel.* New York: Charles Scribner's Sons, 1970. Pp. xiv + 219.

Narrative of Herschel's scientific life and work. Covers his observations and experiences during his years at Cape Town in South Africa and his continuation of William Herschel's observational programs; his contributions to optical theory and practical photography; and his service as Master of the Mint. 236 citations, selected John Herschel bibliography, and short bibliography of secondary works. Reviewed in: *Isis*, 61 (1970), 415-417.

*   Calinger, Ronald. "Kant and Newtonian Science: The Pre-Critical Period."

Cited herein as item 754.

1258.   Campbell, W.W. "Simon Newcomb." *Memoirs of the National Academy of Sciences*, 17 (1924), 1-69.

Extensive biographical sketch, presented at the NAS

in 1916, and based largely upon Newcomb's autobiography
(item 1346). Includes an extensive partially annotated
bibliography by R.C. Archibald of Newcomb's writings in
astronomy, economics, mathematics, and his many political
writings on education, labor, the role of government
in science, and psychic phenomena. Bibliography includes
genealogical studies and obituary notices.

1259. Cannon, Walter F. "John Herschel and the Idea of Science."
    Journal of the History of Ideas, 22 (1961), 215–239.

    Examines John Herschel's role in influencing scientific
    attitudes in early Victorian England circa the 1830s.
    Notes especially his significance during the period when
    "In the England of the 1830s, 'to be scientific' meant
    'to be like physical astronomy.'" Notes too the growing
    influence of an evolutionary cosmology both in astronomy
    and geology. 82 citations.

1260. Caspar, Max. *Johannes Kepler*. Stuttgart: Kohlhammer,
    1948. Pp. 479.

    Definitive biography in German providing background
    on the general state of astronomy during Kepler's time
    as well as the general political and intellectual climate.
    Provides chronological development of Kepler's life and
    work. Reviewed in: *Isis*, 41 (1950), 216–219. Translated
    into English and edited by C. Doris Hellman as: *Kepler*.
    (London: Abelard-Schuman, 1959). Pp. 401.

*   Chapin, Seymour L. "The Astronomical Activities of Nicolas
    Claude Fabri de Peiresc."

    Cited herein as item 188.

1261. Clerke, Agnes. *The Herschels and Modern Astronomy*. New
    York: Macmillan, 1895. Pp. 224.

    Brief account of the lives and work of the Herschel
    family and their role in the development of astronomy.
    Brief bibliographical preface. No direct citations.

1262. Cortie, A.L. *Father Perry, The Jesuit Astronomer*. London:
    The Catholic Truth Society, 1890. Pp. 113.

    The life of Stephen Joseph Perry (1833–1889), astronomer
    and long-time director of the Stonyhurst Observatory in
    England. Reviews his studies in meteorology and magnetic
    phenomena, and his studies of the Sun, solar phenomena
    and the transits of Venus. Includes a bibliography of
    Perry's works.

* Cotter, Charles H. "George Biddell Airy and his Mechanical Correction of the Magnetic Compass."

   Cited herein as item 514.

1263. Crowther, J.G. *British Scientists of the Twentieth Century*. London: Routledge and Kegan Paul, 1952. Pp. xiii + 320.

   Includes biographical sketches of James Jeans and Arthur Stanley Eddington, including analyses of their major works. Provides short lists of their major works.

1264. Cudworth, William. *Life and Correspondence of Abraham Sharp*. London: Sampson Low, et al., 1889. Pp. xvi + 342.

   A detailed popular biography of this Yorkshire astronomer and mathematician based upon his correspondence and largely upon his contacts with J. Flamsteed, for whom Sharp served as assistant at Greenwich.

1265. Darwin, G.H. *Scientific Papers*. 5 volumes. Cambridge: Cambridge University Press, 1907-1916. Pp. 2200+.

   Collected works organized by topics which include: tides, periodic orbits, figures of equilibrium, geo-physics, tidal friction and cosmogony. Volume five contains biographical memoirs of Darwin by Francis Darwin and E.W. Brown, and G.H. Darwin's lectures on Hill's *Lunar Theory*, edited by F.J.M. Stratton and J. Jackson.

1266. Davis, Herman S. "David Rittenhouse." *Popular Astronomy*, 4 (1896), 1-12.

   Brief review of the life of this early American astronomer, concentrating upon his researches.

1267. Dijksterhuis, E.J. "Christian Huygens." *Centaurus*, 2 (1953), 265-282.

   Describes completion of Huygens' *Collected Works* (item 1318) in 22 volumes. Includes subject analysis of various volumes.

1268. Dingle, Herbert. "Thomas Wright's Astronomical Heritage." *Annals of Science*, 6 (1950), 404-415.

   Address delivered at the University of Durham in 1950 for bicentenary of Wright's *Original Theory or New Hypothesis of the Universe*. Reviews the history of cosmological thinking from the Greeks through post-Newtonian period to provide background for intellectual climate within which Wright worked.

1269.   Douglas, A. Vibert. *The Life of Arthur Stanley Edding-
          ton*. London: Thomas Nelson, 1957. Pp. xi + 207.

          Scientific biography based upon Eddington's letters,
          early student journals, and records collected by a long-
          time personal friend, C.J.A. Trimble. Chapters include
          his Greenwich years, work on stellar atmospheres and
          interiors, relativity, and his "fundamental theory."
          Numerous citations.

1270.   Dreyer, J.L.E. *Tycho Brahe, A Picture of Scientific
          Life and Work in the Sixteenth Century*. London: Adam
          and Charles Black, 1890. Reprinted, New York: Dover,
          1963. Pp. xvi + 405.

          Authoritative account of Brahe's life and empire of
          scientific research at Uraniborg on his island granted
          by King Frederick II. In addition to biographical chap-
          ters, extensive commentary is provided on the revival
          of astronomy in Europe, the Nova of 1572 and Comet of
          1577--events examined with greater accuracy than was
          possible previously--and their role in shaping Tycho's
          system of the world.

1271.   Dreyer, J.L.E. "Descriptive Catalogue of a Collection
          of William Herschel Papers Presented to the Royal
          Astronomical Society by the Late Sir W.J. Herschel,
          Bart." *Monthly Notices of the Royal Astronomical
          Society*, 78 (1918), 547-554.

          Describes contents, including Herschel's chronological
          observing journals and "reviews of the heavens," (1774-
          1819); his general observations with the 40-foot tele-
          scope; observations of stars, variables, the Sun, Moon,
          planets and comets; notes on the construction of tele-
          scopes; and a correspondence collection including 190
          outgoing and 700 incoming letters.

1272.   Dreyer, J.L.E. "Flamsteed's Letters to Richard Towneley."
          The Observatory, 45 (1922), 280-294.

          Describes some 70 letters recently uncovered that
          provide insight into Flamsteed's life and work, especially
          into his relations with Halley and Newton. Numerous
          citations.

1273.   Dunkin, Edwin. *Obituary Notices of Astronomers*. Edinburgh:
          1879. Pp. viii + 257.

          Provides useful information on many obscure 19th-century
          astronomers.

1274.   Dunnington, G. Waldo. *Carl Friedrich Gauss: Titan of Science. A Study of His Life and Work*. New York: Exposition Press, 1955. Pp. xvi + 479.

General account of Gauss' contributions to the theory of numbers, the co-invention of the electromagnetic telegraph, theoretical and observational astronomy, optics, non-Euclidean geometry, and geodesy. Describes Gauss' life and intellectual environment in Göttingen including the international recognition afforded him after his development of an efficient method to determine the orbits of asteroids after the first one was discovered by Piazzi in 1801. Extensive documentation. Reviewed in: *Isis*, 50 (1959), 285-286.

1275.   Eddington, A.S. "Obituary Notice of Robert Emden." *Monthly Notices of the Royal Astronomical Society*, 102 (1942), 77.

Provides commentary on Emden's book *Gaskugeln* (item 916).

1276.   Eggen, Olin J. "Sherburne Wesley Burnham and his Double Star Catalogue." *Astronomical Society of the Pacific Leaflet* No. 295 (November, 1953). Pp. 8.

Brief biographical sketch of Burnham, the growth of his self-taught interests in astronomy, his tenure at the Lick and Yerkes Observatories and the production of his catalogue of visual binary stars.

1277.   Euler, Leonhard. *Manuscripta Euleriana Archivi Academiae Scientiarum URSS*. Vol. 1. Descriptio Scientifica. J. C. Kopelevič, et al., eds. Moscow: Izdatel'stvo Akademiia Nauk, 1962. Pp. 428.

Attempts a complete description, in Russian, of Euler's manuscripts in the Russian Academy library. German preface. Volume 2, *Opera Mechanica*, is edited by G.K. Mikhailov. Reviewed in: *Isis*, 58 (1967), 271-274.

*   Euler, Leonhard. *Opera Omnia*.

Cited herein as item 762.

*   Evans, D.S., T.J. Deeming, B.H. Evans, and S. Goldfarb, eds. *Herschel at the Cape, Diaries and Correspondence, 1834-1838*.

Cited herein as item 281.

1278.   Evershed, J. "Recollections of Seventy Years of Scientific
        Work." *Vistas in Astronomy*, 1 (1955), 33-40.

        Records his contribution to various aspects of observa-
        tional and experimental solar physics including eclipse
        expeditions, prominence observations, pressure effects
        in sunspots, measurement of the relativistic deflection
        of starlight during solar eclipses, and solar spectroscopy.
        14 citations.

*   Flamsteed, John. *The Gresham Lectures of John Flamsteed*.

        Cited herein as item 540.

1279.   Fleming, Donald. *John William Draper and the Religion
        of Science*. Philadelphia: University of Pennsylvania
        Press, 1950. Pp. ix + 235.

        Intellectual biography of a person who made broad
        contributions to many aspects of science, technology
        and philosophy including early studies in the application
        of spectroscopy to astronomy in the latter half of the
        19th century. Reviewed in: *Isis*, 42 (1951), 256-257.

1280.   Forbes, Eric G. "The Life and Work of Tobias Mayer
        (1723-62)." *Quarterly Journal of the Royal Astronomical
        Society*, 8 (1967), 227-251.

        Biographical study examining Mayer's training, early
        interests in cartography, his switch to a study of lunar
        topography in 1747, and his eventual researches on the
        theory of the motion of the Moon, "stimulated by his
        astronomical work at Nuremberg and especially by his
        careful study of the causes of the Moon's libration."
        Euler's influence upon Mayer is noted, as is the stimula-
        tion of the prize by British Parliament for a useful
        method of longitude determination at sea. 109 citations.

1281.   Forbes, Eric G. "Tobias Mayer (1723-62): A Case of For-
        gotten Genius." *British Journal for the History of
        Science*, 5 (1970), 1-20.

        Provides a biographical sketch of the scientific life
        and training of Mayer, an important contributer to the
        18th-century problem of navigation through the use of
        tables of the motion of the Moon. Reviews Mayer's youth
        in Esslingen, his early employment in Augsburg as a
        cartographer, his lunar studies in Nuremberg and his
        growing interest in improving knowledge of geographical
        longitude through observations of the position of the
        Moon. 103 citations.

1282.   Forbes, Eric G., ed. *The Unpublished Writings of Tobias Mayer*. 3 volumes. Göttingen: Vandenhoeck and Ruprecht, 1972. Pp. viii + 227; 136; 104.

Volume I covers astronomy and geography; Volume II covers artillery and mechanics; Volume III covers the theory of the magnet and terrestrial magnetism. Includes detailed introduction and notations on bibliographical sources. Volume I contains seven manuscripts which deal with the determination of latitude and longitude, cartography, the history and contemporary status of astronomy, the theory of the Earth's figure and planetary motion, especially the motion of Mars. Abstract presented at the 1968 Paris meetings of the Congrès International d'Histoire des Sciences, Tome XII "Monographies de Savants" (Paris: Blanchard, 1971), pp. 19–25. Reviewed in: *Annals of Science*, 32 (1975), 180–181; *Isis*, 66 (1975), 279–280.

1283.   Forbes, Eric G. "The Maskelyne Manuscripts at the Royal Greenwich Observatory." *Journal for the History of Astronomy*, 5 (1974), 67–69.

Sketch of Maskelyne's life and work as revealed in his rich manuscript collection held in the Records room of the Royal Greenwich Observatory.

1284.   Forbes, Eric G. "Early Astronomical Researches of John Flamsteed." *Journal for the History of Astronomy*, 7 (1976), 124–138.

Provides commentary on the various researches of Flamsteed including his search for stellar parallax; the physical constitution of celestial bodies wherein Flamsteed believed that the fixed stars were similar to the Sun; his theory of cometary motion; and his work on telescopic optics. Provides some background to Flamsteed's *Gresham Lectures*. 44 citations.

1285.   Forbes, Eric G. "Tobias Mayer's Contributions to Observational Astronomy." *Journal for the History of Astronomy*, 11 (1980), 28–49.

Shows that Mayer's contributions were primarily observational rather than theoretical, especially in his development of accurate lunar tables that were capable of predicting lunar parallax from ships at sea. Examines Mayer's instruments, his formulae for atmospheric refraction and other rectification procedures he maintained. Notes Mayer's detection of proper motions for many of his standard

stars. Provides general commentary on progress of
theoretical celestial mechanics during Mayer's time.
112 citations.

1286. Forbes, G. *David Gill: Man and Astronomer*. London: John
Murray, 1916. Pp. xi + 418.

Chronological narrative of Gill's personal and scientific life based largely upon Gill's correspondence.
Highlights Gill's work at the Cape Observatory, his
researches on the solar parallax, and his many contacts
and collaborations, especially with J.C. Kapteyn, G.B.
Airy, A. Auwers, W.L. Elkin, Agnes Clerke and others.
Includes a Gill bibliography.

1287. Fowler, Alfred. "Memories of Sir Norman Lockyer." *Life
and Work of Sir Norman Lockyer* (item 1332), 453-462.

Reviews his long association with Lockyer, beginning
in 1883 and continuing through 1900 as his chief assistant.
No direct citations.

1288. Frost, Edwin Brant. *An Astronomer's Life*. Boston: Houghton
Mifflin, 1933. Pp. xi + 300.

Autobiography of the second director of the Yerkes
Observatory (1904-1932) including background on his
training at Dartmouth and Potsdam, and life at Yerkes.
No direct citations.

1289. Gade, John Allyne. *The Life and Times of Tycho Brahe*.
Princeton: Princeton University Press, 1947. Pp. xii +
209.

A quadricentennial tribute emphasizing the cultural
and personal background to Tycho's work. 8-page bibliography.

1290. Gillispie, Charles Coulston, ed. *Dictionary of Scientific
Biography*. 16 volumes. New York: Scribner's, 1970s-
1980. Pp. 10,000+.

Encyclopedic dictionary covering all periods and areas
of science. Includes over 5000 biographies, each with
bibliographical addenda. Volume 15, a supplement, includes biographies not found in earlier volumes, and
more detailed biographies of a few scientists, especially
a 130-page study of S.P. Laplace. Volume 16 provides a
comprehensive name and subject index identifying over
75,000 topics. Numerous reviews have appeared of this
major reference work. Review symposium by Donald Fleming,

Joseph Needham, Edward Grant and Jacques Roger in: *Isis*, 71 (1980), 633-652.

1291.   Graves, Robert Perceval. *Life of Sir William Rowan Hamilton*. 2 volumes. Dublin: Hodges, Figges and Co., 1882-1885. Pp. xix + 698; xvi + 719.

The life of the mathematician who was professor of astronomy in the University of Dublin and Astronomer Royal of Ireland. Includes detailed extracts from Hamilton's correspondence and extensive commentary on his scientific and institutional work. A full biography by Thomas L. Hankins has just been published: *Sir William Rowan Hamilton* (Baltimore: Johns Hopkins, 1980). Pp. xxi + 474.

*   Gray, George J. *A Bibliography of the Works of Sir Isaac Newton*.

Cited herein as item 16.

1292.   Greenstein, Jesse L. "The Seventieth Anniversary of Professor Joel Stebbins and of the Washburn Observatory." *Popular Astronomy*, 56 (1948), 283-299.

Addresses given at the Washburn Observatory in April 1948 covering the history of the observatory and the scientific life of its long-time director, Joel Stebbins, a pioneer in photoelectric photometry. No direct citations.

1293.   Grimsley, Ronald. *Jean d'Alembert, 1717-1783*. Oxford: Clarendon Press, 1963. Pp. 316.

Traces d'Alembert's intellectual life, accomplishments, and identifies influences upon his philosophy. Reviewed in: *Isis*, 56 (1965), 242-243.

1294.   Gushee, Vera. "Thomas Wright of Durham, Astronomer." *Isis*, 33 (1941), 197-218.

Biographical notes and narrative on aspects of Wright's life and work. Includes extensive citations to reviews of Wright's speculations on the structure of the Universe. 17 citations.

*   Hagar, Charles F. "Through the Eyes of Zeiss."

Cited herein as item 397.

*   Hahn, Roger. *Laplace as a Newtonian Scientist*.

Cited herein as item 774.

1295. Hall, Tord. *Carl Friedrich Gauss. A Biography.* Cambridge, Mass.: M.I.T. Press, 1970. Pp. viii + 173.

A translation of Hall's *Gauss, Matematikernas Konung* by Albert Froderberg. Analyzes Gauss' contributions to mathematics, astronomy and physics, especially his mathematical techniques applied to problems in physics and astronomy. Includes a bibliography of secondary sources in addition to Gauss' collected works. Reviewed in: *Isis* 62 (1971), 260-261.

1296. Halley, E. *Correspondence and Papers of Edmond Halley. Preceded by an Unpublished Memoir of His Life by One of his Contemporaries and the "Eloge" by D'Ortous de Mairan.* Eugene Fairfield Mac Pike, editor and arranger. Oxford: Clarendon Press, 1932. Pp. ix + 300.

Collection of 100 letters written by Halley with listings of his general correspondence identified to date. Includes thirty unpublished Halley manuscripts on his physical experimentation and an outline of Halley's life. Appendices include various Halleiana, notably an annotated bibliography of his published writings and grouped references to Halley's life and work. Reviewed in: *Isis*, 20 (1934), 470-472.

*   Hammer, Franz. "Problems and Difficulties in Editing Kepler's Collected Works."

Cited herein as item 168.

*   Hamor, W.A. "David Alter and the Discovery of Spectrochemical Analysis."

Cited herein as item 481.

1297. Hankins, Thomas L. *Jean d'Alembert. Science and the Enlightenment.* Oxford: Clarendon Press, 1970. Pp. xii + 260.

Biography of d'Alembert centering upon his philosophy regarding the science of mechanics. Examines his training and his various debates with D. Bernoulli, Clairaut, Euler, and others in the *vis viva* controversy and touches upon other matters, especially a conflict over the nature of lunar tables and how best to construct them. A "critique" to Hankins' work, and a rebuttal, appear in *Isis*, 67 (1976), 274-278, and center around the degree to which d'Alembert was Cartesian or Newtonian. Reviewed in: *Isis*, 62 (1971), 255-257; *Centaurus*, 16 (1971-1972), 56-59; *British Journal for the History of Science*, 6 (1972-73), 327-329.

1298.   Hardin, Clyde L. "The Scientific Work of the Reverend
         John Michell." *Annals of Science*, 22 (1966), 27-47.

         Surveys the life and scientific work of Michell,
         emphasizing his contributions to astronomy in the late
         18th century but covering his wide interests in magnetism,
         experimental mechanics, and seismology. Reviews Michell's
         early interest in the physical study of the stars and
         his statistical analysis of star groups that led to
         the conclusion that double stars exist as physical
         systems, and not as chance superpositions, as William
         Herschel argued. 65 citations.

1299.   Hellman, C.D. "George Graham: Maker of Horological and
         Astronomical Instruments." *Popular Astronomy*, 39
         (1931), 186-199.

         Describes the life and work of this early 18th-century
         instrument maker who was responsible for Edmond Halley's
         and James Bradley's instruments. Includes partial
         bibliography of secondary works dealing with Graham's
         instruments.

1300.   Hellman, C.D. "John Bird (1709-1776). Mathematical
         Instrument Maker in the Strand." *Isis*, 17 (1932),
         127-153.

         Traces the life of Bird, who was responsible for the
         design and production of many important astronomical
         instruments, as well as being an ardent astronomical
         observer. Includes information on Bird's instruments
         constructed for James Bradley which replaced those by
         Graham built for Halley, and Bradley's observations
         made with them; instrumentation for Tobias Mayer,
         Maskelyne and others are also noted. A general discussion
         is added on the reception and influence of Bird's publi-
         cations on techniques of dividing scales on astronomical
         instruments and constructing mural circles. 105 citations.

1301.   Henderson, E. *Life of James Ferguson in a Brief Auto-
         biographical Account, and Further Extended Memoir.*
         Edinburgh: Fullarton, 1867. Pp. xiii + 503.

         Edited and extended autobiography in two chronological
         sections reviewing his life and works. Ferguson's many
         publications and instruments are identified and discussed,
         with over 100 illustrations of his various planetaria
         and mechanical devices.

Illustration from Ferguson's autobiographical recollections of observing the stars with a beaded string. (*From: E. Henderson*, Life of James Ferguson *(Item 1301), p. 14.)*

1302.   Herrmann, D.B. "Karl Friedrich Zöllner und sein Beitrag
        zur Reception der naturwissenschaftlichen Schriften
        Immanuel Kants." *NTM--Schriftens. Gesch. Naturwiss.
        Technik und Med.*, 13 (1976), 50-53.

        Analyzes the role of Fredrick Engels' philosophy upon
        Zöllner's reception of Kant at a time when Kant's cos-
        mogony was being confirmed by spectroscopic astronomy.
        Notes Zöllner's tendency toward "spiritism." 18 citations.

1303.   Herschel, Mrs. John. *Memoirs and Correspondence of
        Caroline Herschel*. London: John Murray, 1876. Pp.
        xii + 355.

        Chronological narrative derived from Caroline Her-
        schel's *Journal* including detailed background on the
        life of her brother William together with their lifelong
        collaboration in astronomy. An appendix describes the
        inventory of books and pictures in the Herschel family.

1304.   Herschel, W. *Scientific Papers of William Herschel*.
        2 volumes. J.L.E. Dreyer, editor. London: Royal Society,
        1912.

        Includes previously unpublished manuscripts and publica-
        tions; with a biographical introduction by Dreyer.

1305.   Herschel, W., and J. Herschel. "Microfilm of Herschel
        Archives." *Memoirs of the Royal Astronomical Society*,
        85 (1978); *Quarterly Journal of the Royal Astronomical
        Society*, 18 (1977), 459-463.

        The first citation constitutes a catalogue of the
        Herschel Papers kept at the Royal Astronomical Society,
        and available on microfilm. The second citation describes
        the general collection prepared by J.A. Bennett. See
        also: *JHA*, 7 (1976), 75-108.

1306.   Hertzsprung-Kapteyn, H. *J.C. Kapteyn, Zijn Leven en
        Werken*. Groningen: P. Noordhoff, 1928. Pp. 176.

        Chronological narrative (in Dutch) of Kapteyn's personal
        life written by his daughter. No citations or index.

*   Hetherington, Norriss S. "Cleveland Abbe and a View of
        Science in Mid-Nineteenth Century America."

        Cited herein as item 203.

1307.   Hevelius, Johannes. *Johannes Hevelius and His Catalog
        of Stars*. Provo, Utah: Brigham Young University Press,
        1971. Pp. 89.

Edited by A.A. von Hohenlohe, I. Volkoff and others, this facsimile reprint includes a brief biographical study, descriptions of his observatory, instruments and observations, and an analysis of differences between the manuscript and printed versions of the catalog. Bibliography of Hevelius' works and secondary literature. Reviewed in: *Isis*, 63 (1972), 284-285.

1308.   Hill, George William. *The Collected Mathematical Works of George William Hill*. 4 volumes. Washington, D.C.: Carnegie Institution, 1905-1907. Pp. 363; 339; 577; 460.

Includes an introduction written by Henri Poincaré that constitutes a short biography and critical appraisal of Hill's contributions to astronomy and mathematics. Hill was a major figure in late 19th-century theoretical celestial mechanics. Comprehensive index in Volume 4. See also: F.R. Moulton. *Popular Astronomy*, 22 (1914), 391.

1309.   Hindle, Brooke. *David Rittenhouse*. Princeton: Princeton University Press, 1964. Pp. ix + 394.

Traces in detail Rittenhouse's life as a craftsman-instrument maker, astronomer, and notable personage in 18th-century Philadelphia. Provides a general study of Rittenhouse's intellectual and social environment including his devotion to the American Philosophical Society and his friendships with Franklin and Jefferson. His many astronomical endeavors are chronicled, including the design and construction of clocks, surveying devices, telescopes, and mechanical orreries, and his observation of the 1769 transit of Venus and its reduction for the solar parallax. Extensive citations and bibliography. Reviewed in: *Tech. & Culture*, 6 (1965), 465-468; *Isis*, 57 (1966), 282-284.

*   Hiscock, W.G., ed. *David Gregory, Isaac Newton and Their Circle*.

Cited herein as item 724.

1310.   History of Science Society. *Johann Kepler, 1571-1630, A Tercentary Commemoration of His Life and Work*. Baltimore: Williams & Wilkins Co., 1931. Pp. xii + 133.

Collection of papers with an introduction by A.S. Eddington and with contributions by W. Carl Rufus ("Kepler as a Mathematician"), E.H. Johnson ("Kepler and Mysticism")

and F.E. Brasch ("Bibliography of Kepler's Works").
Reviewed in: *Isis*, 18 (1932), 197-200.

1311.   Hodghead, Beverly L. "Address of the Retiring President
of the Society, in Awarding the Bruce Medal to Ernest
W. Brown, Professor of Mathematics at Yale University."
Publications of the Astronomical Society of the Pacific,
32 (1920), 85-92.

Brief review of Brown's life and career in celestial
mechanics and his singular interest in the theory of the
Moon's motion and the construction of lunar tables. Notes
the influences upon Brown, including his Cambridge
professor G.H. Darwin and the works of G.W. Hill. Notes
that advance in accuracy of knowledge of the lunar motion
obtained through Brown's theory is two orders of magnitude: from greater than 1 second of arc derived by
Newcomb from the work of Hansen and Delaunay, to one
hundredth of a second derivable from Brown.

1312.   Holden, Edward S. *Sir William Herschel, His Life and
Works*. New York: Charles Scribner's Sons, 1881. Pp.
vii + 238.

Chronological narrative and topical review of Herschel's
scientific work. Includes a 10-page Herschel bibliography and secondary works relating to Herschel's life
and science. Reviewed by Simon Newcomb in: *Nation*, 32
(February 17, 1881), 118-119.

1313.   Holden, Edward S. *Memorials of William Cranch Bond and
of His Son George Phillips Bond*. New York: Murdock,
1897. Pp. 291.

Dual biography of the founders and directors of the
Harvard College Observatory. Notable for the letters of
G.P. Bond written during his European travels to observatories. The Bonds exemplified the observational and experimental side of mid-19th-century American astronomy.

1314.   Holden, Edward S., and C.S. Hastings. *A Synopsis of the
Scientific Writings of Sir William Herschel*. Smith-
sonian Report for 1880. Washington: U.S. Government
Printing Office, 1881. Pp. 114.

Includes secondary works on Herschel's life and contributions, and an analytical summary of his papers arranged
chronologically. Subject index.

1315.   Hoskin, Michael A. *William Herschel, Pioneer of Sidereal Astronomy*. New York: Sheed and Ward, 1959. Pp. 79.

Short biography of Herschel's life, and an analysis of his major works including: recognizing the orbital motions of binary stars as due to their mutual gravitational attraction; analyzing the motion of the Sun through space; determining the structure of the Milky Way; and promoting the concept of a universe evolving with time. Brief bibliography and numerous citations.

1316.   Huggins, William, ed. *The Scientific Papers of Sir William Huggins*. London: William Wesley, 1909. Pp. xii + 539.

Reprints, with some commentary, the published papers of work done at Huggins' Observatory since its inception in 1856. Papers are arranged by topic including descriptions of the Observatory and its equipment, the spectra of stars, nebulae, novae, comets, the Sun, Moon, planets and aurorae, as well as radial velocity observations and examination of chemical spectra. Includes Huggins' popular lectures and major addresses.

1317.   Hughes, Edward. "The Early Journal of Thomas Wright of Durham." *Annals of Science*, 7 (1951), 1-24.

Annotated reprint of Wright's complete journal, which constituted a chronicle of his life from 1711 through 1746. Describes the contents of Wright's manuscript collection at Newcastle Central Library. 77 citations.

1318.   Huygens, Christiaan. *Oeuvres complètes de Christiaan Huygens*. 22 volumes. La Haye: Martinus Nijhoff and others, 1885-1950.

General direction provided by the Dutch Society of Sciences and edited by a number of scholars. Volume XV (1925) includes astronomy, as does XVII (1932). Volume XXI (1944) includes cosmology. Reviewed in: *Isis*, 42 (1951), 56-57.

1319.   Irons, James Campbell. *Autobiographical Sketch of James Croll with [a] Memoir of his Life and Work*. London: Edward Stanford, 1896. Pp. 553.

Expansion of an autobiography of the important geologist and cosmologist, including a detailed analysis of his publications and correspondence. Many letters transcribed verbatim, especially his correspondence with Charles and George Darwin, A.R. Wallace, Charles Lyell, J.D. Hooker,

Herbert Spencer and others on the probable age and origin of the Sun and related topics.

*   Jones, Bessie Zaban, and Lyle Gifford Boyd. *The Harvard College Observatory: The First Four Directorships, 1839-1919.*

Cited herein as item 305.

1320.   Kellner, L. *Alexander von Humboldt.* Oxford: Oxford University Press, 1963. Pp. 247.

Biography of Humboldt and a discussion of the manuscript notes for his preparation of *Cosmos*, regarded by the biographer as "an unparalleled source of information on the knowledge of his time in geophysics and astronomy." Reviewed in: *British Journal for the History of Science*, 2 (1964), 171.

1321.   Kepler, J. *Opera Omnia.* Charles Frisch, ed. 8 volumes. Frankfurt and Erlangen: Heyder and Zimmer, 1858-1871. Pp. 6350+.

Kepler's collected works including notes, manuscripts and letters. Volume 8, part 2 includes a general biography review of 16th-century astronomy, and a detailed vita of Kepler's activities, all in Latin. Superseded by the comprehensive *Gesammelte Werke*.

1322.   Kepler, J. *Johannes Kepler Gesammelte Werke.* 19 volumes. München: C.H. Beck, 1888-1975. 5 linear feet.

Edited by a succession of historians including Walther von Dyck, Max Caspar, Franz Hammer and Martha List, under the auspices of the Kepler-Kommission of the Bavarian Academy of Science. Notation and discussion in German. Project reviewed in: *Vistas in Astronomy*, 9 (1967), 261-264.

1323.   Kilmister, C.W. *Sir Arthur Eddington.* Oxford: Pergamon Press, 1966. Pp. 279.

Selected reprints of papers and abstracts from Eddington's work introduced by Kilmister with extensive annotations and a critical introduction to Eddington's life and work. Reviewed in: *Annals of Science*, 23 (1967), 246-247.

1324.   Koestler, Arthur. *The Watershed.* New York: Anchor Books, 1960. Pp. 280.

Excerpted biographical chapter from Koestler's provoca-
tive *The Sleepwalkers* (1959). Explores Kepler's character and times with detailed commentary on his association with Tycho Brahe. Extensive citations and annotations.

1325. Krafft, Fritz, Karl Meyer, and Bernhard Sticker, eds.
Internationales Kepler-Symposium. Weil der Stadt,
1971. Referate und Diskussion. Hildesheim: Gerstenberg,
1973. Pp. xii + 490.

Collection of papers, with discussion, of various
aspects of Kepler's scientific life including papers by
Krafft on Kepler's celestial physics. Other contributors
comment on his optics, philosophy, mathematics and poetry.
Reviewed in: *British Journal for the History of Science*,
7 (1974), 294-295.

* Kraus, John. *Big Ear.*

Cited herein as item 489.

1326. Kulikovsky, Piotr G. *M.V. Lomonosov, Astronom i Astro-*
fizik. Moscow: Gosudarstvennoe Izdatelstvo Fiziko-
matematicheskoi Literatury, 1961. Pp. 103.

Examines works of Lomonosov (1711-1765) that deal with
astronomy including tides, magnetism, gravity, optics,
navigation and cartography. Attempts to place Lomonosov
within the astronomical community then found in Russia
and reviews his observations of the 1761 transit of
Venus and his defense of Copernicanism. Reviewed in:
Isis, 56 (1965), 386-387.

* Lankford, John. "A Note on T.J.J. See's Observations of
Craters on Mercury."

Cited herein as item 652.

1327. Laplace, S.P. *Oeuvres complètes de Laplace.* 14 volumes.
Paris: Gautier-Villars, 1878-1912. Pp. 5600+.

Prepared and published under the auspices of the Academy
of Sciences, arranged mainly in chronological order.
Volumes 1-5 include his *Traité de mécanique céleste*,
and Volume 6 his *Exposition du système du monde*.

1328. Laurent, Fernand. *Jean Sylvain Bailly, premier maire de*
Paris. Paris: Boivin et Cie, 1927. Pp. viii + 460.

The life of Bailly, an important intellectual personage
of late 18th-century Paris whose astronomical work
brought him into close contact with Lacaille, Clairaut,
Laplace and others. Examines his political life as well
as his astronomical life, which included a number of
major historical reviews of astronomy. Three quarters
of the work is devoted to Bailly's political life in
Paris during the Revolution. Reviewed in: *Isis*, 11
(1928), 393-395. See also item 1371.

1329.   Lenzen, Victor F. "Charles S. Peirce as Astronomer."
        Studies in the Philosophy of Charles Sanders Peirce,
        Second Series. E.C. Moor and R.S. Robin, eds. Amherst:
        University of Massachusetts Press, 1964. Pp. 33-50.

        Reviews Peirce's work while on the staff of the
        Harvard College Observatory working as an observer,
        principally in Harvard's program in stellar photometry,
        but including some other areas such as the measurement
        of gravity, standards of measure, and spectroscopy.
        Includes commentary on Peirce's work in the U.S. Coast
        Survey in the 1860s. Numerous textual citations.

1330.   Lindsay, Robert Bruce. *Julius Robert Mayer, Prophet of*
        Energy. New York: Pergamon Press, 1973. Pp. vii + 238.

        Includes a biographical sketch, a short study providing
        a critique of Mayer's work, and translations of five
        papers centering on the mechanical equivalence of heat
        including "Celestial Dynamics" (1848), and others specu-
        lating on the gravitational source of the Sun's heat.
        Reviewed in: *Isis*, 66 (1975), 145-146.

1331.   Livingston, Dorothy Michelson. *The Master of Light: A*
        Biography of Albert A. Michelson. New York: Charles
        Scribner's Sons, 1973. Pp. xi + 376.

        Personal accounting and family recollections of
        Michelson's life as well as a general review of his
        scientific career in experimental physics. Reviews his
        measurements of the speed of light; rigidity of the
        Earth; the standard meter in terms of spectral standards;
        the angular diameters of the Jovian satellites and giant
        stars; and his long search for the Earth's drift through
        the ether. Numerous citations. Reviewed in: *Isis*, 66
        (1975), 432-434.

*   Lockyer, J. Norman. *Contributions to Solar Physics*.

        Cited herein as item 1014.

1332. Lockyer, T. Mary, and Winifred L. Lockyer, eds. *Life and Work of Sir Norman Lockyer*. London: Macmillan, 1928. Pp. xii + 474.

After a general biography provided by the Lockyer sisters (Chapters I-XXVI), 13 chapters are contributed by scientists evaluating aspects of Lockyer's work in light of contemporary research. Topics include the constitution of the Sun; the discovery of celestial helium and its detection in terrestrial sources; Lockyer's theory of dissociation as a precursor to ionization equilibrium; Lockyer's Meteoritic Hypothesis and contemporary theories of stellar evolution. Contributors include Herbert Dingle, Megh Nad Saha, C.E. St. John, H.N. Russell and A. Fowler. Provides an important topical review of state of knowledge of solar physics and stellar astronomy circa 1927.
Contains items 929, 1001, 1029, 1109, 1138, 1146, 1287.

1333. Lohne, J. "The Fair Fame of Thomas Harriott." *Centaurus*, 8 (1963), 69-84.

Examines Baron von Zach's quest to bring Harriott's work to wider attention after he discovered Harriott's papers in 1784, and the bitter opposition to this project by the English historian of science S.P. Rigaud. Recounts von Zach's discovery of Harriott's papers and his arguments on their merit, and Rigaud's unfortunately successful counterarguments. See also *Centaurus*, 7 (1960-61), 220-221. 5 citations.

1334. Loria, Gino. "Nel secondo centenario della nascita di G.L. Lagrange, 1736-1936." *Isis*, 28 (1938), 366-375.

Biographical sketch of Lagrange, who was born in Italy but spent most of his professional life in France. Examines his professional career, years in the army, and his founding of the Academy of Science in Torino. Notes relation with D'Alembert, and Lagrange's visits to Paris circa 1763-64 and his move to Berlin to assume position vacated by Euler. Reviews the first centenary of his death and the production of two volumes of memorial papers. Identifies rivalry between France and Italy for the nationality of Lagrange. 8 citations and bibliography of papers reviewing Lagrange's letters and works.

1335. Lubbock, Constance Ann [Herschel]. *The Herschel Chronicle*. Cambridge: Cambridge University Press, 1933. Pp. x + 388.

Exposition, by a granddaughter, of the personal lives of William Herschel and his sister Caroline based upon documentation from Caroline's "Journals" and brief biographical notes by William.

1336. MacPherson, H. *Makers of Astronomy*. Oxford: Clarendon Press, 1933. Pp. 240.

Brief sketches of the lives and works of 58 astronomers arranged in chronological order from Copernicus through to the modern era. No direct citations.

1337. MacPike, E.F. "Doctor Edmond Halley." *Popular Astronomy*, 17 (1909), 408-412.

Examines extant literature dealing with Halley's life. Numerous citations.

1338. MacPike, E.F. *Hevelius, Flamsteed, Halley: Three Contemporary Astronomers and Their Mutual Relations*. London: Taylor and Francis, 1937. Pp. x + 140.

The first three chapters, providing brief biographical sketches of these three astronomers, are followed by a comparative study of their work and interaction, partly as representatives of Greenwich and Paris Observatories. Useful as a supplement to MacPike's *Correspondence and Papers of Edmond Halley* (item 1296). Reviewed in: *Annals of Science*, 4 (1939), 110-111; *Isis*, 29 (1938), 433-434.

* Mayall, N.U. "Bernhard Schmidt and his Coma-Free Reflector."

Cited herein as item 443.

1339. McCarthy, Martin F. "Fr. Secchi and Stellar Spectra." *Popular Astronomy*, 58 (1950), 153-169.

Centers upon influences that led Angelo Secchi to studies of stellar spectra. Details the development of his classification scheme for spectra based upon the colors and line spectra of stars. Notes Secchi's use of line structure in developing a classification by temperature; his interest in peculiar spectra, notably the correlation of variations in line spectra with brightness in variable stars and spectroscopic distinctions between eclipsing binaries and intrinsic variables; and the limitations upon his work imposed by visual techniques before the application of photography to spectroscopy. 47 citations.

1340.   McCormmach, Russell. "John Michell and Henry Cavendish: Weighing the Stars." *British Journal for the History of Science*, 4 (1968), 126-155.

Examines the personal relationship of the late 18th-century natural philosopher John Michell with Henry Cavendish, their scientific collaboration, and Michell's 1784 work suggesting that stars could be weighed by measuring the gravitational retardation of their light. Commentary is also provided on Michell's and Cavendish's project of weighing the world using a delicate torsion balance. Emphasis is upon their Newtonian philosophies. 114 citations.

1341.   Meadows, A.J. *Science and Controversy: A Biography of Sir Norman Lockyer*. Cambridge: M.I.T. Press, 1972. Pp. ix + 331.

Detailed description and analysis of the scientific, personal and public life of the pioneer Victorian solar and stellar astrophysicist. Lockyer's theories of stellar evolution and the temperature-dependent constitution of matter (his "dissociation hypothesis") are examined in detail and discussed within the scientific context of late 19th- century physics and astronomy. Extensive citations and select bibliography. Essay review in: *JHA*, 4 (1973), 131-133.

1342.   Millburn, John R. *Benjamin Martin, Author, Instrument-Maker, and "Country Showman."* Leyden: Nordhoff, 1976. Pp. xii + 233.

General biography emphasizing Martin's role in the popularization of science and his contributions to scientific instrument development in the 18th century. Reviewed in: *Tech. & Culture*, 19 (1978), 114-116.

1343.   Milne, E.A. *Sir James Jeans, A Biography*. Cambridge: Cambridge University Press, 1952. Pp. xvi + 176.

Short scientific biography edited, after Milne's death, by S.C. Roberts and G.J. Whitrow. After a chronological narrative, chapters on aspects of Jeans' science are included on the partition of energy, rotating fluid masses, star clusters, equilibrium configurations for stars, and on his philosophy. Few direct citations; Jeans bibliography.

*   Minnaert, M. "Forty Years of Solar Spectroscopy."

Cited herein as item 1019.

1344.   Moigno, Abbe. *Le Reverend Père Secchi*. Paris: Gautier-
        Villars, 1879. Pp. xxvii + 276.

        Biography of Angelo Secchi and a review of the astronomy
        of his time, including his observatory and observations.
        In 6 parts: Part 1 constitutes the biography and part
        2 a review of Secchi's publications. The remaining parts
        include statements on Secchi's work by various astronomers

*   Morgan, Julie. "The Huggins Archives at Wellesley College."

        Cited herein as item 43.

*   Multhauf, Robert, ed. "Holcomb, Fitz and Peate: Three
        19th-Century American Telescope Makers."

        Cited herein as item 447.

1345.   Nassau, J.J. "Ambrose Swasey, Builder of Machines, Tele-
        scopes, and Men." *Popular Astronomy*, 45 (1937), 407-
        418.

        Brief biographical sketch centering upon the origins
        and development of the firm of Warner and Swasey. No
        direct citations.

1346.   Newcomb, Simon. *Reminiscences of an Astronomer*. Boston:
        Houghton Mifflin, 1903. Pp. viii + 424.

        Autobiographical narrative and general observations on
        progress in astronomy and scientific life in Washington
        during the second half of the 19th century. Includes
        chapters on large telescopes, their construction and
        use; the Lick Observatory; science in England and Europe;
        the transits of Venus and the solar parallax; the Naval
        Observatory and Almanac Office. No direct citations.
        Reviewed in: *The Observatory*, 27 (1904); 216-219, 294;
        Science, 22 (Dec. 8, 1905), 748-750.

1347.   Newton, I. *Isaac Newton's Papers and Letters on Natural
        Philosophy*. I.B. Cohen, ed. Cambridge, Mass.: Harvard
        University Press, 1958. Pp. xiii + 501.

        Selected reprints with annotations and introduction.
        Includes Newton's work on physical optics, the improve-
        ment of the telescope, correspondence with the Rev.
        Bentley on the stability of the Universe, relations with
        Halley and the development of the *Principia*.

*   Nininger, Harvey H. *Find a Falling Star*.

        Cited herein as item 984.

1348. Norberg, Arthur L. "Simon Newcomb's Early Astronomical
       Career." *Isis*, 69 (1978), 209-225.

       Traces Newcomb's life and career to his attainment of
       the directorship of the Almanac Office in 1877. Reviews
       Newcomb's early training; the scientific life of mid-
       19th-century Washington; his tenure as professor at
       the U.S. Naval Observatory; and Newcomb's developing
       studies of the motions of the Moon and planets. 58 cita-
       tions.

1349. Nordenmark, N.V.E. *Pehr Wilhelm Wargentin*. Uppsala:
       Almquist & Wiksells, 1939. Pp. 464.

       Biography, in Swedish with French summary, of an
       18th-century astronomer who specialized in the produc-
       tion of tables of the motions of the Jovian satellites.
       Examines the influence of Anders Celsius and the impor-
       tance of Jovian satellite tables for the determination
       of terrestrial longitude. Biography also treats Wargentin's
       institutional activities, including his founding of the
       Swedish Royal Observatory in 1753, and the introduction
       of the Gregorian Calendar into Sweden. Reviewed in:
       Isis, 32 (1940), 361-363.

1350. Ogilvie, Marilyn Bailey. "Caroline Herschel's Contribu-
       tions to Astronomy." *Annals of Science*, 32 (1975),
       149-161.

       Reviews the scientific work of Caroline Herschel and
       how her personality, perseverance and position motivated
       her attention to accuracy. 59 citations.

1351. Osterbrock, Donald E. "James E. Keeler, Pioneer Astro-
       physicist." *Physics Today*, 32 (February 1979), 40-47.

       Biographical sketch of Keeler's scientific life at
       the Allegheny and Lick Observatories, his early studies
       and education at Johns Hopkins, and contact with S.P.
       Langley. 23 citations.

1352. Palmer, Charles Skeele. "Two Hours with Alvan Clark,
       Sr." *Popular Astronomy*, 35 (1927), 143-145.

       Personal account of memories of Clark during the 1880s.

1353. Palter, Robert, ed. *The 'Annus Mirabilis' of Sir Isaac
       Newton, 1666-1966*. Cambridge, Mass.: M.I.T. Press,
       1971. Pp. viii + 351.

       Includes papers on Newton's life and milieu, his philos-
       ophy, aspects of his science, and his influence upon his

and subsequent times. Reviewed in: *British Journal for the History of Science*, 6 (1972-73), 322-323.

1354.   Patterson, Elizabeth C. "Mary Somerville." *British Journal for the History of Science*, 4 (1969), 311-339.

Traces the development of Mary Somerville's life and interest in science and her writings on science, especially *The Mechanism of the Heavens* (1831) and *On the Connexion of the Physical Sciences* (1834) which went through 10 editions in the next 40 years. Examines impact of her writings on 19th-century development of science. Provides a description of her surviving letters. 164 citations. See items 809 and 1407.

1355.   Pelseneer, Jean. "La Correspondence de N. Bowditch. Opinions de Le Gendre." *Isis*, 14 (1930), 227-228.

Provides short note on Bowditch's three volumes of correspondence held at the Boston Public Library. Identifies the correspondents, noting Berzelius, de Lalande, Encke, J. Herschel, Legendre and Quetelet. Illustrates the value of the collection by reprinting excerpts of letters with Legendre. 1 citation.

1356.   Philip, A.G. Davis, and David H. DeVorkin, eds. *In Memory of Henry Norris Russell*. Albany: Dudley Observatory, 1977. Pp. 181.

Proceedings of historical sessions of IAU Symposium 80 on the Hertzsprung-Russell Diagram in honor of the centenary of Russell's birth. Articles include reminiscences of Russell by students and colleagues and short historical reviews of Russell's early career and the creation of the relationship that bears his name. Contains items 841, 933, 1084, 1087, 1111, 1123, 1129, 1132, 1148.

1357.   Quill, Humphrey. *John Harrison: The Man Who Found Longitude*. London: John Baker, 1966. Pp. xiv + 255.

Examines the life of John Harrison, the first to produce a sufficiently accurate and reliable marine chronometer and thereby the first to provide a reliable means for the determination of longitude at sea. Reviewed in: *Isis*, 59 (1968), 117-118.

1358.   Ronan, Colin A. *Edmond Halley, Genius in Eclipse*. New York: Doubleday, 1969. Pp. 251.

Popular biography of Halley's scientific life and
accomplishments. Reviewed in: *Isis*, 61 (1970), 547-548.

1359. [Rosse] William Parsons. *The Scientific Papers of
William Parsons, Third Earl of Rosse*. London: Percy
Lund, Humphries & Co., 1926. Pp. 221.

Collected by Charles Parsons, these 24 papers are
arranged in chronological order by journal and are re-
produced without alteration or annotation. Includes
descriptions of Lord Rosse's telescopes and observations
of nebulae.

1360. Roth, Günter D. *Joseph von Fraunhofer: Handwerker,
Forscher, Akademiemitgleid, 1787-1826*. Stuttgart:
Wissenschaftliche Verlagsgesellschaft, 1976. Pp. 167.

General narrative of Fraunhofer's life touching
briefly upon his scientific accomplishments including
his optical work and instruments responsible for the
measurements of stellar parallax. Notes Fraunhofer's
pioneering examination of solar and stellar spectra.
Reviewed in: *Annals of Science*, 34 (1977), 72-73; *Isis*,
70 (1979), 478.

1361. Roule, L. *Buffon et la description de la nature*. Paris:
E. Flammarion, 1930. Pp. 248.

Biography of George Louis Leclerc, Count Buffon
(1707-1788). Includes the development of his cosmological
speculations, which were of great influence and of
historical significance in that he envisioned the development of the Earth as a continuing process in time.
Reviewed in: *Isis*, 23 (1935), 264-265.

1362. Rumrill, H.B. "The Rittenhouse Bicentenary Celebration."
Popular Astronomy, 41 (1933), 84-92.

Description of meeting held in April, 1932 in Philadelphia together with a brief sketch of aspects of
Rittenhouse's career.

* Rybka, Eugeniusz. *Four Hundred Years of the Copernican
Heritage*.

Cited herein as item 587.

* Sarton, George. "Laplace's Religion."

Cited herein as item 805.

1363.   Scaife, W. Lucien, ed. *John A. Brashear, The Autobiog-
        raphy of a Man Who Loved the Stars*. New York: American
        Society of Mechanical Engineers, 1924. Pp. xxii + 262.

        Romantic narrative of his life in Pittsburgh, the
        development of the Brashear Optical Works in the late
        19th century, and contact with S.P. Langley and the
        Allegheny Observatory. Includes a listing of major
        Brashear instrumentation.

*   Schur, Wilhelm. *Beiträge zur Geschichte der Astronomie
        in Hannover*.

        Cited herein as item 254.

1364.   Scott, J.F. *The Scientific Work of René Descartes
        (1596-1605)*. London: Taylor and Francis, 1952. Pp.
        x + 211.

        Contains a forward by H.W. Turnbull, several chapters
        on the life of Descartes, and an exposition of his
        1637 work *Discours de la méthode* with its associated
        works on optics, atmospheric color, and geometry. Chap-
        ters are included on the *Principia Philosophiae* which
        contains Descartes' mature cosmological thinking. Re-
        viewed in: *Annals of Science*, 8 (1952), 282-283.

1365.   Seares, F.H. "George Ellery Hale. The Scientist Afield."
        Isis, 30 (1939), 241-267.

        Brief biographical sketch centering upon his insti-
        tutional work. Includes general citations to Hale
        obituaries.

1366.   See, T.J.J. "The Services of Benjamin Peirce to American
        Mathematics and Astronomy." *Popular Astronomy*, 3
        (1895), 49-57.

        Reviews Peirce's training under Bowditch and his
        subsequent researches in celestial mechanics. No direct
        citations.

*   See, T.J.J. "The Services of Nathaniel Bowditch to
        American Astronomy."

        Cited herein as item 234.

1367.   See, T.J.J., and Hector MacPherson. "Tribute to the
        Memory of William Huggins." *Popular Astronomy*, 18
        (1910), 387-398; 398-401.

        Descriptions of Huggins' scientific work.

1368.   Shapiro, Barbara J. *John Wilkins 1614-1672: An Intellectual
        Biography*. Berkeley: University of California Press,

1969. Pp. 333.

Includes a chapter on his scientific interests including his writing of *Discovery of a World in the Moone* in 1638. Reviewed in: *Isis*, 61 (1970), 413–414.

1369. Shapley, Harlow S. *Through Rugged Ways to the Stars*. New York: Scribner's, 1969. Pp. 180.

Candid personal reminiscences of his long career in galactic research and his relations with others in the field, especially with A. van Maanen and E.P. Hubble. Provides recollections of his political and humanistic ventures during World War II, after the War, and through the McCarthy era. Based upon tape-recorded interviews. Reviewed in: *Isis*, 61 (1970), 148–149.

1370. Sidgwick, J.B. *William Herschel*. London: Faber & Faber, 1955. Pp. 228.

Evaluates Herschel's researches in the light of modern 20th-century astronomy. Reviewed in: *Isis*, 47 (1956), 88–89.

1371. Smith, Edwin Burrows. *Jean-Sylvain Bailly, Astronomer, Mystic, Revolutionary, 1736-1793. (Transactions of the American Philosophical Society, N.S.*, 44 [1954], 427–538). Philadelphia: The American Philosophical Society, 1954. Pp. iv + 112.

Biographical study of Bailly's life within the context of late 18th-century France. Emphasizes Bailly's studies of the history of astronomy which resulted in his publication of five major volumes between 1775 and 1787. See items 76 and 77. Numerous footnote citations and extensive bibliography. Reviewed in: *Isis*, 50 (1959), 182–183.

1372. Stetson, Harlan True. "Elihu Thomson: His Interest in Astronomy." *Popular Astronomy*, 48 (1940), 470–479.

Examines the astronomical interests of this engineer who made many contributions to astronomical instrumentation, including his advice on the first casting of a large glass disk in the United States, in 1929 at the National Bureau of Standards.

1373. Szanser, Adam J. "F.G.W. Struve (1793-1864). Astronomer at the Pulkovo Observatory." *Annals of Science*, 28 (1972), 327–346.

General biographical study centering on Struve's work at Dorpat, with a description of his development of the Observatory's major astronomical instrumentation and his move to help found the Pulkovo Observatory in 1834–1835 and his eventual transfer there. Reviews Struve's many interests, from geodesy, double stars, and the absorption of light in space, to stellar proper motions and parallaxes. 33 citations.

1374.   Szanser, Adam J. "Johannes Hevelius (1611-1687)--Astronomer of Polish Kings." *Quarterly Journal of the Royal Astronomical Society*, 16 (1976), 488-498.

Describes the life of Hevelius, his observatory and instruments including his aerial telescopes, and the observations he made with them of the Moon, planets and positions of stars. Examines his relationship with the Polish royal court and contacts with Edmond Halley. 4 citations.

1375.   Thomas, Shirley. *Men of Space: Profiles of the Leaders of Space Research, Development and Exploration.* 8 volumes. Philadelphia: Chilton, 1960-1969. Pp. 1800+.

Biographical series includes scientists, administrators and engineers prominent in space history. About 10 biographical appreciations appear in each of the first seven volumes, while the last volume is reserved for general discussion, and profiles of the series advisory committee as well as of major international programs in space exploration. Some names associated with space astronomy and space science include: James van Allen, Fred L. Whipple, Samuel Herrick, Frank D. Drake, Carl Sagan, Harold Urey, Herbert Friedman. Volumes 1 and 2 reviewed in: *Tech. and Culture*, 3 (1962), 212-214.

1376.   Trier, Betty, tr. "Sayings of Gauss and Bessel." *Popular Astronomy*, 13 (1905), 231-235.

Offers excerpted translations from the correspondence of Gauss and Bessel based upon their "Briefwechsel" (Leipzig, 1880). Centers upon opinions of each other, the role of observations and instrumentation in astronomy, the quality of students and other general topics. 25 citations.

1377.   Turner, A.J. *Science and Music in Eighteenth Century Bath.* Bath: University Press, 1977. Pp. viii + 131.

Exhibition catalogue of William Herschel's life and work at Bath including bibliography and transcriptions from his writings.

1378.   Van Biesbroeck, G., ed. "E.E. Barnard's Visit to G. Schiaparelli." *Popular Astronomy*, 42 (1934), 553-558.

Transcription of notes recording a conversation between the two astronomers when Barnard visited Schiaparelli in Milan in the Summer of 1893. The conversation centers upon observing styles, telescopes, and Mars.

1379. Van Dyck, Walther. *Georg von Reichenbach*. München:
       Deutschen Museums, 1912. Pp. 140.

       Biography of noted astronomical instrument-maker
       and civil engineer of the early 19th century.

*   van Woerden, Hugo, Willem N. Brouw, and Henk C. van de
       Hulst, eds. *Oort and the Universe: A Sketch of Oort's
       Research and Person.*

       Cited herein as item 1180a.

1380. Volkmann, Harald. *Carl Zeiss und Ernst Abbe: Ihr Leben
       und ihr Werk*. Munich: R. Oldenbourg, 1966. Pp. 46.

       Traces the origins and development of the Zeiss Optical
       Works through a biographical sketch of its founder and
       his associate, the respected optical designer Ernst
       Abbe, who became his successor. Continues the historical
       account through World War II.

*   Wallis, Peter, and Ruth Wallis. *Newton and Newtoniana,
       1672-1974.*

       Cited herein as item 62.

1381. Warner, Deborah Jean. *Alvan Clark & Sons: Artists in
       Optics*. Washington, D.C.: Smithsonian Inst. Press,
       1968. Pp. 120.

       Short biography of Alvan Clark, George Bassett Clark
       and Alvan Graham Clark emphasizing the development of
       their optical firm. A detailed, annotated catalogue of
       the astronomical instruments produced by the Clarks
       between 1844-1897 forms the bulk of this work. Includes
       an appendix listing the paintings, mostly miniature
       portraits, by Alvan Clark. No index. 391 citations.
       Reviewed in: *JHA*, 1 (1970), 158; *Isis*, 60 (1969), 261-
       262; *Annals of Science*, 25 (1969), 366-368.

*   Warner, Deborah Jean. "Lewis Morris Rutherfurd: Pioneer
       Astronomical Photographer and Spectroscopist."

       Cited herein as item 409.

1382. Wattenberg, Diedrich. *Johann Gottfried Galle, 1812-1910.*
       Leipzig, 1963. Pp. 162.

       Life and work of the German astronomer including his
       studies of meteor orbits, stellar positions and motions.
       Reviewed in: *Centaurus*, 10 (1964-65), 55.

1383. Webb, W.L. *Brief Biography and Popular Account of the
       Unparalleled Discoveries of T.J.J. See*. Boston, Mass.:
       Nichols, 1913. Pp. xii + 298.

       Biography of Thomas Jefferson Jackson See, an out-

spoken and highly controversial turn-of-the-century
figure in American astronomy. Written under the guidance
of See, and with many statements of questionable authority, it is still of interest as a document contemporary
to his times.

1384. Weiner, Philip P. "The Peirce-Langley Correspondence
    and Peirce's Manuscript on Hume and the Laws of Nature."
    Proceedings of the American Philosophical Society,
    91 (1947), 201-228.

    Transcription of 13 letters between C.S. Peirce and
    S.P. Langley circa 1901-1902 that provide insight into
    Peirce's philosophy and perceived state of science. In-
    troduced by brief biography of Peirce contrasting his
    philosophy with that of Langley. 13 citations.

*   Whatton, A.B., ed. *The Transit of Venus Across the Sun*.

    Cited herein as item 889.

1385.   Whiston, William. *Astronomical Lectures* [1728]. I.B.
    Cohen, ed. New York: Johnson Reprint Corporation,
    1972. Pp. xl + 368 + 136.

    Reprint of 1728 original edition that provides impor-
    tant commentary on the state of astronomy in the immedi-
    ate post-Newtonian era. Reviewed in: *JHA*, 4 (1973), 204-
    205.

*   Whiteside, D.T. "The Expanding World of Newtonian Re-
    search."

    Cited herein as item 735.

1386.   Whiteside, D.T., ed. *The Mathematical Papers of Isaac
    Newton*. 7 volumes. Cambridge: Cambridge University
    Press, 1967-1976. 1.5 linear feet of text.

    Chronologically ordered collection, heavily annotated
    and analyzed by Whiteside. Constitutes a standard work
    for interpreting Newton's work during the period 1664-
    1695. Other volumes in preparation. Newton's correspondence
    has been organized in a parallel work by A.R. Hall and
    L. Tilling. Extensively reviewed in journals during the
    period.

*   Wilson, David B. "Herschel and Whewell's version of
    Newtonianism."

    Cited herein as item 740.

1387.   Wilson, Margaret. *Ninth Astronomer Royal, The Life of
    Frank Watson Dyson*. Cambridge: Heffer, 1951. Pp. xiv +
    294.

Personal account by his daughter, with a general accounting of his scientific life. Foreword by H. Spencer Jones. No citations.

1388.   Woolley, Richard. "James Bradley, Third Astronomer Royal." *Quarterly Journal of the Royal Astronomical Society*, 4 (1963), 47-52.

Reviews Bradley's career, including his observing techniques; his ability to understand the interrelated roles of solar motion, proper motion, refraction and aberration; and his operation of the Royal Observatory.

1389.   Wright, Helen. *Sweeper in the Sky, the Life of Maria Mitchell, First Woman Astronomer in America*. New York: Macmillan, 1949. Pp. vii + 253.

Popular narrative of her life on the island of Nantucket, her early contact with astronomy and discovery of a comet in 1847, and her later directorship of the Vassar College Observatory, founded in 1865. Bibliographical note, no direct citations.

1390.   Wright, Helen. *Explorer of the Universe: A Biography of George Ellery Hale*. New York: Dutton, 1966. Pp. 480.

Chronicles Hale's scientific, entrepreneurial, and organizational career providing extensive background both on his personal life and the scientific world he moved in and helped to create. Traces in detail Hale's creation of his private Kenwood Observatory in Chicago; the founding of the *Astrophysical Journal*; the founding in succession of the Yerkes Observatory and the Mount Wilson Solar Observatory; and Hale's long efforts at creating the 200-inch Palomar telescope. Extensive citations to manuscript sources and primary papers. Reviewed in: *Isis*, 58 (1967), 124-125.

1391.   Wright, Helen, J.N. Warnow, and Charles Weiner, eds. *The Legacy of George Ellery Hale*. Evolution of Astronomy and Scientific Institutions, in Pictures and Documents. Cambridge: MIT Press, 1972. Pp. viii + 293.

Three-part illustrated work beginning with biographical essay of Hale's life by Wright; a collection of Hale's writings; and a collection of papers by astronomers reviewing areas of Hale's accomplishments, and developments since his death. Essay review, including bibliography of secondary works on Hale: *JHA*, 5 (1974), 61-63. See also: *Isis*, 64 (1973), 138-139.
Contains items 367, 391, 452, 1008.

*Lord Rosse's Reflecting Telescope,
Birr Castle, Ireland. (from the S.W.)
6 ft. Aperture, 54 feet focus.*

Lord Rosse's 6-foot speculum reflector, the largest of its kind ever con-·
structed; built in the 1840s. *(From Helen Wright, Joan Warnow and Charles
Weiner, eds.,* The Legacy of George Ellery Hale *(Item 1391), p. 221; photograph courtesy of Lick Observatory Archives.)*

TEXTBOOKS AND POPULAR WORKS

18TH CENTURY

*   Costard, George. *The History of Astronomy*.

        Cited herein as item 84.

1392.   Ferguson, James. *Astronomy explained upon Sir Isaac Newton's Principles* [1756]. London: T. Longman, et al., 1794 (9th Edition). Pp. 503.

        One of the earliest and most popular textbooks in English to include detailed account of the 1761 transit of Venus discussed in contrast to Horrox's observations of the 1639 transit. Text is famous for the many highly-detailed woodcut foldouts of Ferguson's mechanical devices to describe celestial motions. Textbook went through 17 editions in England (the first in 1756) and America, with one extensive revision in 1773. First American edition in 1806 revised and annotated by Robert Patterson. Numerous contemporary citations.

*   Gorman, Mel. "Gassendi in America."

        Cited herein as item 198.

1393.   Strong, Nehemiah. *Astronomy Improved*. New Haven: Thomas and Samuel Green, 1784. Pp. 52.

        One of the first textbooks on astronomy in the United States. Strong was a well-known producer of 18th-century almanacs, and was Professor of Natural Philosophy at Yale in the period 1756-1759.

19TH CENTURY

*   Airy, G.B. *Gravitation: An Elementary Explanation of
        the Principal Perturbations in the Solar System.*

        Cited herein as item 743.

1394.   Airy, G.B. *Six Lectures on Astronomy.* London: Simpkin
        and Marshall, 1856 (3rd Edition). Pp. viii + 222.

        General lectures on practical astronomy delivered under
        the auspices of the Ipswich Museum in March, 1848. Of
        interest for his descriptive analysis of aspects of
        practical astronomy including methods of parallax deter-
        minations; theories of motion of the planets and their
        satellites; and the calculation of the figure of the
        Earth from pendulum experiments. No direct citations or
        index.

*   Arago, François. *Popular Astronomy.*

        Cited herein as item 74.

1395.   Ball, Robert S. *The Story of the Heavens.* New Edition.
        London: Cassell, 1893. Pp. xix + 556.

        Frequently revised popular review first published in
        1886 covering all aspects of astronomy. No direct cita-
        tions.

1395a. Biot, J.B. *Traité élémentaire d'astronomie physique.*
        Paris: Bachelier, 1841-1857. 5 volumes.

        Comprehensive technical review of all aspects of prac-
        tical astronomy, celestial mechanics, and techniques for
        the reduction of physical observations. Some citations.

1396.   Chambers, George F. *A Handbook of Descriptive and Practical
        Astronomy.* 2 volumes. Oxford: Clarendon Press, 1890.
        (4th Edition). Pp. xxxii + 676; xix + 558.

        A popular, comprehensive review. Volume 1 treats the
        physical observations of the Sun, planets and comets, and
        Volume 2 reviews instruments and the techniques of practical
        astronomy. Chapters on spectroscopy are written by Maunder.
        Profusely illustrated with a chapter (in Volume 2) on the
        history of astronomy, a chronology, and a detailed, 28-
        page bibliography.

*   Chauvenet, W. *Manual of Spherical and Practical Astronomy*.
        Cited herein as item 820.

*   Clerke, Agnes M., Alfred Fowler, and J. Ellard Gore.
        The Concise Knowledge Astronomy.

        Cited herein as item 126.

1397. Dick, Thomas. *The Practical Astronomer*. New York: Harper,
        1846. Pp. xiv + 437.

        Well-illustrated popular review emphasizing telescopic
        and descriptive astronomy.

1398. Dick, Thomas. *The Complete Works of Thomas Dick*. 1st
        American Edition. 9 volumes bound in 3. Hartford, 1849.
        Pp. 1600.

        Well-illustrated compendium of popular reviews represent-
        ing early to mid-19th-century astronomy. Includes sections
        on telescopes and observatories highlighting Lord Rosse's
        observatory.

1399. Guillemin, Amedee. *The Heavens, An Illustrated Handbook*.
        2nd Edition. N. Lockyer, ed. London: Richard Bentley,
        1966. Pp. xv + 524.

        Heavily-illustrated review of astronomy written at the
        outset of the application of spectroscopy.

*   Herschel, John F.W. *A Treatise on Astronomy*.
        Cited herein as item 95.

*   Herschel, John F.W. *Outlines of Astronomy*.
        Cited herein as item 96.

1400. Humboldt, Alexander von. *Cosmos: A Sketch of A Physical
        Description of the Universe*. 4 volumes. E.C. Otte,
        trans. London: George Bell, 1871-1876; and numerous
        other publishers.

        Includes considerable review of early to mid-19th-cen-
        tury astronomy.

*   Langley, Samuel Pierpont. *The New Astronomy*.
        Cited herein as item 1010.

1401. Mitchell, O.M. *The Orbs of Heaven, or, The Planetary and
        Stellar Worlds*. 4th Edition. London: Ingram and Cooke,
        1853. Pp. viii + 300.

Popular study of recent astronomy based upon a series
of public lectures at Cincinnati College through which
the author raised support for the building of an observatory in that city in the early 1840s. A 60-page Appendix
on telescopes, observatories, and major figures in history
by Denison Olmsted is included. Well-illustrated. No
direct citations or index.

1402. Newcomb, Simon. *Popular Astronomy* [1878]. 2nd, Revised
Edition. London: Macmillan, 1890. Pp. xx + 579.

Important secondary work first published in 1878 and
reprinted and revised many times. This general view of
late 19th-century astronomy also provides a brief introduction to the development of the Copernican system
and Newtonian celestial mechanics. Concentrates on elements of practical astronomy, instrumentation, stellar
distances, the Sun and solar system, and stellar astronomy,
including contemporary problems of cosmology and cosmogony.
Appendices include listing of major telescopes and observatories, lists of double stars, nebulae, star clusters,
comets, and elements of the planetary orbits. Provides
extensive bibliography on the solar parallax, 1854-1881,
a short general bibliography, and a glossary of technical
terms.

*   Newcomb, Simon. *The Stars*.

Cited herein as item 1063.

1403. Newcomb, Simon. *Side-Lights on Astronomy and Kindred
Fields of Popular Science*. New York: Harper and Bros.,
1906. Pp. vii + 349.

Collected popular reviews and commentary. Consists of
21 papers on various topics, including reviews of recent
problems of astronomy, aspects of American Astronomy,
the organization of scientific research, the evolution
of astronomical knowledge. No direct citations.

*   Nichol, J.P. *The Architecture of the Heavens*.

Cited herein as item 1156.

1404. Norton, William A. *A Treatise on Astronomy, Spherical
and Practical* [1867]. 4th Edition. New York: John
Wiley, 1872. Pp. xiv + 443 + 115 (tables).

Revised edition of original 1867 text including extensively rewritten chapter on the physical constitution
of the Sun, and descriptions of instruments, comets,

stars, tides, etc. Includes extensive chapter on the
solution of practical problems in astronomy.

1405.   Olmsted, Denison. *An Introduction to Astronomy*. New York:
        Collins, Brother & Co., 1839; 1846 (5th Edition.).
        Pp. xv + 288.

        Standard mid-19th-century textbook emphasizing practical
        aspects but with new sections on descriptive astronomy,
        instruments, the Sun, physical aspects of the planets
        and satellites, and stellar astronomy including a short
        chapter, "Of the System of the World," an argument for
        the Copernican system and the structure of the material
        universe. Brief citations, no index.

1406.   Proctor, Richard A., and A. Cowper Ranyard. *Old and
        New Astronomy*. London: Longmans, Green and Co., 1892.
        Pp. viii + 816.

        Proctor's magnum opus left incomplete at the time of
        his death (1888) and completed by Ranyard. Large scale,
        comprehensive review of descriptive and practical astron-
        omy. Proctor completed sections on the Earth and planetary
        system. Ranyard added the sections on the stellar uni-
        verse. In many areas--especially in those regarding the
        statistics of stellar motions--in which Proctor had
        provided important studies, Ranyard attempts to reproduce
        Proctor's methodology and conclusions. Profusely illus-
        trated. Detailed annotations and some citations.

*   Schellen, H. *Spectrum Analysis*.

        Cited herein as item 931.

*   Somerville, Mary. *The Mechanism of the Heavens*.

        Cited herein as item 809.

1407.   Somerville, Mary. *The Connexion of the Physical Sciences*.
        London: John Murray, 1834; 1836. Pp. 458.

        Classic review of astronomy and physics at mid-century
        revised many times through the 1840s and 50s. Intended
        for popular audiences, with detailed notes but few direct
        citations.

1408.   Todd, David P. *Stars and Telescopes*. Boston: Little,
        Brown and Co., 1899. Pp. xvi + 419.

        Popular survey of special interest for its many portraits
        and illustrations and extensive bibliographies at the end

of each chapter. Based upon a text entitled *Celestial Motions* by William Thynne Lynn.

1408a. Williams, L. Pearce. *Album of Science: The Nineteenth Century*. New York: Charles Scribner's Sons, 1978. Pp. xiv + 413.

A pictorial history of science portraying the visual sources created for popular consumption, and thereby revealing popular images of science. Includes over 60 pages of astronomical instruments and celestial scenes. Numerous citations.

1409.   Young, Charles A. *A Text-Book of General Astronomy*. Boston: Ginn, 1888; 1889; 1895; 1898; (Revised) 1904. Pp. ix + 630.

The most complete and general of Young's many texts (in order of sophistication: *Lessons in Astronomy* 1890-1903; *Elements of Astronomy* 1889-1903; *Manual of Astronomy* 1902-1904 and later; *General Astronomy*). Provides comprehensive background on the state of astronomy circa 1900 including many detailed discussions of technique, both historical and contemporary, and the beginnings of the applications of spectroscopic methods and the applicability of the laws of modern physics. Extensive indirect citations.

20TH CENTURY

1410.   Berendzen, R. "The Case Studies Project on the Development of Modern Astronomy." *Education in and History of Modern Astronomy*. (item 125), 146-154.

Development of a set of curricular materials which resulted in the textbook *Man Discovers the Galaxies* (item 1183). Includes discussion of reception of materials by students. 40 citations, discussion.

1411.   Davidson, Martin. *The Stars and the Mind*. London: The Scientific Book Club, 1947. Pp. x + 210.

Popular introduction to the history of astronomy including commentary on the impact of astronomy on Christian faith; cosmogony and religion; thermodynamics and cosmology No direct citations.

*   Dickson, F.P. *The Bowl of Night*.

         Cited herein as item 1198.

*   Dingle, Herbert. *Modern Astrophysics*.

         Cited herein as item 913.

1412.   Eddington, Arthur Stanley. *Stars and Atoms*. Oxford:
        Oxford University Press, 1927. Pp. 131.

        Discursive synthesis of a number of popular lectures
        by Eddington circa 1926 reviewing highlights of his major
        work: *The Internal Constitution of the Stars*. Includes
        three lectures covering stellar interiors, the mass/
        luminosity relation, recent studies of Algol, Sirius B,
        the solar chromosphere, Betelgeuse, pulsating variables,
        subatomic energy and stellar evolution. The 1928 edition
        adds material on the recent identification by I.S. Bowen
        of the elements responsible for the chief spectral lines
        in gaseous nebulae. No direct citations or index.

1413.   Flammarion, Camille. *Popular Astronomy* [1894]. 2nd Edition.
        J. Ellard Gore, translator. London: Chatto & Windus,
        1907. Pp. xix + 696.

        Second edition of the English translation, originally
        published in 1894, of Flammarion's famous 1880 text
        Astronomie Populaire. Includes six sections covering all
        aspects of current astronomy and in many cases provides
        historical commentary. Throughout the text Flammarion's
        belief in the existence of habitable worlds is quite
        evident, especially in his chapter on Mars. An appendix
        by Gore reviews principal advances between 1894 and 1907,
        and various parenthetic annotations are provided in
        Flammarion's text to indicate advances between the latest
        French edition and the first English translation. No
        direct citations.

1414.   Hale, George Ellery. *The Depths of the Universe*. New York:
        Charles Scribner's Sons, 1924. Pp. 23.

        Reprint of popular articles in *Scribner's Magazine* re-
        viewing recent studies by Shapley on the scale of the
        Universe, E.E. Barnard's studies of nebulae, and Hale's
        own work on solar magnetism. Reviewed in: *Nature*, 114
        (1924), 3-4; *Isis*, 7 (1925), 152-153.

1415.   Jeans, James H. *The Astronomical Horizon*. Oxford: Oxford
        University Press, 1945. Pp. 23.

Based upon the 1944 Deneke lectures, this popular exposition traces the "receding" horizons in astronomical knowledge beginning with Greek thought, then moving on to the scientific revolution and Halley's detection of stellar proper motions and the recognition that stars are suns. The bulk of the narrative begins with William Herschel's studies of the structure of the Universe and centers upon contemporary extra-galactic research and cosmology. No direct citations.

*   Macpherson, Hector. *Modern Astronomy*.

Cited herein as item 105.

*   Russell, Henry Norris. *The Solar System and Its Origin*.

Cited herein as item 1052.

*   Russell, Henry Norris. "The Time Scale of the Universe."

Cited herein as item 1219.

1416.   Russell, Henry Norris, Raymond Smith Dugan, and John Q. Stewart. *Astronomy*. 2 volumes. Boston: Ginn, 1926. Pp. xi + 470 + xxi; xii + 471-932 + xxx. Twice reprinted.

Standard general textbook prepared as a revision and extension of C.A. Young's texts from the turn of the century. Volume 1 includes astronomical instruments, practical astronomy, a general survey of the Solar System, and introductory celestial mechanics. Volume II covers all aspects of astrophysics: spectroscopy, atomic theory, stellar and galactic astronomy. Chapter on stellar constitution and evolution provides important primary source information on the state of knowledge at the time. Selected chapter bibliographies.

1416a. Rutherford, F.J., Gerald Holton, and Fletcher G. Watson, editors and directors. *The Project Physics Course*. New York: Holt, Rinehart and Winston, 1970. Pp. xviii + 790.

Developed as a unified introduction to physics with a strong humanistic orientation. Significant portions of the text and associated handbooks provide historical commentary through selections from historical literature and excerpts from primary sources. In six units; unit 2, "Motions in the Heavens," examines theories of motion through Newton. Numerous citations.

*   Smart, William M. *Text-Book on Spherical Astronomy*.

     Cited herein as item 686.

1417. Unsöld, Albrecht. *The New Cosmos*. New York: Springer
     Verlag, 1969. Pp. xii + 373.

     Comprehensive mathematical introduction to modern
     astronomy providing historical chapters on the develop-
     ment of modern astrophysics. Translation by W.H. McCrea
     from the first German edition (1967) titled "Der neue
     Kosmos." Includes a general bibliography of texts, mono-
     graphs and journal articles.

*   Woodbury, David, O. *The Glass Giant of Palomar*.

     Cited herein as item 343.

INDEX

Authors are indicated by an *; in cases where an individual is both author and subject of the same entry, the entry number is listed twice.

Bell, Trudy E.   *0265, *0354, *0464, *1246
Bennett, J.A.   *0005, *0422, *0423, *0424, 1305
Bentley, Rev.   1347
Berendzen, Richard   *0125, *0163, *0182, *0214, *0355, *1183, *1410
Berlin Observatory   1255
Berman, L.   0352
Bernoulli, D.   0605, 0716, 1297
Bernoulli, J.   0277, 0897
Berry, Arthur   *0078
Berry, Robert Elton   *1247
Bessel, F.W.   0081, 0284, 0823, 0839, 0840, 0872, 0897, 0970, 1247a, 1255, 1376
Besterman, T.   *0006
Beta Lyrae   1083
Betelgeuse   1057
Bethe, H.   1133, 1135, 1137
Bienkowska, Barbara   *0532
Biermann, Kurt R.   *0164, *0604, *0942, *1248
Bigourdan, Guillaume   *0079, *0183
Biot, J.B.   0608, *1395a
Bird, J.   0403, 1300
Birkenmajer, Alexander   *0425
Bishop, Jeanne E.   *1133
Black Holes   0806
Blackwell, D.E.   *0868, *0969
Blair, G. Bruce   *0184
Blancanus, J.   0444
Blunck, Jurgen   *0643
Blunt   0457
Board of Longitude   0285
Bobrovnikoff, N.T.   *0666, *0970
Bochner, Salomon   *0699
Bode, J.E.   0277, 0384, 0781
Bode's Law   0790, 0797
Bogota Observatory   0263
Bohnenberger   0897
Bohr, N.   0917
Bok, Bart J.   *0150, *1163
Bolograph   1011
Bolometer   0512
Bond, G.P.   0334, 1313
Bond, W.C.   *0266, 0334, 1313
Bondi, H.   *0151, 1196
Bonner Durchmusterung   0829, 0835
Bonnet, C.   0781
Boquet, F.   *0080
Borel, P.   0535
Borelli, G.A.   0698, 0707